Analytical Methods of Food Authentication

VISIT OUR FOOD SCIENCE SITE ON THE WEB

http://www.foodsci.com

e-mail orders: direct.orders@tps.co.uk

Analytical Methods of Food Authentication

Edited by

P.R. ASHURST

Principal of Dr P.R. Ashurst & Associates
Consulting Chemists
Hereford
UK

and

M.J. DENNIS

Head of Food Authenticity and Statutory Methods
CSL Food Science Laboratory
Norwich
UK

BLACKIE ACADEMIC & PROFESSIONAL
An Imprint of Chapman & Hall
London · Weinheim · New York · Tokyo · Melbourne · Madras

Published by Blackie Academic and Professional,
an imprint of Thomson Science, 2–6 Boundary Row, London SE1 8HN, UK

Thomson Science, 2–6 Boundary Row, London SE1 8HN, UK

Thomson Science, 115 Fifth Avenue, New York, NY 10003, USA

Thomson Science, Suite 750, 400 Market Street, Philadelphia, PA 19106, USA

Thomson Science, Pappelallee 3, 69469 Weinheim, Germany

First edition 1998

© 1998 Thomson Science

Thomson Science is a division of International Thomson Publishing I(T)P

Typeset in 10/12 Times by Blackpool Typesetting Services Ltd
Printed in Great Britain by St Edmundsbury Press Ltd, Bury St Edmunds, Suffolk

ISBN 0 7514 0426 8

All rights reserved. No part of this publication may be reproduced, stored in a retrieval system or transmitted in any form or by any means, electronic, mechanical, photocopying, recording or otherwise, without the prior written permission of the publishers. Applications for permission should be addressed to the rights manager at the London address of the publisher.
  The publisher makes no representation, express or implied, with regard to the accuracy of the information contained in this book and cannot accept any legal responsibility or liability for any errors or omissions that may be made.

A catalogue record for this book is available from the British Library

Library of Congress Catalog Card Number: 97-77015

∞ Printed on acid-free text paper, manufactured in accordance with ANSI/NISO Z39.48-1992 (Permanence of Paper)

Contents

List of Contributors		xi
Preface		xiii

1 Introduction to methods for food authentication   1
M.J. DENNIS and P.R. ASHURST

1.1	Aims of food authentication	1
	1.1.1 Food producing companies	1
	1.1.2 Enforcement authorities	1
	1.1.3 National governments	2
1.2	Networks	2
1.3	Sampling	3
	1.3.1 National surveillance	3
	1.3.2 Food companies	4
1.4	Interpreting databases	4
1.5	Influences on data evaluation	6
	1.5.1 Legislation	6
	1.5.2 Historical changes	7
1.6	New technology	7
	1.6.1 Traditional techniques	7
	1.6.2 Food irradiation	8
	1.6.3 Genetically modified organisms	11
1.7	Methods of analysis	11
References		13

2 Stable isotope analysis by mass spectrometry   14
D.A. KRUEGER

2.1	Introduction	14
2.2	Carbon isotopes	16
2.3	Oxygen isotopes	23
2.4	Hydrogen isotopes	26
References		29

3 Nuclear Magnetic Resonce spectroscopy   36
I.J. COLQUHOUN and M. LEES

3.1	Introduction	36
3.2	Practical considerations	37
	3.2.1 Background	37
	3.2.2 Sample preparation	37
	3.2.3 Acquisition time	38
	3.2.4 Quantitative NMR	39
	3.2.5 Dynamic range and water suppression	39
	3.2.6 Detection limits	40
	3.2.7 Signal assignments	40
	3.2.8 Data preparation for multivariate analysis	41
3.3	Applications	42
	3.3.1 Fruit juices and fruits	42
	3.3.2 Wines	51
	3.3.3 Vegetable oils	52
	3.3.4 Fish and meat	58

	3.4	Site-specific natural isotope fractionation by NMR	59
		3.4.1 Introduction	59
		3.4.2 Natural isotope fractionation	60
		3.4.3 Determining site-specific ratios by NMR	60
		3.4.4 ^{2}H-NMR for quantitative determinations of site-specific ratios	61
		3.4.5 Choice of isotopic probve for SNIF-NMR® analysis	62
		3.4.6 Ethanol as isotopic probe	63
		3.4.7 Wine authentication	64
		3.4.8 Fruit juice authentication	65
		3.4.9 Sample treatment: the SNIF-NMR® concept	67
		3.4.10 Authentication of other sugar-rich products	69
		3.4.11 Origin determination of alcoholic beverages	69
		3.4.12 Acetic acid as isotopic probe	70
		3.4.13 SNIF-NMR® applied to other isotopic probes	70
		3.4.14 The future	71
	References		72

4 Infrared spectroscopy   76
C.N.G. SCOTTER and R. WILSON

	4.1	Theory	76
	4.2	Mid-infrared spectroscopy: equipment and methods	79
	4.3	Near-infrared instrumentation and methods	82
		4.3.1 Instrumentation	82
		4.3.2 NIR methods	82
	4.4	Applications of mid-infrared spectroscopy	83
		4.4.1 Fruit products	83
		4.4.2 Coffee	84
		4.4.3 Edible oils	87
	4.5	Applications of NIR spectroscopy	88
		4.5.1 Cereals	88
		4.5.2 Fruit products	89
		4.5.3 Food-grade oils – substituting in part or wholly inferior oils for more expensive types	90
		4.5.4 Vegetable fat adulteration of dairy products	91
		4.5.5 Animal fat substitutes in sausages	91
		4.5.6 Meat species discrimination	91
		4.5.7 Classification of Iberian pig carcass quality	91
		4.5.8 Discrimination of fresh from frozen-thawed meat	92
		4.5.9 Tea – inappropriately described type	92
		4.5.10 Coffee – type and coffee content	92
	Acknowledgements		93
	Appendix 4A		93
	References		94

5 Oligosaccharide analysis   97
N. LOW

	5.1	Introduction	97
		5.1.1 Adulterants	98
		5.1.2 Carbohydrate classification	99
	5.2	Oligosaccharide formation in foods and inexpensive sweeteners	99
		5.2.1 Carbohydrate metabolism in plants	99
		5.2.2 Oligosaccharide synthesis in plants	100
		5.2.3 Oligosaccharide formation in inexpensive sweeteners	100
	5.3	Principle of the fingerprint oligosaccharide method	104
	5.4	Sample preparation	105
		5.4.1 Sample preparation method 1 (SPM1)	105

		5.4.2 Sample preparation method 2 (SPM2)	106
5.5	Oligosaccharide fingerprinting employing high performance anion exchange liquid chromatography with pulsed amperometric detection (HPAE-PAD)		106
	5.5.1	Sample analysis by high performance anion exchange liquid chromatography	106
	5.5.2	Interpretation of HPAE-PAD results	110
	5.5.3	HPAE-PAD conclusions	119
5.6	Oligosaccharide fingerprinting employing capillary gas chromatography with flame ionization detection (CGC-FID)		120
	5.6.1	Sample analysis by capillary gas chromatography with flame ionization detection	120
	5.6.2	Interpretation of CGC-FID results	121
	5.6.3	Other applications of oligosaccharide fingerprinting by CGC-FID	129
	5.6.4	CGC-FID conclusions	129
5.7	Oligosaccharide fingerprinting employing capillary zone electrophoresis		131
	5.7.1	Introduction	131
	5.7.2	Sample preparation for CZE	132
	5.7.3	Interpretation of CZE results	133
	5.7.4	CZE conclusions	133
5.8	Conclusions		133
References			135

6 Enzymic methods of food analysis   137
G. HENNIGER

6.1	Introduction		137
6.2	Specificity of enzymatic methods		141
	6.2.1	Highly ('absolutely') specific enzymes	141
	6.2.2	Less specific enzymes	142
	6.2.3	Relative specificity of enzymatic procedures	142
	6.2.4	Enzymes with slow side activity	143
	6.2.5	Group-specific enzymes	143
	6.2.6	Stereo-specific enzymes	144
6.3	The analytes		146
	6.3.1	Acetaldehyde	146
	6.3.2	Acetic acid	147
	6.3.3	Ammonia	147
	6.3.4	L-Ascorbic acid ('vitamin C')	148
	6.3.5	L-Aspartic acid	148
	6.3.6	Cholesterol	149
	6.3.7	Citric acid	149
	6.3.8	Creatine/creatinine	150
	6.3.9	Ethanol	150
	6.3.10	Formic acid	151
	6.3.11	D-Fructose ('fruit sugar')	152
	6.3.12	D-Gluconic acid/D-glucono-δ-lactone	152
	6.3.13	D-Glucose ('grape sugar')	153
	6.3.14	L-Glutamic acid	153
	6.3.15	Glycerol	154
	6.3.16	Guanosine-5'-mono-phosphate	154
	6.3.17	D-3-Hydroxybutyric acid	154
	6.3.18	D-Isocitric acid	155
	6.3.19	D-Lactic acid	155
	6.3.20	L-Lactic acid	156
	6.3.21	Lactose	156
	6.3.22	Lactulose	157
	6.3.23	L-α-Lecithin	158
	6.3.24	D-Malic acid	158

		6.3.25 L-Malic acid	158
		6.3.26 Maltose	159
		6.3.27 Nitrate	159
		6.3.28 Oxalic acid	160
		6.3.29 Pyruvate	160
		6.3.30 Raffinose	161
		6.3.31 D-Sorbitol/Xylitol	161
		6.3.32 Starch	162
		6.3.33 Succinic acid	162
		6.3.34 Sucrose ('sugar')	163
		6.3.35 Sulphite	163
		6.3.36 Urea	164
	6.4	Sample preparation	164
		6.4.1 Sample preparation methods	164
		6.4.2 Special sample preparations	165
	6.5	Principles and measuring technique	167
		6.5.1 Reaction mechanism	167
		6.5.2 Reaction equilibrium	170
		6.5.3 Measurement of enzymatic reactions	171
		6.5.4 Measurement techniques	172
		6.5.5 The reaction	173
		6.5.6 Calculation	175
	6.6	Interferences with the determination	176
		6.6.1 The nature of interference	176
		6.6.2 Recognizing interferences	177
	6.7	Sensitivity	177
	6.8	Conclusion	180
		Further reading	181

7 DNA/PCR techniques — 182
W.S. DAVIDSON

7.1	The problem: species identification of raw and processed meat		182
7.2	Protein-based methods		182
	7.2.1	Gel electrophoresis	182
	7.2.2	Immunology	183
	7.2.3	Liquid chromatography	184
	7.2.4	Need for alternative approach	184
7.3	DNA-based methods		185
	7.3.1	Hybridization	185
	7.3.2	PCR direct sequence analysis (FINS)	187
	7.3.3	Species specific primers	193
	7.3.4	Sex specific primers	194
	7.3.5	Amplification of restriction site polymorphisms	195
	7.3.6	PCR-single strand conformational polymorphisms (PCR-SSCP)	195
	7.3.7	Arbitrary primers (AP-PCR and RAPD)	196
	7.3.8	Microsatellites	198
7.4	Summary and future trends		199
	Acknowledgements		200
	References		200

8 Electrophoretic methods — 204
A.T. ANDREWS

8.1	Introduction		204
8.2	Gel-based methods		205
	8.2.1	Polyacrylamide gel electrophoresis (PAGE)	205
	8.2.2	SDS-PAGE	211

		8.2.3	Agarose gel electrophoresis	212
		8.2.4	Isoelectric focusing (IEF)	214
		8.2.5	Immunoelectrophoresis	218
		8.2.6	Two-dimensional methods	220
	8.3	Detection of separated components		222
		8.3.1	Staining procedures	222
		8.3.2	Blotting of gel patterns	227
	8.4	Capillary electrophoresis (CE)		228
		8.4.1	Capillary zone electrophoresis (CZE)	230
		8.4.2	Micellar electrokinetic capillary chromatography (MECC)	230
		8.4.3	CE in polymer solutions	231
		8.4.4	CE in gel-filled capillaries	232
		8.4.5	Capillary isoelectric focusing	232
	8.5	Some applications of electrophoretic analysis		233
	References			238

9 Antibody techniques 241
E. MÄRTLBAUER

	9.1	Introduction		241
	9.2	Principles		244
		9.2.1	Marker-free methods	244
		9.2.2	Enzyme immunoassay	247
		9.2.3	Sensitivity	253
		9.2.4	Specificity	254
	9.3	Specific applications		256
		9.3.1	Milk and milk products	256
		9.3.2	Meat and meat products	259
	9.4	Statistical parameters for quantitative evaluation of EIA		263
	9.5	Conclusions		265
	References			266

10 Trace element analysis for food authenticity studies 270
H.M. CREWS

10.1	Introduction	270
10.2	General considerations	272
10.3	Flame emission and atomic absorption spectrometry (FES and AAS)	273
	10.3.1 Applications using FAAS, FES, ETAAS and AFS	275
10.4	Atomic emission spectrometry (AES)	278
	10.4.1 Applications using DCP-AES and ICP-AES	280
10.5	Neutron activation analysis (NAA)	281
	10.5.1 Applications of NAA	282
10.6	Mass spectrometry (MS)	283
	10.6.1 Applications using thermal ionization MS (TIMS) and inductively coupled plasma-MS (ICP-MS)	287
References		288

11 Pyrolysis mass spectrometry in food analysis and related fields: principles and application 292
M. LIPP and E. ANKLAM

11.1	Introduction	292
11.2	Principles	293
	11.2.1 Pyrolysis techniques	293
	11.2.2 Ion generation	295
	11.2.3 Mass spectrometer	295

11.3 Data evaluation		296
11.4 Food analysis by Py-MS		297
11.4.1 Application to food without special emphasis on proof of authenticity		297
11.4.2 Application to food authentication		298
References		304

12 The principles of multivariate data analysis — **308**
M.J. ADAMS

12.1 Introduction	308
12.2 Univariate statistics	308
12.3 Multivariate statistics	310
12.4 Data reduction	313
12.5 Pattern recognition	319
12.5.1 Cluster analysis	319
12.5.2 Discriminant analysis	323
12.5.3 Neural networks	325
12.6 Calibration	326
12.7 Conclusions	335
References	336

Index   **337**

Contributors

Professor M.J. Adams	School of Applied Sciences, University of Wolverhampton, Wolverhampton, UK.
Professor A.T. Andrews	Institute of Cardiff, University of Wales, Cardiff, UK.
Professor E. Anklam	European Commission, Directorate General Joint Research Centre, Environmental Institute, TP 740, Ispra, Italy.
Dr P.R. Ashurst	Dr P.R. Ashurst & Associates, Kingstone, Hereford, UK.
Dr I.J. Colquhoun	Institute of Food Research, Norwich, UK.
Dr H.M. Crews	CSL Food Science Laboratory, Norwich, UK.
Professor W.S. Davidson	Department of Biochemistry, Memorial University of Newfoundland, St Johns, Newfoundland, Canada A1B 3X9.
Dr M.J. Dennis	CSL Food Science Laboratory, Norwich, UK.
Dr G. Henniger	Biochemical Research Division, Boehringer-Mannheim, D-82372 Penzberg, Germany.
Mr D.A. Krueger	Krueger Food Laboratories Inc, Cambridge MA 02139 3170, USA.
Dr M. Lees	Eurofins SA, 44073 Nantes, Cedex 03, France.
Dr M. Lipp	European Commission, Directorate General Joint Research Centre, Environmental Institute, TP 740, Ispra, Italy
Professor N. Low	Department of Applied Microbiology and Food Science, University of Saskatchewan, Saskatoon, Canada S7N 5A8.
Professor E. Märtlbauer	Tierärztliche Fakultät, Lehrstuhl für Hygiene und Technologie der Milch, D 80539 München, Germany.
Mr C.N.G. Scotter	Campden and Chorleywood Food Research Association, Chipping Campden, Gloucestershire, UK.
Dr R. Wilson	Institute of Food Research, Norwich, UK.

Preface

As we stated in the earlier volume *Food Authentication* the adulteration of food is not a new issue. It has, within the past decade become much sharper, partly because of the trend in Europe and other parts of the world to move from compositional legislation to label declaration and partly because of the development of analytical methods. The legal issues surrounding label declarations are pursued with differing vigour by the various enforcement bodies, but this volume is not concerned as such with that aspect.

Having examined in some detail the main food commodity groups, for it is within commodities that the problem mainly exists, this volume surveys the range of analytical methods that are today applied to the investigation and confirmation of authenticity or adulteration. At the outset it must be recognized that for any food commodity it is invariably the case that there is no single test that will confirm the authenticity of a product. As will be seen almost all types of analysis, both complex and simple, are used in one or another commodity. Two common strands are evident in all authenticity work. One is the creation and maintenance of databases and the other is the handling of the large amounts of data generated.

There is deliberately no chapter in this volume dealing with databases because these will vary to meet the special needs of each commodity. We have, however, included a chapter on the chemometric methods that are now widely used for data processing and evaluation – a direct result of the computing resources that are available universally at relatively low cost.

Stable isotope analyses have become an important tool in the detection of adulteration and chapters are included on both mass spectrometry and nuclear magnetic resonance as alternative ways of obtaining ratios of the key stable isotopes. Mass spectrometry linked to pyrolysis is dealt with in a separate chapter and the significance of both mid- and near-infrared spectroscopy are recognized. Other analytical techniques that each have their own chapters include oligosaccharide analysis, trace elements analysis and the important biochemical techniques such as enzyme based methods, analyses linked to DNA identification and comparison, antibody techniques and electrophoresis. Finally, a short miscellaneous chapter examines some of the simpler but none the less effective tools such as microscopy.

We have also included a chapter on the detection of irradiated foodstuffs which although perhaps not strictly an authenticity issue does rate highly in consensus of consumers and enforcement bodies.

This book is the work of its authors, all eminent and respected specialists in their own fields and the editors are very appreciative of their contributions. We take full responsibility for the overall work 'warts and all' and it is hoped the reader will find this a worthy successor to the original volume.

1 Introduction to methods for food authentication
M.J. DENNIS and P.R. ASHURST

1.1 Aims of food authentication

When food analysts carry out food authenticity tests, what are they trying to achieve? The answer depends very much on the requirements of the customer. Food producers, food enforcement agencies and national governments each have subtly different needs and these must be taken into account when the choice of method of analysis is made.

1.1.1 *Food producing companies*

Reputable food producers want to ensure the products they manufacture meet the demands of legislation in the countries where they are sold. They also wish to avoid paying a premium price for an inferior commodity. Equally the costs of the authenticity test are important and, in some manufacturing situations, the speed with which the test can be accomplished may be very important. Although food companies will wish to use soundly based methodology to meet their obligations under due diligence legislation, they do not necessarily need the degree of method validation which is appropriate in other situations. Thus, when purchasing raw materials, they are in a position to specify the composition of the material they are buying. They do not necessarily have to justify this specification; they can simply say 'this is not the product I want'. One consequence of this is that the larger, more scientifically aware companies, can be among the first to adopt new technologies in food authentication when these meet a perceived need.

1.1.2 *Enforcement authorities*

In contrast, method validity is of crucial importance to enforcement authorities. It is vital that they can show that the methods used are defensible in a court of law. Thus, the methods must be rigorously defined and their validity established either through long usage, or preferably, through collaborative testing. Enforcement authorities will therefore tend to concentrate on validating and applying methodologies developed by universities and research institutes rather than undertaking cutting-edge research. In this, their position is similar to that of the food companies, who do not consider it their role to devote resources to research but are content to apply the best methods currently available.

1.1.3 *National governments*

The organization of the enforcement of food law differs between countries. However, it is not unusual for law enforcement to be undertaken at a local or district level. This is the case in the UK where enforcement is generally under the direction of local authorities. Nevertheless, national governments must have an interest in co-ordinating and directing these activities. Only at the national or international level can the combination be found of the appreciation of the need to invest in authenticity research and the resources to carry out such work. Governments need to ensure that sound, well-validated methods are available to meet the needs of both industry and enforcement in order to protect the public from misleading or fraudulent labelling. In this they seek to meet the needs of the public, as voters and consumers, and the food industry as an engine of the economy. There is no conflict of interest between these parties. Each needs to know that when they buy a premium product, they get what they pay for.

Governments have a difficult task in deciding where best to direct resources for food authentication. Where existing methods of analysis have not been fully validated or usage has indicated problems, then it is appropriate to fund method development and validation work. Equally, where new technologies arise and have potential application to food authentication (e.g. DNA biotechnology) then further research to develop these applications is merited; particularly where current methods need to be improved. It is also important to maintain an awareness of the market place to ensure that important authenticity issues are not being overlooked.

1.2 Networks

Regular contacts of scientists for the discussion and debate of issues has always been fundamental to the development of the scientific method of study. However, the pace of change in food authentication is unrelenting and it is necessary for all parties to maintain close awareness of both scientific developments, and developments in the market place. In the UK for example, the Ministry of Agriculture's Food Authenticity Working Party allows representatives of consumers, food manufacturers and retailers, food law enforcement co-ordinators and scientists to contribute their views as to how the authenticity of foods should be ensured.

In the United States, the Food and Drug Administration (FDA) is seeking a closer working relationship with the food industry and inviting them to become partners and customers of the planned Joint Institute for Food Safety and Applied Nutrition (JIFSAN) which is being set up in co-operation with the University of Maryland (Anon, 1997). The FDA emphasized the need for information sharing between state and federal regulators, food manufacturers and international agencies. It was recognized that industry is often in a unique position to provide information on dishonest competitors, significant changes in commodity markets and new technologies. One proposal considered was that US Customs services should co-operate with other countries to prevent exporters seeking alternative ports of entry when adulterated products were rejected at the border. The European Commission has recognized the importance of providing support for food authenticity networks by funding Food Authenticity Issues and Methodologies (FAIM). This group brings together over 40 food scientists to review authenticity issues and analytical methodology in fruit-based products, honey, meat and fish, dairy products, oils and fats, cereals, tea and coffee.

Thus there is considerable international recognition of the need to encourage information exchange in food authenticity. These more formal schemes together with the individual initiatives of scientists working in key commodity areas ensure that resources are targeted towards the most deserving issues. Nevertheless, a clear awareness of progress in ensuring the authenticity of food commodities is most commonly achieved through well-targeted surveys.

1.3 Sampling

1.3.1 *National surveillance*

There are a number of issues which should be considered when a national authenticity surveillance of a specific food (or group of foods) is contemplated. It is necessary to find a basis for determining priorities and the relative importance of the many possible issues which could be studied. Some basis for weighting the various factors must be established. It is, in principle, possible to consider any claim which is made on a food label as an issue requiring testing, but some issues will be more important than others. Many factors may influence this judgement. For instance, is this a problem which consumers are aware of? Or will feel particularly strongly about? (e.g. kosher or vegetarian labels); will the authenticity issue have an effect on the health of a sub-group of the population? (e.g. those allergic to cow's milk might be severely affected if goat's milk yoghurt or cheese contained this ingredient). Is the food commodity sold to a large percentage of the population or will only a few people be affected? Is the issue a recently discovered problem or have manufacturers and retailers known about the problem for sufficient time to put adequate monitoring systems in place? Are suitable methods of analysis available?

If it is decided on the basis of these considerations that a surveillance exercise is appropriate, then a detailed sampling plan, incorporating any constraints which the analytical method imposes, should be prepared. The type of sales outlet for the food may be significant; supermarket, small shop, restaurant or mobile sales van. Will the surveillance cover the whole country or just selected areas? Will it cover premium brands, budget brands and own label brands? How much will the analysis cost? Would it be more effective to evaluate a large number of samples with, perhaps, a cheap but less sensitive method, to gain a good idea of how widespread a problem may be? Alternatively, is a highly accurate but expensive test required in order to detect a sophisticated adulteration issue? Has a cut-off limit been established from authentic samples by reputable laboratories using the test in question, and will these authentic data cover the range of food commodities and countries of origin that it is proposed to investigate? Have any new processing methods been carefully evaluated and shown not to affect the proposed tests? Is a confirmatory test available which will add validity to the proposed survey? All of these factors and possibly others need to be taken into account when planning a surveillance exercise.

1.3.2 *Food companies*

For the food industry, the emphasis on where authenticity testing should be directed is rather different. Lees (1996) outlined some of the considerations which needed to be taken into account. The analysis of every batch of raw materials before incorporation into the food product is termed positive release. This approach is very effective in detecting adulteration but it carries a high cost. It is appropriate for high risk situations such as when a new supplier is being used, where a regular supplier consistently delivers a product close to an authenticity limit value (and a Gaussian distribution might be expected) or where a previously good supplier has had a batch fail an authenticity test. In situations where a supplier has a good track record of providing authentic materials, then other approaches can be considered. Screening tests which are simple, fast and cheap may help to maintain confidence in a product but are unlikely to detect sophisticated adulterations. The most cost effective approach is to undertake random sampling using the best tests currently available. By using statistics, it is possible to calculate the minimum number of analyses required to ensure, for a given confidence interval (a%) that a p% non-conformity will be detected. So, for 100 sample batches in one year, if it is assumed that 1 in 4 batches are not authentic (p 25%) then only 10 batches need to be analysed to give a 95% probability (a%) of identifying the problem. This level of confidence can be increased by ensuring co-ordination of surveillance activities with retailers, trade associations, importers and producers.

Having established that analysis is required to investigate sample authenticity, it is necessary to ensure that appropriate and sufficient authentic data are available and to specify how subsequent data from unknown samples will be interpreted.

1.4 Interpreting databases

When carrying out food authentication studies, it is best to establish a firm cut-off point at which the sample will be deemed to have failed the test before any tests are carried on unknown samples. This is particularly beneficial when the data from the sample under test are found to be on the borderline between pass and fail. In setting such cut-off points, two quite different situations may apply. In the first case (type A) a fixed value may have been established in legislation, or by specification, for an analytical parameter beyond which the sample is considered not to comply with the description of the product, e.g. virgin olive oil must have a stigmastadiene concentration below 0.15 mg/l. (EC Commission Regulation 2568/91). The second case (type B) is where a database of authentic values has been built up and is used to assess the authenticity of a sample.

For the type A situation, it is not sufficient to measure the analytical parameter and make a simple comparison with the limiting value, e.g. parameter x should be not more than 3.0 mg/l, therefore 2.99 mg/l passes, 3.01 mg/l fails. Experienced analysts are aware that analytical measurements are not perfectly precise and hence some allowance must be made for measurement precision. For well-established methods, the repeatability (r) and reproducibility (R) are available from collaborative trial data. Repeatability (r, within laboratory variation) is the value below which the absolute difference between two single test results obtained with the same method on identical test material under the **same** conditions may be expected to lie with 95% probability. Reproducibility (R, between laboratory variation) is the value below which the absolute difference between two single test results obtained with the same method on identical test material under **different** conditions may be expected to lie with 95% probability.

Thus R represents the maximum difference which two experienced laboratories would be expected to achieve if they both analysed the same sample. It might seem, at first glance, that by adding the value R to the limit value, a realistic cut-off value would be defined. However this would be too generous an approach. The R value relates to the difference between two measurements but a limit value, fixed by legislation or contract, contains no variability. It is therefore appropriate to reduce the allowance for measurement precision to allow for the fact that only one measurement is undertaken. Thus $R(\sqrt{2})^{-1}$ is the appropriate factor rather than R. However, a limit value provides for a one-sided (as opposed to two-sided) statistical test. The question at issue is whether the measured value from the sample exceeds the defined limit value. Thus a further reduction is required to take this into account and the factor becomes $0.84R(\sqrt{2})^{-1}$. This factor applies to single analyses. However, a further reduction in this factor is appropriate if a number of repeats are carried out. The appropriate factor depends on whether the repeats are carried out in the same or at a different laboratory and the theoretical justification for these has been discussed by Martin and Martin (1996).

Thus for n repeats within a single laboratory, the allowance needed to take account of measurement precision from the limit value, L_A, becomes:

$$L_A + 0.84R'(\sqrt{2})^{-1} \qquad (1.1)$$

where

$$R' = \sqrt{R^2 - r^2\left(1 - \frac{1}{n}\right)}$$

Similarly where a repeat analysis is carried out by several laboratories, then the acceptable interval is reduced further. For n measurement by p laboratories, the factor becomes:

$$L_A + 0.84R''(\sqrt{2})^{-1} \qquad (1.2)$$

where

$$R'' = \frac{1}{\sqrt{p}}\sqrt{R^2 - r^2\left(1 - \frac{1}{n}\right)}$$

By using these revised cut-off limits it is possible to prevent an unnecessarily large allowance for measurement precision being used which could have the effect of permitting a small degree of adulteration that could otherwise be detected. However, analysts should be aware that there is a possibility that samples passing the test with values which are close to the limit value within the limiting ranges defined above ($L_A \pm 0.84R'(\sqrt{2})^{-1}$) have a probability that the true value exceeds the limit value and these results should therefore be treated with caution and the position explained to the customer.

The situation is much simpler where analytical values can be compared directly with a database (type B). In this case, the measurements made for the database already contain a contribution from measurement precision and it is not necessary to apply further factors to take this into account. Now the most precise evaluation is obtained when the sample is compared against a well-defined database which is closely related to the sample. Thus in evaluating the authenticity of an orange juice concentrate from Brazil it is preferable to compare analytical data with a database of Brazilian concentrate rather than with samples of concentrate from anywhere in the world.

1.5 Influences on data evaluation

1.5.1 *Legislation*

For many commodities, legislation or codes of practice are in place which govern the processes which may be applied to the production of a particular food commodity. These can be complex and the treatment of different commodities is not necessarily logical. For instance, the use of enzymes to assist the expression of juice is permitted in Europe for apples but not oranges. Different countries and even different wine regions have very different rules about how wine should be made. A clear understanding of the legislation relating to particular food commodities is therefore essential when deciding on their authenticity because changes in processing procedures can influence analytical database values. Equally, legislators need to take into account the effect that changes in legislation will have on authenticity methods. Otherwise, new legislation on permitted manufacturing procedures can create a situation where it is no longer possible to enforce proper food commodity labelling.

1.5.2 Historical changes

Some methods for food authentication have been available for many years and still remain valid. However, the database of authentic values may not always remain valid even though the analytical method is still appropriate. A surprising example of this effect are changes in the nitrogen factors of meat. Nitrogen in meat is traditionally measured by its conversion to ammonia, which is removed by distillation and determined by titration. A factor is then used to convert nitrogen to protein and hence to meat content which in turn can be used to measure the amount of added water in meat products. This procedure was first described around 80 years ago (Stubbs and More, 1919) and data were assembled on the appropriate conversion factors for different meat animals and even different cuts of meat. Over time, new husbandry practices, or possibly the development of new breeds have led to changes in nitrogen factors which required that this database be updated. This provides a good example of the need to maintain an awareness of the potential for changes in databases even when such changes would appear to be unlikely.

This need is well understood where isotopic methods are regularly used. Chapter 2 (Krueger) outlines the environmental and biochemical effects which alter isotopic enrichments both between plant species and through climatological factors. The latter can mean that annual adjustments are required to some authenticity databases for regions or countries when unusual weather conditions are experienced.

1.6 New technology

1.6.1 Traditional techniques

Consumers appear to have an ingrained preference for traditional methods for preparing food and a dislike of new food production techniques even when these may have well-established benefits. However, in the case of cold pressed oils and particularly olive oil, the traditional techniques have established a product of unique quality. Hence methods which are able to detect the incorporation of refined oils in cold pressed oils are an essential feature of authenticating virgin olive oil. This is readily achieved because the bleaching process used in refining converts sterols present in the oil into sterenes by a dehydration reaction. The introduction of such marker chemicals into potential adulterants is a feature of many authenticity tests. Another good example is the presence of specific oligosaccharides in hydrolysed syrups (Chapter 5, Low). The difficulty for researchers who are developing authenticity tests is to be fully aware of the technical developments in the food industry which might enable food products to be used as adulterants and this further emphasizes the need for efficient information networks outlined earlier.

1.6.2 *Food irradiation*

Food irradiation may be the first in what could become an increasingly important class of authenticity issues. It is an example of new technology which can provide benefits to the consumer (in terms of longer shelf-life and lower bacterial counts) but one which many consumers do not want. Although the process is considered to have no adverse effects on food safety, the consumer demands the right to choose whether or not to purchase products prepared in this way. A number of methods have now been well validated to detect the use of this technology and these have recently been reviewed (McMurray *et al.*, 1996). There is no single test which can be used to detect irradiated foods. Irradiation leads to small changes in the composition of foods but which of these changes are most useful for detecting the process depends on what the major components of the food are.

The irradiation of herbs and spices was considered an important development by much of the food industry because of concerns raised over the safety of existing chemical procedures used for reducing bacterial loads. The most effective method for detecting this treatment is thermoluminescence (Sanderson, 1989; MAFF validated method V27, 1993). The principle depends on the energy from irradiation being trapped in microscopic particles of grit accompanying the sample. This energy can be released as photons by heating the sample and it is usual to do this as part of a controlled study. A portion of the sample is irradiated and data compared with that from the unirradiated sample. This comparative test is able to give a clear indication of whether the sample has been irradiated and is the method of choice. However, it depends on having available a suitable radiation source which means that the technique is not widely available. Photostimulated luminescence is able to overcome this disadvantage for many samples. In this technique the energy trapped in the grit is released by illuminating the sample with a suitable light source. The light released is of lower wavelength (and hence higher energy) than the incident light so that fluorescence from samples is not measured. The technique is significantly cheaper and quicker than thermoluminescence but is probably best considered as a screening technique since for some samples calibration by treating the sample with radiation is desirable. These methods are applicable to many foods where grit is present – even if only in very small amounts. Figure 1.1 (taken from Sanderson *et al.*, 1995 with permission) shows the application of photostimulated luminescence to the detection of irradiated Brown Shrimp.

INTRODUCTION TO METHODS FOR FOOD AUTHENTICATION 9

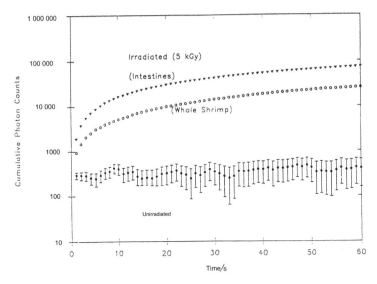

Figure 1.1 PSL detection of trapped intestinal grits from Brown Shrimp. Anti-Stokes luminescence can be stimulated from trapped silicates directly through the membranes of dissected intestines (triangles), and through the whole shrimp (circles). In both cases the PSL detected from irradiated samples is between two and three orders of magnitude greater than the background signals associated from unirradiated controls.

Electron Spin Resonance (ESR) is a second technique which is applicable to detection of food irradiation. In this case the radiation is trapped as an unpaired electron which is detected spectroscopically. The major difficulty with the technique is that unpaired electrons are usually unstable and are easily quenched by water which is, of course, present in many foods of interest. The method is well developed for chicken bones where it provides a rapid, sensitive and inexpensive test (MAFF validated method V28, 1993). Figure 1.2 indicates an ESR spectrum of chicken bone unirradiated (a) and irradiated (b) at 0.2 kGy (about 20% of the dose usual for chicken). The signal produced by irradiation is clear cut. This figure was provided by Dr E. Stewart, Agriculture and Food Science Centre, Belfast whose permission for its reproduction is gratefully acknowledged.

A second method of analysis for detecting irradiation was also developed on chicken. The lipids in chicken undergo a number of fragmentation reactions when irradiated. Some non-specific fragments (such as alkanes) are produced when lipid is irradiated and attempts have been made to use these as detection tests. However, Boyd *et al.* (1991) discovered in irradiated chicken an unusual compound, 2-dodecylcyclobutanone, which was not present in fresh birds. The methodology required for this analysis is familiar to many analytical chemists as it involves solvent extraction, clean-up and detection by GC-MS. It is therefore possible, in principle, for the method to be applied in many laboratories. Further studies have shown that the reaction appears to be a general one for fats and hence is applicable to a wide range of fat-containing foods including vegetables such as avocado. In some ways, this method shows parallels with that of detecting the presence of refined oils in cold pressed oils by sterene analysis.

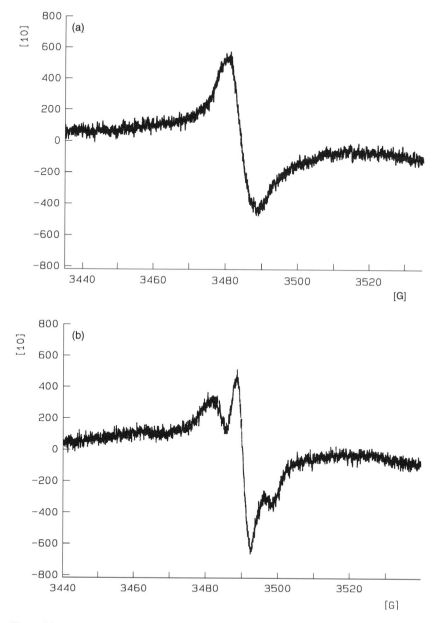

Figure 1.2 (a) Control chicken bone fragments unirradiated. (b) A sliver of irradiated chicken bone, 0.2 kGy, in a glass tube.

1.6.3 *Genetically modified organisms*

The appropriate labelling of foods containing genetically modified organisms (GMOs) shares a number of similarities with the food irradiation issue. Again this is new technology which has the potential to provide benefits to the consumer but is eyed with suspicion by many consumers who have doubts about its safety. An issue of current interest concerns genetically modified (GM) soya. This material contains a gene which provides resistance to the weedkiller Roundup. It therefore permits the use of this weedkiller to enhance crop yields and hence produce cheaper soya. Safety concerns have been focused on the incorporation, as part of the biotechnological procedure, of a second gene which confers antibiotic resistance. This, it is claimed, could be passed on to hazardous micro-organisms although experts consider this unlikely.

Perhaps more importantly, some consumers have a philosophical dislike of the concept of manipulating genes. They consider it their right to be able, through appropriate labelling, to choose not to purchase such products. This necessarily leads to the need to have in place appropriate methodology for ensuring that any claims that a product does not contain GMOs can be substantiated. Advances in DNA technology (Chapter 7, Davidson) mean that small changes in genes of 1 or 2 base pairs can be readily detected so that detecting the considerable changes to the genome which occur in GMOs should not pose a problem where sufficient genetic material remains. However, many of these commodities may be used for the preparation of food chemicals, e.g. lecithin is made from soya beans. Detecting whether a pure compound came from ordinary or GMO soya beans is likely to prove extremely difficult unless isotopic differences between the two can be found. These are unlikely unless there are major changes to metabolic pathways which lead to the product in question. Thus it is likely that ensuring that food ingredients came from non-GMO material will depend on appropriate DNA testing of the unprocessed commodity followed by a chain of documentation to guarantee the authenticity of ingredients.

1.7 Methods of analysis

Man receives information from the environment via five senses but we place an overwhelming emphasis on sight. Smell and taste are such important functions of food yet they are used remarkably rarely for food authentication because they are considered subjectively. **Extra** virgin olive oil is so designated because of its flavour and the origin of fine wines can often be determined by experts but by and large these remain notable exceptions. Sensations of taste and smell have the fundamental problem that they are not quantitative. This potential problem can be overcome to some extent by using taste panels. Indeed, in some ways a taste panel can be considered a scientific instrument in that it seeks to translate sensory data into numerical data. The statistical procedures used in this application bear a resemblance to those used for the treatment of data from more conventional instruments. Indeed, pattern recognition, principal components analysis and other statistical data reduction and classification techniques now play a central role in many instrumental analyses. However, the basic data generated by a taste panel are essentially imprecise and electronic sensor technology (the electronic nose) seeks to redress this deficit. It does so by converting an aroma into quantitative electronic information and differentiating foodstuffs by using many of the statistical concepts mentioned above. As yet, sound methods for food authentication have not been established with these instruments but undoubtedly they have a role to play in the future for ensuring the consistency of food products.

Appearance defines the origin of many primary food commodities. For example, there is no problem in identifying a salmon until it is processed by filleting etc. The visual triggers which are important in identifying food material present the simplest means of establishing authenticity. Therefore it is not surprising that one of the most important instruments for this purpose is the microscope. Microscopy is particularly valuable for detecting the presence of offals in comminuted meat – a topical issue at the moment. It still plays a major role in establishing the unifloral nature of honeys by pollen analysis (mellisopalynology). Figure 1.3 provides an example of some typical pollen grains which can be used to characterize the floral origins of imported honeys. Microscopy is also valuable for identifying fruit and vegetable mixtures. Figure 1.4 shows an example of how microscopy can be used to detect the presence of potato or parsnip in horse-radish sauce. Differentiating these starches by microscopy is trivial but how easy would it be using chemical methods?

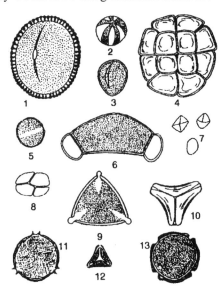

Figure 1.3 Characteristic pollen grains found in imported honey, original magnification ×1000. (From R.W. Sawyer, *J.A.P.A.* (1975), Vol. 13.)

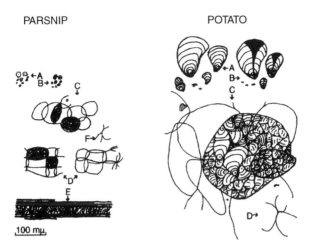

Figure 1.4 Ground parsnip. (Reprinted from the *Journal of AOAC International*, 1989, **72**(4), 620. Copyright 1989 by AOAC International, Inc.)

References

Anon (1997) Food adulteration seen needing new enforcement paradigm. *Inside Laboratory Management*, March 9.
Boyd, D.R., Crone, A.V.J., Hamilton, J.T.G., Hand, M.V., Stevenson, M.H. and Stevenson, P.J. (1991) Synthesis, characterisation and potential use of 2-dodecylcyclobutanone as a marker for irradiated chicken. *J. Agric. Food Chem.*, **39**, 789–92.
Lees, M. Authenticity testing of food and beverages – risk versus cost. *Food Authenticity 96*, 1–3 September 1996, University of East Anglia, Norwich, UK.
McMurray, C.H., Stewart, E., Gray, R. and Pearce, J. (eds) (1996) *Detection Methods for Irradiated Foods*, Royal Society of Chemistry.
MAFF validated method V27 (1993) Detection of irradiated herbs and spices. *J. Assoc. Publ. Analysts*, **29**(3), 189–200.
MAFF validated method V28 (1993) Detection of irradiated meats which contain bone fragments. *J. Assoc. Publ. Analysts*, **29**(3), 201–208.
Martin, G.G. and Martin, Y.-L. (1996) Discussion of some statistical concepts applied to fruit juice analysis. *Food Processing*, **6**(5), 186–94.
Sanderson, D.C.W., Slater, C. and Cairns, K.J. (1989) Detection of irradiated food. *Nature*, **340**, 23–24.
Sanderson, D.C.W., Carmichael, L.A., Riain, N.I., Naylor, J. and Spencer J.Q. (1995) Luminescence studies to identify irradiated food. *Food Science and Technology Today*, **8**(2), 93–96.
Stubbs, G.O. and More, A. (1919) The estimation of the approximate quantity of meat in sausages and meat pastes. *Analyst*, **44**, 125–27.

2 Stable isotope analysis by mass spectrometry
D.A. KRUEGER

2.1 Introduction

In the last twenty years, high precision stable isotope ratio analysis (SIRA) has become an important tool in the food quality control laboratory. The characteristic isotope signatures of materials from different origins has allowed for the detection of various types of fraud and misbranding that would be difficult or impossible to detect by conventional chemical means. Problems as diverse as determining the origin of the carbonation in sparkling wine, the sugar in maple syrup and the vanillin in vanilla extract have been solved by this technique. This chapter will discuss the use of isotope ratio mass spectrometry for the authentication of foodstuffs, and the detection of economic adulteration.

Following the discovery of isotopes by Thompson in 1913, a great deal of effort was made to establish the existence and abundance of the isotopes of the various elements. Most of this work was performed by mass spectrometry, particularly for the stable, non-radioactive isotopes. Conventional mass spectrometry was used to establish the average natural abundances of the isotopes and to measure some of the larger natural variations in their abundance. However, for the stable isotopes of most heavy elements, variations in natural abundance were too small to be accurately measured in this way. Precise measurement of the natural variability of most stable isotopes was not possible until the development of the ratio mass spectrometer (McKinney et al., 1950). The improved measurement precision of instruments of this general design produced another burst of measurement activity. Results from these studies produced a wide variety of practical applications of stable isotope ratio measurements, ranging from establishing the planetary origin of meteorites, to demonstrating the dietary habits of ancient cultures and determining the temperature of the oceans during different geological epochs. In the 1970s, this technique began to be applied to problems of food analysis.

The ratio mass spectrometer is a magnetic sector type mass spectrometer with two important modifications. First, the ion detector consists of two or more independent detector elements. These detector elements are precisely aligned to simultaneously collect ions of two or more masses associated with two or more of the isotopomers of the ionized analyte. This simultaneous measurement of the mass fragments associated with the different isotopes of the analyte allows for a very precise measurement of the ratio of the different isotopic species. The precision of the ratio measurement is considerably higher than that obtainable by mass scanning in a single collector mass spectrometer.

The second key feature is the dual sample inlet device in the ion source.

Analyte is introduced into the ion source as a gas, usually carbon dioxide for carbon isotope ratio measurement, and hydrogen gas for hydrogen isotope ratio measurement. The gas is introduced dynamically through a precise capillary tube leak from the sample reservoir, while the source is under active vacuum. The inlet device contains two such leaks, which are very carefully matched in their flow characteristics. From one capillary, the sample gas is allowed into the mass spectrometer source. From the second capillary, a well-characterized reference gas of known isotope ratio is allowed into the source. A set of switching valves alternates which of the two gases is allowed in at any given time. A series of ratio measurements is then made, alternating between the sample and reference gases. This arrangement allows for very precise comparison of the isotope abundances of the two gases, and from the known value of the reference gas, to an accurate determination of the isotopic abundances of the sample gas.

The results of SIRA measurements are by convention expressed not in terms of atom percent or mole fraction, but in terms of differences between the analyte isotopic ratio and an arbitrarily chosen reference standard ratio. These differences are expressed as parts per thousand (‰, permil) difference between the sample gas ratio and the reference gas ratio. The symbol of this ratio is δ, and it is expressed according to the following formula:

$$\delta(‰) = 1000 \times \left[\frac{R_s}{R_r} - 1\right] \quad (2.1)$$

where R_s and R_r are the ratios of the analyte isotopomers of the sample and the reference. The nominal isotope ratios, analytes and measured mass ratios for the commonly analysed stable isotopes are listed in Table 2.1. The table also lists the reference values normally used for the computation of δ values.

The natural variations in SIRA values are caused by a variety of physical effects such as evaporation and diffusion, as well as kinetic and equilibrium isotope effects associated with chemical and biological reactions. Measurements of these variations were first applied to problems of food quality control by Bricout (Bricout and Merlivat, 1971), who used differences in the deuterium

Table 2.1 Major stable isotopes and their standard reference values

Element	Isotope ratio	Analyte	Mass ratio (R)	Reference	Reference R
Hydrogen	$^2H/^1H$	H_2	3/2	SMOW	0.000316*
Carbon	$^{13}C/^{12}C$	CO_2	45/44	PDB	0.011237
Oxygen	$^{18}O/^{16}O$	CO_2	46/44	SMOW	0.0039948*
Nitrogen	$^{15}N/^{14}N$	N_2	29/28	Air	0.007353*
Sulphur	$^{34}S/^{32}S$	SO_2	66/64	Troilite	0.0450045

SMOW, Standard Mean Ocean Water; PDB, Pee Dee Belemnite; Air, Atmospheric Air Nitrogen; Troilite, Canyon Diablo Troilite.
*Double atom ratios; the analyte contains two atoms of the element of interest.

content of juice water to differentiate between fresh and reconstituted orange juice. Several reviews of the earlier literature on the use of isotopic analysis in food control have been published (Winkler and Schmidt, 1980; Krueger and Reesman, 1982; Bricout, 1982; Winkler, 1984; Schmidt, 1986; Krueger, 1988b).

2.2 Carbon isotopes

One important category of isotopic analysis in food control is carbon stable isotope analysis. Carbon 13 is a minor isotope of carbon with a natural abundance of approximately 1%. In terrestrial plant tissue, the principal source of variation in $^{13}C/^{12}C$ ratios derives from the different photosynethetic pathways for carbon dioxide fixation. Plants incorporate carbon dioxide photosynthetically by three different mechanisms: the Calvin cycle (C-3) pathway, the Hatch-Slack (C-4) pathway and the Crassulacean acid metabolism (CAM) pathway. It was observed that the C-3 pathway results in a large carbon isotope fractionation relative to atmospheric carbon dioxide. The C-4 pathway produces a much smaller fractionation. Consequently, carbon SIRA values from plant derived materials tend to cluster into two ranges of values associated with the major photosynthetic pathways (Smith and Epstein, 1971). CAM plants, which are relatively rare, give carbon SIRA values intermediate between the C-3 and C-4 ranges.

Smaller isotopic fractionations are association with enzymatic and chemical effects resulting from various physiological, geological and manufacturing processes. Thus small variations in $^{13}C/^{12}C$ ratios are observed between different types of molecules within a tissue and between products in a commercial process. Differences are also observed between different positions within a particular molecule.

$^{13}C/^{12}C$ ratios are normally measured using an isotope ratio mass spectrometer with the sample carbon in the form of carbon dioxide. Carbon dioxide can sometimes be isolated from the sample directly, but is generally obtained indirectly by combustion. The sample may either be combusted by oxygen or hot metal oxides, usually copper oxide. Carbon dioxide from combustion is purified cryogenically and the ratio of the masses 46, 45 and 44 are measured with the ratio mass spectrometer. The ratio of mass 45 to mass 44 is a measure of the $^{13}C/^{12}C$ ratio. The ratio of mass 46 to mass 44 is also measured to assess oxygen isotope fractionation in order to correct for the small contribution of the oxygen isotope ^{17}O the mass 45 signal. The corrected 45/44 ratio is compared to that of CO_2 obtained from a standard carbon source of known carbon SIRA value. $^{13}C/^{12}C$ ratios are normally reported as parts per thousand (permil) difference between the sample ratio and the ratio of the arbitrary international standard PDB (Pee Dee Belemnite) – a fossil calcium carbonate from South Carolina USA.

$$\delta^{13}C\ (\text{‰}) = 1000 \times \left[\frac{\left(\frac{^{13}C^{16}O_2}{^{12}C^{16}O_2}\right)_s}{\left(\frac{^{13}C^{16}O_2}{^{12}C^{16}O_2}\right)_r} - 1 \right] \qquad (2.2)$$

$$\text{RPDB} = 0.011237$$

One of the first applications of carbon isotope ratio analysis to food analysis involved sugar. It was observed that the sugar cane and sugar beet plants had different $^{13}C/^{12}C$ ratios, and it was suggested that the $^{13}C/^{12}C$ ratio could differentiate between cane and sugar beet sugar (Smith and Epstein, 1971). This suggestion was subsequently demonstrated analytically (Bricout and Fontes, 1974; Smith, 1975). Cane sucrose had $\delta^{13}C$ values between -10 and -12‰, while sugar beet sucrose had $\delta^{13}C$ values near -25‰.

Carbon SIRA was used to differentiate maple syrup from cane sugar and corn syrup (Hillaire-Marcel et al., 1977; Carro et al., 1980; Parker, 1982). The $\delta^{13}C$ value of maple syrup ranged from -22.4 to -24.8‰, while the results from cane sugar and corn syrup ranged from -10 to -12‰. Maple syrup was not well differentiated by carbon SIRA from sugar beet sugar, which averaged -26‰. The results from Canadian syrups were approximately 0.5‰ more positive than those from Vermont. The method was collaboratively studied and approved as an AOAC International official method for detecting cane sugar and corn syrup in maple syrup (Morselli and Baggett, 1984). Malic acid isolated from maple syrup was used as an internal standard to refine the detectability of added sugar (Paquin et al., 1994). The malic acid was isolated by HPLC and $^{13}C/^{12}C$ ratios determined for both the acid and the whole syrup. Malic acid yielded results slightly more negative than the whole syrup. Differences between the values of greater than 2‰ were suggested as a criterion for detecting cane sugar or corn syrup addition.

Carbon SIRA has proved useful in the detection of adulteration of honey. Most honeys are derived from Calvin cycle plants and have relatively uniform $\delta^{13}C$ values near -25‰. The method has proved useful in detecting additions of cane sugar and corn syrup to honey (Ziegler et al., 1977; Doner and White, 1978; White and Doner, 1978b; Doner et al., 1979a). Regional studies have been published regarding the $\delta^{13}C$ values of honeys from Spain (Serra-Bonheri and Ventura-Coll, 1995) and from Israel (Lindner et al., 1996). The method has since been collaboratively studied and is now an AOAC International official method (White and Doner, 1978a).

Honeys from some sources have relatively more positive $\delta^{13}C$ values than most honeys, in some cases considerably more so (Ziegler et al., 1979a, b). Citrus honeys have been shown to be consistently about 2‰ more positive than other honeys and require a different approach to interpretation (White, 1980; White and Robinson, 1983). A method for identifying pure honeys which have unusually positive $\delta^{13}C$ values has been developed. A fraction from the honey which is mostly protein is isolated by precipitation with sodium tungstate. Comparison of the carbon SIRA value of the protein fraction with that of the total honey provides an indication of cane or corn syrup addition. Pure honeys have only a small difference in the measured ratios of the two fractions, while adulterated honeys have a protein SIRA value >1‰ more negative than the total honey SIRA value (White and Winters, 1989; Rossmann et al., 1992). The method has since been collaboratively studied and is now an AOAC International official method (White, 1992).

Sample preparation of honey for carbon isotope ratio analysis is normally conducted by combustion of the honey and analysis of the resulting CO_2. It has been shown that two of the commonly employed combustion techniques do not produce concordant values, and this bias must be taken into account in the interpretation of honey SIRA values (Krueger, 1993). An automated procedure for honey SIRA has been developed, which uses an elemental analyser configured on-line with a single inlet ratio mass spectrometer (Brookes et al., 1991). Carbon SIRA has been used to evaluate candied fruit processed with honey. The sweetener was isolated by water extraction, and carbon SIRA of the sweetener allowed for the differentiation of products sweetened with honey and those sweetened with cane sugar and corn syrup sweeteners (Doner et al., 1979b).

The carbon SIRA value of wheat has been determined in the course of studies of atmospheric ^{14}C (Baxter and Farmer, 1973). The carbon SIRA values of other grain carbohydrates, such as barley, oats and potatoes were measured in conjunction with studies of distilled spirits (Simpkins and Rigby, 1982). The detection of corn syrup addition to barley malt extract by carbon isotope analysis has also been studied (Krueger, 1982).

Carbon SIRA values of fruit juices were used to detect additions of cane sugar and corn syrup. It was first applied to orange and grapefruit juices (Nissenbaum et al., 1974), and subsequently to apple juice (Doner et al., 1980), grape juice (Krueger and Reesman, 1982) and to various other C-3 fruit juices as shown in Tables 2.2 and 2.3. The carbon SIRA method has been collaboratively studied, and has been adopted as an official method by AOAC International for apple and orange juices (Doner and Bills, 1982), and as a CEN method for analysis of fruit juices (Koziet et al., 1993).

The $\delta^{13}C$ values of fruit juice pulp from pulpy juices was used to refine the interpretation of carbon SIRA results (Parker, 1982; Bricout and Koziet, 1987). The pulp was found to have similar $\delta^{13}C$ values to the juice, therefore, the difference between the juice and pulp SIRA values can be used as an indicator of sugar addition. Examination of both the $^{13}C/^{12}C$ and $^{15}N/^{14}N$ ratios of citrus pulp allowed an improvement in the accuracy of origin assignment of citrus juices (Kornexl et al., 1996). The carbon SIRA value of pineapple juice was found to be different from that of most fruits (Doner et al., 1979a; Parker, 1982; Krueger et al., 1992). The Crassulacean Acid Metabolism (CAM) photosynthesis of pineapple juice results in $\delta^{13}C$ values near $-12‰$. Consequently, cane and corn sugar cannot be detected in pineapple juice, but

STABLE ISOTOPE ANALYSIS BY MASS SPECTROMETRY

Table 2.2 $^{13}C/^{12}C$ ratio of sugars and major commercial fruits

Sugar source	$^{13}C/^{12}C$ (‰)	Range	Reference
Corn syrup	−12.3	−12.2 to −12.4	Bender (1968)
	−10.4		Carro et al. (1980)
	−9.7	−9.5 to −9.8	Doner and Bills (1982)
		−9.5 to −10.7	Parker (1982)
	−13.5		Bricout and Koziet (1985)
	−10.4		Rossmann et al. (1990)
Cane sugar	−12.2		Nissenbaum et al. (1974)
	−11.1	−10.3 to −12.2	Bricout and Fontes (1974)
	−11.6	−11.1 to −12.1	Smith (1975)
	−11.2	−10.5 to −11.7	Carro et al. (1980)
	−11.4	−11.2 to −11.7	Bricout and Koziet (1985)
Apple juice	−25.4	−22.5 to −27.9	Doner et al. (1980)
	−25.6	−22.5 to −27.2	A'brams (1982)
	−24.2	−23.4 to −25.5	Lee and Wrolstad (1988)
	−25.9	−25.5 to −26.8	Rossman et al. (1990)
	−25.3	−22.7 to −27.1	Krueger (1993)
Cranberry juice	−25.0		Parker (1982)
	−26.3		Krueger et al. (1986)
	−24.8	−24.1 to −25.5	Hong and Wrolstadt (1986)
Orange juice	−24.6	−24.3 to −25.0	Nissenbaum et al. (1974)
	−24.5	−23.4 to −25.6	Doner and Bills (1981)
	−25.7	−23.5 to −26.5	Krueger and Reesman (1982)
	−25.1	−23.5 to −28.1	Bricout and Koziet (1985)
	−25.4	−24.0 to −26.2	Rossmann et al. (1990)
	−25.5	−24.6 to −26.1	Krueger (1995)
	−27.2	−27.0 to −27.4	Krueger (1995)
Grape juice	−26.7	−23.5 to −30.5	Krueger and Reesman (1982)
	−26.2		Parker (1982)
Grapefruit juice	−24.6		Nissenbaum et al. (1974)
	−25.8		Parker (1982)
	−26.1	−24.9 to −26.8	Krueger et al. (1986)
	−26.6 [sic]	−26.8 to −28.0	Rossmann et al. (1990)
	−26.1	−24.6 to −27.5	Krueger (1995)
Lemon juice	−24.1		Parker (1982)
	−25.6	−25.1 to −26.2	Krueger et al. (1986)
	−24.1	−23.6 to −25.1	Doner (1985)
	−25.4	−23.6 to −26.3	Rossmann et al. (1990)
Pineapple juice	−12.9		Doner et al. (1979)
	−12.5		Parker (1982)
	−12.2	−11.2 to −13.5	Krueger et al. (1992)
	−12.9	−11.9 to −14.3	Low et al. (1994)
Raspberry juice	−25.4	−24.6 to −26.1	Krueger et al. (1986)
	−24.1	−23.2 to −24.7	Spanos and Wrolstad (1987)
	−24.0	−21.5 to −26.1	Durst et al. (1995)
Strawberry juice	−24.3		Parker (1982)
	−24.2	−24.0 to −24.5	Krueger et al. (1986)
	−24.3	−23.1 to −25.7	Krueger (1995)

Table 2.3 Carbon SIRA values of various fruits[a]

Fruit	$\delta^{13}C$ (‰)	Fruit	$\delta^{13}C$ (‰)
Apricot	−24.1[b]	Papaya	−26.2[c]
Avocado[d]	−31.8	Papaya	−25.8
Banana[d]	−24.0	Passion fruit	−27.1
Blackberry	−26.0	Peach	−26.6
Blueberry	−26.4	Pear	−25.4
Cantaloupe	−25.5	Pear	−25.2
Carambola	−25.4	Persimmon	−26.6
Cherry	−26.3	Plum	−25.0[b]
Coconut	−21.1	Plum	−23.8
Feijoa	−25.1	Plum	−25.5
Guava	−25.9	Pomegranate	−27.8
Honeydew melon	−26.3	Prickly pear	−12.8
Kiwi fruit	−24.7	Prune	−26.0[b]
Kumquat	−26.3	Sapodilla plum	−24.3
Lime	−24.4	Tamorillo	−25.7
Lime	−23.7	Tangerine	−24.1
Lime	−24.7	Tomato	−25.4
Mango	−25.5	Watermelon	−24.6
Nectarine	−26.5		

[a] Data from Krueger et al. (1986) except and noted.
[b] Data from Parker (1982).
[c] Data from Doner et al. (1979a).
[d] Fruit flesh was analysed.

sugar beet sugar can. In addition, pineapple juice can be detected by this method in other fruits.

The carbon SIRA value of organic acids isolated from fruit juices has been used to detect additions of acidulants such as citric and malic acid to fruit juices. Early studies reported upon the $\delta^{13}C$ values of malic and tartaric acid in grapes (Di Marco et al., 1977) and malic acid in apples (Lee and Wrolstad, 1988). Isolation of citric acid from lemon juice by precipitation as calcium citrate was used to detect additions of synthetic citric acid in lemon juice (Doner, 1985). SIRA analysis of malic and citric acids isolated by HPLC from fruit juices was used to detect additions of these acids (Schmidt et al., 1993). Additions of synthetic ascorbic acid have also been detected in this manner (Gensler et al., 1995).

Carbon SIRA values have been measured of the ethanol from a variety of malt whiskies and wines in conjunction with atmospheric ^{14}C studies (Baxter et al., 1969; Baxter and Walton, 1970; Farmer et al., 1972; Baxter and Farmer, 1973). Carbon SIRA was used to detect ethanol derived from sugar cane and corn in a variety of distilled spirits (Simpkins and Rigby, 1982) and in wine (Bricout et al., 1975; Dunbar, 1982a, d). The $\delta^{13}C$ value of ethanol from beer has been reported (Schmidt, 1986). The carbon SIRA value of tequila and mescal ethanols, derived from the succulent agave, a CAM plant, was found to be near −12‰ (Winckler and Schmidt, 1980).

It was observed that the $\delta^{13}C$ value of ethanol was less useful in detecting additions of synthetic alcohol to alcoholic beverages from Calvin cycle plant sources (Rauschenbach et al., 1979). Measurement of the difference in the $\delta^{13}C$ values of the methylene and methyl groups of ethanol was more effective at identifying the presence of synthetic ethanol (Rossmann and Schmidt, 1989). $\delta^{13}C$ value of ethanol prepared by fermentation of fruit juices has been used to detect added cane and corn sugar in those juices (Martin et al., 1991). The correlation of $\delta^{13}C$ values of ethanols from various sources with deuterium NMR measurements has also been reported (Martin et al., 1991, 1996). Measurement of the $\delta^{13}C$ value of dissolved CO_2 in sparkling wines was found to be effective at detecting artificial carbonation (Dunbar, 1982a).

Carbon SIRA of vinegar has been used to differentiate wine and cider vinegars from vinegars derived from corn and synthetic alcohol sources (Schmid et al., 1978; Krueger and Krueger, 1985b; Remaud et al., 1992). The $\delta^{13}C$ results from corn-derived vinegar are considerably more positive than those of apple and grape sources, while the $\delta^{13}C$ results of many acetic acids from synthetic sources are more negative than any of the agricultural sources. The use of this method has been collaboratively studied and is now an AOAC International method for the detection of corn-derived vinegar in apple cider vinegar (Krueger, 1992).

The distribution of carbon isotopes between the two positions, methyl and carboxyl, of vinegar has been studied. Pyrolysis of sodium or calcium salts of acetic acid results in cleavage of the acetate molecule, allowing for separate measurement of the methyl and carboxyl $\delta^{13}C$ values (Meinschein et al., 1974; Schmid et al., 1981; Krueger and Krueger, 1984; Rossmann and Schmidt, 1989). This information provides an additional piece of information that assists in clarifying the origin of vinegars.

Carbon SIRA analysis of vegetable oils showed that corn oil has a $\delta^{13}C$ value near $-15‰$, while most other oils have $\delta^{13}C$ values near -26 to $-30‰$ (Gaffney et al., 1979; Winkler and Schmidt, 1980). This allows for detection of corn oil in other vegetable oils. The detection of other oils in corn oil has also been evaluated using carbon SIRA (Rossell, 1994).

The $\delta^{13}C$ value of meat was found to reflect the diet of the animals; for example, corn fed beef and pork was found to have a $\delta^{13}C$ value near $-12‰$. The $\delta^{13}C$ values of soybean and various grain protein meals were also determined. These data indicate that in some circumstances carbon SIRA can be used to differentiate the range fed from feed lot fed meat, and to detect soybean protein in meat (Gaffney et al., 1986).

Carbon SIRA of vanillin isolated from vanilla extract has been used to differentiate natural and synthetic vanillin. Vanillin is isolated from the extract by solvent extraction and purified by adsorption chromatography (Bricout et al., 1974), gas chromatography (Hoffman and Salb, 1979; Krueger and Krueger, 1983), thin layer chromatography (Martin et al., 1981), recrystallization (Krueger and Krueger, 1985a) or HPLC (Lamprecht et al., 1994). The isolated vanillin is combusted and the $^{13}C/^{12}C$ ratio of the recovered CO_2 is determined.

Vanillin from natural vanilla yields $\delta^{13}C$ values near $-20‰$, while synthetics derived from wood lignin, guaiacol and clove eugenol yielded results from -27 to $-31‰$. The results of SIRA performed on vanillin isolated from vanilla extract by different procedures have been compared (Guarino, 1982); the results from the different procedures were substantially similar.

It was observed that since natural vanillin is slightly depleted in ^{13}C relative to synthetic vanillin, a synthetic vanillin might be adjusted by the addition of ^{13}C enriched vanillin. This would allow the synthetic to mimic the $\delta^{13}C$ value of natural vanillin. The two most readily available ^{13}C enriched vanillins are enriched in the methyl and carbonyl carbons respectively. Ring labelled and uniformly labelled vanillins are more expensive to prepare.

Adjusted vanillin prepared from methyl ^{13}C enriched vanillin has been detected by cleaving the methyl group from the isolated vanillin with boron trifluoride-dimethyl sulphide complex and determining the $\delta^{13}C$ value of the resulting protocatechualdehyde (Bricout et al., 1981). The demethylated vanillins yield results similar to vanillins without isotopic adjustment. Methyl ^{13}C adjusted vanillin has also been detected by demethylation of the vanillin with hydroiodic acid and determining $\delta^{13}C$ value of the resulting methyl iodide (Krueger and Krueger, 1983). Natural vanillin yielded methyl $\delta^{13}C$ values between -23 and $-27‰$, while a methyl ^{13}C adjusted vanillin yielded a methyl $\delta^{13}C$ value of $+26‰$. Adjusted vanillin prepared from carbonyl ^{13}C enriched vanillin has been detected by first oxidizing the vanillin to vanillic acid with sodium chlorite, then decarboxylation of the vanillic acid with bromine, yielding CO_2 derived from the vanillin carbonyl position (Krueger and Krueger, 1985a). Natural vanillin yielded carbonyl $\delta^{13}C$ values between -23 and $-30‰$, while carbonyl ^{13}C adjusted vanillin yielded a carbonyl $\delta^{13}C$ value of $+17‰$.

Carbon SIRA values have been used in the authentication of natural esters used for flavour formulations. Natural ethyl butyrate from orange juice was shown to yield different results from ethyl butyrate derived from fermentation sources. The fermentation ethyl butyrate was also differentiable from synthetic sources (Byrne et al., 1986). Similar results have been reported for a variety of other natural ester materials (Culp and Noakes, 1982).

Carbon SIRA values of natural and synthetic cinnamaldehyde and oil of cassia have been reported (Hoffman and Salb, 1980; Culp and Noakes, 1990), as have those of natural and synthetic benzaldehyde and oil of bitter almond (Krueger, 1987; Culp and Noakes, 1990). $\delta^{13}C$ values of other aromatics such as methyl salicylate, anethole and vanillin have also been reported (Culp and Noakes, 1992). The determination of the origin of caffeine from various sources was accomplished through the use of carbon isotope ratios, in combination with the isotope ratios of oxygen and hydrogen (Dunbar and Wilson, 1983a). A detailed study of the $\delta^{13}C$ values of linalool and linalyl acetate from various sources has been reported (Hanneguelle et al., 1992). The carbon isotope ratios of various components isolated from orange oil have also been reported (Braunsdorf et al., 1993a).

Carbon isotope ratio measurements of individual components in complex mixtures have been accomplished through the development of a gas chromatographic interface to a ratio mass spectrometer (GC-IRMS) (Barrie et al., 1984). The mixture is separated by gas chromatography, and the column effluent directed to a furnace where the components are combusted. Carbon dioxide derived from the sample is separated from the other combustion gases in the interface, and is continuously directed into the mass spectrometer inlet. The signal peaks associated with ^{12}C and ^{13}C carbon dioxide are separately integrated over the whole peak, and the ratio of the peak areas is taken. An aliquot of a reference carbon dioxide sample is periodically allowed into the inlet as well for calibration.

The results obtained are only slightly less precise that those obtained by conventional dual inlet methods. It has also been established that methylene chloride extraction of a number of flavour compounds from fruit juice matrix followed by evaporative concentration does not result in any significant carbon isotope fractionation when the extract is subsequently analysed by GC-IRMS. This is true even for relatively volatile substances which are incompletely recovered (Dautraix et al., 1995). The GC-IRMS approach to flavour isotope ratio analysis has been reviewed (Mosandl, 1992, 1995a, b; Braunsdorf et al., 1993b).

GC-IRMS has been used for the determination of the carbon SIRA values of lactones (Bernreuther et al., 1990) and ionones (Braunsdorf et al., 1991) in flavours. The δ^{13}C values of aldehydes in orange oil have been determined by this approach (Braunsdorf et al., 1992). The method has also been used for evaluating the authenticity of peppermint oil (Faber et al., 1995) and the essential oil of *Coriandrum sativum* L. (Frank et al., 1995). GC-IRMS of the headspace vapours of alcoholic beverages has been used for the measurement of alcohol carbon SIRA values (Hener et al., 1995). The internal pattern of carbon SIRA values of components in an essential oil has been used to detect addition of foreign components to the oil (Martin, 1995; Mossandl et al., 1995).

The carbon isotope ratio of the enantiomers of chiral substances has been determined by GC-IRMS through the use of chiral capillary gc columns. The method has been used to measure the isotope ratio of linalool enantiomers (Hener et al., 1992). A multi-dimensional gas chromatographic system has been interfaced to a ratio mass spectrometer to further improve the preseparation of flavour mixtures prior to carbon isotope ratio analysis (Nitz et al., 1992).

2.3 Oxygen isotopes

Measurement of the ratio of the two oxygen stable isotopes ^{18}O and ^{16}O is another approach which has been utilized to solve some problems in food adulteration. ^{18}O is a rare isotope of oxygen, with a natural abundance of approximately 0.2 atom percent. Minor variations in ^{18}O concentration can yield some information about the origin and processing of foodstuffs. It has proved experimentally difficult to measure the ^{18}O/^{16}O ratio in most organic materials;

thus the use of oxygen isotope ratios is largely restricted to measurement of the $^{18}O/^{16}O$ ratio of water. Variations in the ^{18}O concentration of the water in foods derive largely from three sources.

The water in most foods is derived ultimately from groundwater; hence, the first source of ^{18}O variation is natural groundwater variation. Groundwater $^{18}O/^{16}O$ ratios vary with geography; water from northern and southern latitudes tend to be depleted in ^{18}O relative to more central or equatorial latitudes (Craig, 1961a). This is due to the processes of evaporation and condensation, both of which are fractionating processes. Evaporation of water tends to enrich the ^{16}O concentration in the vapour phase and to enrich the ^{18}O concentration in the residual liquid phase. As ocean water, which is relatively constant in its $^{18}O/^{16}O$ ratio, evaporates and is reprecipitated as rainwater, it tends to become depleted in ^{18}O. There is a significant temperature effect to this fractionation. As the air in the atmosphere circulates to colder northern and southern latitudes, the residual water vapour remaining after previous rainfalls is more and more depleted in ^{18}O. This results in a gradual decrease in $^{18}O/^{16}O$ ratio with increasing latitude. There is also a terrain effect, as the lower temperatures associated with the high altitudes of mountainous areas also tend to reduce the $^{18}O/^{16}O$ ratio.

A second factor is transpiration of water from plant tissue. As water diffuses across plant tissue membranes and evaporates into the air, the oxygen isotopes are fractionated, with water remaining in the tissue becoming enriched in ^{18}O relative to the source groundwater. The extent of this effect will vary with plant physiology and local climatic conditions. Finally, in the course of many commercial processes, water is evaporated from the product. This process also fractionates oxygen isotopes, leading to ^{18}O enrichment in the residual water in food. Oxygen isotope fractionation can result from other physical, chemical and biochemical processes as well, although the above factors account for most of the significant sources of the $^{18}O/^{16}O$ variations in foods.

$^{18}O/^{16}O$ ratios are normally measured using an isotope ratio mass spectrometer with the sample oxygen in the form of carbon dioxide. Water isolated from the sample is equilibrated with the oxygen in tank CO_2 under controlled conditions, and the mass 46/44 ratio of the CO_2 is measured. This ratio is compared to that of CO_2 equilibrated under the same conditions with a standard water sample with a known $^{18}O/^{16}O$ ratio. $^{18}O/^{16}O$ ratios are normally reported as parts per thousand (permil) difference between the sample ratio and the ratio of the arbitrary international standard water SMOW (Standard Mean Ocean Water) (Craig, 1961b).

$$\delta^{18}O\ (‰) = 1000 \times \left[\frac{\left(\frac{^{12}C^{18}O^{16}O}{^{12}C^{16}O_2}\right)_s}{\left(\frac{^{12}C^{18}O^{16}O}{^{12}C^{16}O_2}\right)_r} - 1 \right] \quad (2.2)$$

RSMOW = 0.0039948*.   (*Double atom ratio.)

One area where $^{18}O/^{16}O$ ratio measurement has proved useful is the detection of water added to fruit juices and other beverages. Several groups have studied the $^{18}O/^{16}O$ ratio of the water fraction of various fruits and fruit juices. The conclusions have been similar in each of these studies; water in the fruit is enriched in ^{18}O relative to the local groundwater. Studies of grapes (Dunbar, 1982b), oranges (Nissenbaum et al., 1974; Bricout et al., 1972; Hess, 1974), tomatoes (Lesaint et al., 1974) and apples (Bricout, 1973a, b; Hess, 1974) and various other fruits (Dunbar and Wilson, 1983b) have shown an enrichment in relation to local meteoric water. Since reconstituted juices will reflect the $^{18}O/^{16}O$ of the water used to dilute the concentrate, these products will yield $^{18}O/^{16}O$ values more negative than fresh juice. This concept has also been applied to detecting added water in wines (Dunbar, 1982c).

One problem exists, however, with utilizing ^{18}O data for detecting reconstituted juices. Fruit concentrates are often reconstituted in locations geographically distant from the growing region of the fruit; the relationship between the true fruit value and the water used to reconstitute the product may no longer be the same as that presented in published studies. Further complicating the interpretation of ^{18}O data of fruit juices is the observation that the results obtained for ^{18}O measurement vary with the procedure used to obtain them (Nissenbaum and Feld, 1980; Cohen and Saguy, 1984).

One application of $^{18}O/^{16}O$ ratios is of considerable utility in detecting adulterated fruit juice concentrates. It has been observed that the residual water in orange juice concentrate is greatly enriched in ^{18}O relative to the starting juice, and to the range of meteoric waters (Brause et al., 1984, 1986). Orange juices concentrated to 65° Brix have ^{18}O values in the range +10 to +15‰. Even when rediluted to retail 42° Brix product, the values of ^{18}O are still positive, while meteoric waters are almost universally negative. Since a synthetic juice must be compounded from available water supplies, the ^{18}O value of a synthetic juice will reflect the negative ^{18}O of the starting water. This provides a means of detecting adulterated orange juice concentrates formulated with beet sugar, concentrates which will not be detected by ^{13}C analysis. The physical mechanism of the effect has been studied (Yunianta et al., 1995).

It must be cautioned, however, that contrary to the assertion of the study's author, ^{18}O enriched water can be obtained economically. Discharge waters from the late stages of multistage concentrating or drying plants will be enriched in ^{18}O. Since these waters would ordinarily be regarded as wastewater, it may undoubtedly be obtained inexpensively by those formulating adulterated products. Positive ^{18}O values, although useful, cannot be regarded as proof of authenticity of orange juice concentrate.

Oxygen SIRA values have been determined for a range of organic materials. The organic material is pyrolyzed in a nickel vessel which allows diffusion of hydrogen out of the system, resulting in the sample oxygen in the form of carbon dioxide. This technique has been used to differentiate between natural and synthetic vanillin (Brenninkmeijer and Mook, 1982) and between orange juice solids and sugar beet sugar (Doner et al., 1987).

2.4 Hydrogen isotopes

Another type of isotopic analysis which has proved useful in authenticity evaluation of foodstuffs is hydrogen SIRA, the ratio of the stable isotopes of hydrogen. Natural hydrogen consists of two stable isotopes of masses 1 and 2. The heavy isotope, deuterium, is quite rare; it has a natural abundance of about 0.015 atom percent. Because of their very low mass, hydrogen isotopes are subject to much larger fractionations from a number of physical and biological processes than are the heavier isotopes of carbon or oxygen. This results in a larger diversity of natural variation in deuterium abundance in hydrogen from various sources.

The ultimate source of hydrogen in most plant tissues is groundwater. Most surface and groundwater is depleted in deuterium relative to ocean water, the ultimate source and reservoir of the earth's water. Further, similarly to oxygen isotopes, groundwater deuterium content varies with geographic latitude. Northern and southern latitudes are somewhat depleted in deuterium relative to equatorial latitudes (Craig, 1961a). It is observed that plant tissue deuterium within a species also varies due to this latitude effect.

A second factor affecting deuterium abundance is transpiration of water from plant tissue. As water diffuses across plant tissue membranes and evaporates into the air, the hydrogen isotopes are fractionated, with water remaining in the tissue becoming enriched in deuterium relative to the source groundwater. The extent of this effect will vary with plant physiology and local climatic conditions. A third factor consists of enzymatic or chemical isotopic fractionations associated with various physiological and manufacturing processes. Thus, variations in deuterium content are observed between different types of molecule within a tissue, and between products in a commercial process. Differences are also observed between different positions within a particular molecule.

Finally, the natural variation of deuterium abundance is affected by various chemical exchange processes. Hydrogen is a very labile element in many chemical environments. Oxygen bound hydrogen in particular exchanges very quickly with water. This process is formulated in the following equation:

$$HOH + ROD \underset{\text{very fast}}{\longleftrightarrow} HOD + ROH \qquad (2.3)$$

This effect tends to randomize the hydrogen isotopes within such an exchanging system. From an experimental standpoint, this exchange must be taken into account in any experimental design or theoretical prediction. Oxygen bound hydrogen will tend to mirror the isotopic composition of the last water with which it came into contact. In general, with organic substances, one can only reliably examine the relatively non-labile carbon bound hydrogen with any assurance of the results having coherent meaning.

D/H ratios are normally measured using an isotope ratio mass spectrometer with the sample hydrogen in the form of hydrogen gas. The hydrogen is normally isolated from the sample in the form of water, either directly or indirectly by combustion of the sample or a suitable derivative. The water is then converted to hydrogen by reaction with a hot metal, usually either uranium or zinc. The ratio of DH (mass 3) to HH (mass 2) is then measured with a ratio mass spectrometer. This ratio is compared to that of hydrogen obtained from a standard water sample with a known D/H ratio. D/H ratios are normally reported as parts per thousand (permil) difference between the sample ratio and the ratio of the arbitrary standard water SMOW (Standard Mean Ocean Water) (Craig, 1961b).

$$\delta D(‰) = 1000 \times \left[\frac{(DH/HH)_s}{(DH/HH)_r} - 1\right] \quad (2.4)$$

RSMOW = 0.000316*. (*Double atom ratio.)

D/H ratio measurement has proved useful in the detection of added water in fruit juices and other beverages. Several groups have studied the D/H ratio of the water fraction of various fruits and fruit juices. Similar conclusions have been reached in each of these studies, and parallel the observations regarding $^{18}O/^{16}O$ ratios; water in the fruit is enriched in deuterium relative to the local groundwater. Studies of oranges (Nissenbaum et al., 1974; Bricout et al., 1972), tomatoes (Lesaint et al., 1974) and apples (Bricout, 1973a, b) and various other fruits (Dunbar and Wilson, 1983b) have shown an enrichment in relation to local meteoric water. Since reconstituted juices will reflect the deuterium content of the water used to dilute the concentrate, these products will yield D/H values more negative than fresh juice.

As with $\delta^{18}O$ data, the use of δD values to detect added water in fruit juice is complicated by the fact that fruit juice concentrates are often reconstituted in locations geographically distant from the growing region of the fruit. The relationship between the true fruit value and the water used to reconstitute the product may no longer be the same as that presented in published studies. The D/H results obtained from fruit juice water can also be affected by the procedure used to obtain them (Nissenbaum and Feld, 1980; Cohen and Saguy, 1984).

The D/H ratio of carbon-bound hydrogen of fruit juice sugars has allowed for the detection of added sugar in fruit juices. In one approach, the exchangeable oxygen-bound hydrogen was eliminated by formation of the sugar nitrate ester derivative, replacing all of the hydroxyl groups with nitrate ester groups. Combustion of the nitrate ester yielded water; the hydrogen in this water was derived from the sugar carbon-bound hydrogen (Dunbar and Schmidt, 1984). It was observed that sugar from sugar beets yielded δD values considerably more negative than those of several fruit juices. The method was used for the detection of added beet sugar in wine (Dunbar *et al.*, 1983). The method was particularly effective for detecting beet sugar in orange juice. The mean hydrogen SIRA value for beet sugar was approximately 100‰ more negative than that of orange juice (Bricout and Koziet, 1985, 1987; Doner *et al.*, 1987). The method was also used to determine δD values of the carbon-bound hydrogen of sugars from apple, grape, grapefruit and lemon juices (Rossmann *et al.*, 1990). The method was also applied to the hydrogen SIRA of honey (Rossmann *et al.*, 1992).

Another approach to the determination of the D/H ratio of carbon-bound hydrogen from sugar involved formation of a formic acid derivative (Krueger, 1995). The sugar is selectively oxidized, using ceric ammonium nitrate in perchloric acid, to formic acid, which is isolated by steam distillation and neutralized as its calcium salt. Combustion of the resulting calcium formate yielded water, the hydrogen from which was derived from the sugar carbon-bound hydrogen. The results observed for beet sugar and sugars from orange and grape juices paralleled those of the nitrosugar technique, although the results were shifted approximately 50‰ more positive than those from the nitrosugar technique. The method was shown to be useful in detecting added beet sugar in citrus juices.

It was suggested that the determination of the D/H ratio of the methyl hydrogen of ethanol from fermentation would allow for the detection of added sugar beet sugar to fruit juices. Using chromium trioxide ethanol would be oxidized to acetic acid which would then be converted to a salt and combusted to yield water whose hydrogen derived from the methyl group (Krueger, 1988b). This approach was subsequently employed to detect the addition of beet sugar to wine musts and apple juices (Rossmann and Schmidt, 1989). This method also allowed for the detection of synthetic ethanol in wine and other fermentation alcohols.

The D/H ratio of the water derived from direct combustion of ethanol has also been used for the detection of synthetic alcohol in alcohols from fermentation sources (Bricout *et al.*, 1975; Rossmann and Schmidt, 1989; Rauschenbach *et al.*, 1979). The D/H ratio of vanillin has been used to differentiate natural vanillin from that of synthetic sources (Bricout *et al.*, 1973; Brenninkmeijer and Mook, 1982; Culp and Noakes, 1992). Natural vanillin from vanilla beans yielded δD values ranging from -50 to -115‰ and averaging -80‰, while vanillin from wood pulp lignin sources yielded δD values ranging from -170 to -205‰ and averaging -185‰. Vanillin from guaiacol sources overlapped the range of vanillin from vanilla, ranging from -15 to -90‰.

Hydrogen SIRA has been used for the detection of synthetic benzaldehyde (Krueger, 1988a; Butzenlechner *et al.*, 1989; Culp and Noakes, 1992). Natural benzaldehyde derived from bitter almond oil and apricot kernel oil yielded δD values averaging -110‰. Synthetic benzaldehyde, derived from toluene, produced two clusters of results. Those benzaldehydes produced by air oxidation of toluene yielded δD values averaging $+600$‰. Those derived from benzal chloride hydrolysis yielded results averaging −60‰. The unusually high deuterium content of the more positive synthetic benzaldehyde has been shown to be due to an extraordinary isotopic fractionation during its synthesis. All of the deuterium enrichment relative to the toluene starting material occurs at the one hydrogen position associated with the aldehyde carbon (Hagerdorn, 1992).

Synthetic cinnamaldehyde has been differentiated from cassia oil cinnamaldehyde by hydrogen SIRA (Krueger, 1988a; Culp and Noakes, 1992). Synthetic cinnamaldehyde is derived from benzaldehyde, and the D/H results observed parallel those observed for benzaldehyde. The δD values of natural cinnamaldehydes and cassia oils averaged −120‰. Synthetic samples yielded two distributions of results. One type yielded δD values near +500‰, while the other yielded δD values near −30‰.

The deuterium content of a series of natural and synthetic essential oil components, including estragole, carvacrol, linalool, mint terpenoids and eugenol has been published (Bricout et al., 1973; Bricout and Koziet, 1976). The D/H ratio of a number of linalools and linalyl acetates has been used to differentiate natural from synthetic materials (Hanneguelle et al., 1992; Culp and Noakes, 1992).

The hydrogen isotope ratios of a series of natural and synthetic flavour compounds have been presented, including acetaldehydes, amyl acetates, anetholes, ethyl acetates, ethyl butyrates, ethyl caproates, and methyl salicylates (Culp and Noakes, 1992). The δD values of natural esters and acetaldehydes were generally much more negative than their synthetic counterparts. The D/H ratios have been determined for a series of natural and synthetic phenylethanols and phenethyl acetates (Fronza et al., 1995). Natural cis-3-hexenol can be differentiated from synthetic by hydrogen SIRA (Barbeni et al., 1997).

References

A'brams, P. (1982) Abstract. Presented at AOAC 96th International Meeting, October 25–28, 1982, Washington, DC.

Barbeni, M., Cisero, M. and Fuganti, C. (1977) Natural abundance 2H nuclear magnetic resonance study of the origin of (Z)-3-hexanol. *J. Agr. Food Chem.*, **45**, 237–41.

Barrie, A., Bricout, J. and Koziet, J. (1984) Gas chromatography-stable isotope ratio analysis at natural abundance levels. *Biomed. Mass Spectrum.*, **11**, 583–8.

Baxter, M.S., Ergin, M. and Walton, A. (1969) Glasgow University radiocarbon measurements I. *Radiocarbon*, **11**, 43–52.

Baxter, M.S. and Farmer J.G. (1973) Glasgow University radiocarbon measurements VII. *Radiocarbon*, **15**, 488–92.

Baxter, M.S. and Walton A. (1970) Glasgow University radiocarbon measurements III. *Radiocarbon*, **12**, 496–502.

Bender, M.M. (1968) Mass spectrometric studies of carbon 13 variations in corn and other grasses. *Radiocarbon*, **10**, 468–72.

Bernreuther, A., Koziet, J., Krammer, G., Christoph, N. and Schreier, P. (1990) Chirospecific capillary gas chromatography (HRGC) and on-line HRGC-isotope ratio mass spectrometry of gamma-decalactone from various sources. *Z. Lebensm. Unters. Forsch.*, **191**, 299.

Braunsdorf, R., Hener, U., Lehmann, D. and Mosandl, A. (1991) Analytical differentiation of natural, fermented, and synthetic (nature-identical) aromas. Part 1. Origin specific analysis of (E)-α(β)-ionones. *Dtsch. Lebensm. Rundschau*, **87**, 277–80.

Braunsdorf, R., Hener, U. and Mosandl, A. (1992) Analytical differentiation of natural, fermentatively produced, and synthetic (nature-identical) aroma compounds. Part 2. GC-C-IRMS analysis of important flavor aldehydes – principles and applications. *Z. Lebensm. Unters. Forsch.*, **194**, 426–30.

Braunsdorf, R., Hener, U., Przibilla, G., Piecha, S. and Mosandl, A. (1993a) Anlytische und technologische Einflüsse auf das $^{13}C/^{12}C$-Isotopenverhältnis von Orangenöl-Komponenten. *Z. Lebensm. Unters. Forsch.*, **197**, 24–8.

Braunsdorf, R., Hener, U., Stein, S. and Mosandl, A. (1993b) Comprehensive cGC-IRMS analysis in the authenticity control of flavors and essential oils. *Z. Lebensm. Unters. Forsch.*, **197**, 137–41.

Brause, A.R., Raterman, J.M., Doner, L.W. and Hill, E.C. (1986). Detection of adulteration in apple juice and orange juice by a chemical matrix method. *Fluess. Obst.*, **53**, 15–16, 21–3.

Brause, A.R., Raterman, J.M., Petrux, D.R. and Doner, L.W. (1984) Verification of authenticity of orange juice. *J.A.O.A.C.*, **67**, 535–9.

Brenninkmeijer, C.A.M. and Mook, W.G. (1982) A new method for the determination of the $^{18}O/^{16}O$ ratio in organic compounds, in *Stable Isotopes* (eds H.L. Schmidt, H. Forstel and K. Heinzinger), Elsevier Scientific Publishing, Amsterdam, pp. 661–6.

Bricout, J. (1973a) Composition of stable isotopes in water of fruit juices. *Ann. Fals. Exp. Chim.*, **66**, 195–202.

Bricout, J. (1973b) Control of authenticity of fruit juices by isotopic analysis. *J.A.O.A.C.*, **56**, 739–42.

Bricout, J. (1982) Possibilities of stable isotope analysis in the control of food products, in *Stable Isotopes* (eds H.-L. Schmidt, H. Förstel and K. Heinziger), Elsevier Scientific Publishing, Amsterdam, pp. 483–92.

Bricout, J., Fontes, J.C. and Merlivat, L. (1974) Detection of synthetic vanillin in vanilla extracts by isotopic analysis. *J.A.O.A.C.*, **57**, 713–15.

Bricout, J. and Fontes, J.C. (1974) Analytical distinction between cane and beet sugar. *Ann. Fals. Exp. Chim.*, **67**, 211–15.

Bricout, J., Fontes, J.C., Merlivat, L. and Pusset, M. (1975) Stable isotope composition of ethanol. *Ind. Aliment. Agric.*, **92**, 375–8.

Bricout, J., Fontes, J.C. and Merlivat, L. (1972) Sur la composition en isotopes stables de l'eau des jus d'oranges. *C.R. Acad. Sci. Paris*, Serie D 274, 1803–1806.

Bricout, J. and Koziet, J. (1976) Détermination de l'origine des substances aromatiques par spectrographie de masse isotopique. *Ann. Fals. Exp. Chim.*, **69**, 845–53.

Bricout, J. and Koziet, J. (1985) Detection by isotopic analysis of sugar added to orange juice. *Sci. Aliments.*, **5**, 197–204.

Bricout, J. and Koziet, J. (1987) Control of the authenticity of orange juice by isotopic analysis. *J. Agr. Food Chem.*, **35**, 758–60.

Bricout, J., Koziet, J., Derbesy, M. and Beccat, B. (1981) Nouvelles possibilités de l'analyse des isotopes stables due carbone dans le controle de la qualité des vanilles. *Ann. Fals. Exp. Chim.*, **74**, 691–6.

Bricout, J. and Merlivat, L. (1971) Sur la teneur en deuterium des jus d'oranges. *C.R. Acad. Sci. Paris*, Serie D 273, 1021–3.

Bricout, J., Merlivat, L. and Koziet, J. (1973) Fractionnement des isotopes de l'hydrogène au cours de la biosynthèse de certains constituants aromatiques des végétaux. *C.R. Acad. Sci. Paris*, Serie D 277, 885–8.

Brookes, S. T., Barrie, A. and Davies, J.E. (1991) A rapid $^{13}C/^{12}C$ test for determination of corn syrup in honey. *J.A.O.A.C.*, **74**, 627–9.

Butzenlechner, M., Rossmann, A. and Schmidt, H.-L. (1989) Assignment of bitter almond oil to natural and synthetic sources by stable isotope ratio analyses. *J. Agr. Food Chem.*, **37**, 410.

Byrne, B., Wengenroth, K.J. and Krueger, D.A. (1986) Determination of natural ethyl butyrate by carbon isotopes. *J. Agr. Food Chem.*, **34**, 736–8.

Carro, O., Hillaire-Marcel, C. and Gagnon, M. (1980) Detection of adulterated maple products by stable carbon isotope ratio. *J.A.O.A.C.*, **65**, 840–44.

Cohen, E. and Saguy, I. (1984) Measurement of oxygen-18/oxygen-16 stable isotope ratio in citrus juice: a comparison of preparation methods. *J. Agric. Food Chem.*, **32**, 28–30.

Craig, H. (1961a) Isotopic variations in meteoric water. *Science*, **133**, 1702–3.

Craig, H. (1961b) Standard for reporting concentrations of deuterium and oxygen-18 in natural waters. *Science*, **133**, 1833–4.
Culp, R.A. and Noakes, J.E. (1990) Identification of isotopically manipulated cinnamic aldehyde and benzaldehyde. *J. Agr. Food Chem.*, **38**, 1249–55.
Culp, R.A. and Noakes, J.E. (1992) Determination of synthetic components in flavors by deuterium/ hydrogen isotopic ratios. *J. Agr. Food Chem.*, **40**, 1892–7.
Dautraix, S., Gerola, K., Guilluy, R., Brazier, J.-L., Chateau, A., Guichard, E. and Etievant, P. (1995) Test of isotopic fractionation during liquid-liquid extraction of volatile components from fruits. *J. Agr. Food Chem.*, **43**, 981–3.
Di Marco, G., Grego, S., Tricoli, D. and Turi, B. (1977) Carbon isotope ratios $^{13}C/^{12}C$ in fractions of field grown grape. *Plant Physiol.*, **41**, 139–41.
Doner, L.W. (1985) Carbon isotope ratios in natural and synthetic citric acid as indicators of lemon juice adulteration. *J. Agric. Food Chem.*, **33**, 770–2.
Doner, L.W., Ajie, H.O., Sternberg, L. da S.L., Milburn, J.M., DeNiro, M.J. and Hicks, K.B. (1987) Detecting sugar beet syrups in orange juice by D/H and $^{18}O/^{16}O$ analysis of sucrose. *J. Agric. Food Chem.*, **35**, 610–12.
Doner, L.W. and Bills, D.D. (1981). Stable carbon isotope ratios in orange juice. *J. Agric. Food Chem.*, **29**, 803–4.
Doner, L.W. and Bills, D.D. (1982) Mass spectrometric $^{13}C/^{12}C$ determinations to detect high fructose corn syrup in orange juice: collaborative study. *J.A.O.A.C.*, **65**, 608–610.
Doner, L.W., Chia, D. and White, J.W. (1979a) Mass spectrometric $^{13}C/^{12}C$ determinations to distinguish honey and C3 plant sirups from C4 plant sirups (sugar cane and corn) in candied pineapple and papaya. *J.A.O.A.C.*, **62**, 928–30.
Doner, L.W., Krueger, H.W. and Reesman, R.H. (1980) Isotopic composition of carbon in apple juice. *J. Agric. Food Chem.*, **28**, 362–4.
Doner, L.W., Kushnir, I. and White, J.W. (1979b) Assuring the quality of honey: is it honey or syrup. *Analytical Chemistry*, **51**, 224A–30A.
Doner, L.W. and Phillips, J.G. (1981) Detection of high fructose corn syrup in apple juice by MS $^{13}C/^{12}C$ analysis: collaborative study. *J.A.O.A.C.*, **64**, 85–90.
Doner, L.W. and White, J.W. (1977). Carbon-13/Carbon-12 ratio is relatively uniform among honeys. *Science*, **197**, 891–2.
Dunbar, J. (1982a) Use of $^{13}C/^{12}C$ ratios for studying the origin of CO_2 in sparkling wines. *Fresenius Z. Anal. Chem.*, **311**, 578–80.
Dunbar, J. (1982b) A study of the factors affecting the $^{18}O/^{16}O$ ratio of the water in wine. *Z. Lebensm. Unters. Forsch.*, **174**, 355–9.
Dunbar, J. (1982c) Oxygen isotope studies on some New Zealand grape juices. *Z. Lebensm. Unters. Forsch.*, **175**, 253–7.
Dunbar, J. (1982d) Detection of added water and sugar in New Zealand commercial wines, in *Stable Isotopes*, Analytical Chemistry Symposium Series 11 (eds H.-L. Schmidt, H. Förstel and K. Heizinger), Elsevier Scientific Publishing Company, Amsterdam, pp. 495–501.
Dunbar, J. and Schmidt, H.-L. (1984) Measurement of the 2DH/1H isotope ratios of the carbon bound hydrogen atoms in sugars. *Fresnius Z. Anal. Chem.*, **317**, 853–7.
Dunbar, J., Schmidt, H.-L. and Woller, R. (1983) Möglichkeiten des Nachweises der Zuckerung von Wein über die Bestimmung von Wasserstoffe-Isotopenverhältnissen. *Vitis*, **22**, 375–86.
Dunbar, J. and Wilson, A.T. (1983a) Determination of geographic origin of caffeine by stable isotope analysis. *Anal. Chem.*, **54**, 590–94.
Dunbar, J. and Wilson, A.T. (1983b) Oxygen and hydrogen isotopes in fruit and vegetable juices. *Plant Physiol.*, **72**, 725–7.
Durst, R.W., Wrolstad, R.E. and Krueger, D.A. (1995) Sugar, nonvolatile acid, $^{13}C/^{12}C$ ratio, and mineral analysis for determination of the authenticity and quality of red raspberry juice composition. *J.A.O.A.C.*, **78**, 1195–1204.
Faber, B., Krause, B., Dietrich, A. and Mosandl, A. (1995) Gas chromatography–isotope ratio mass spectrometry in analysis of peppermint oil and its importance in the authenticity control. *J. Essential Oil Res.*, **7**, 123–31.
Farmer, J.G., Syenhouse, M.J. and Baxter, M.S. (1972) Glasgow University radiocarbon measurements VI. *Radiocarbon*, **14**, 326–30.
Frank, C., Dietrich, A., Kremer, U. and Mosandl, A. (1995) GC-IRMS in the authenticity control of the essential oil of *Coriandrum sativum* L. *J. Agr. Food Chem.*, **43**, 1634–7.

Fronza, G., Fuganti, C., Grasselli, P., Servi, S. and Zucchi, G. (1995) Natural abundance 2H nuclear magnetic resonance study of the origin of 2-phenylethanol and 2-phenylethyl acetate. *J. Agr. Food Chem.*, **43**, 439–43.

Gaffney, J., Irsa, I., Friedman, L. and Emken, E. (1979) ^{13}C-^{12}C analysis of vegetable oils, starches, proteins and soya-meat mixtures. *J. Agric. Food Chem.*, **27**, 475–8.

Gensler, M., Rossmann, A. and Schmidt, H.-L. (1995) Detection of added L-ascorbic acid in fruit juices by isotope ratio mas spectrometry. *J. Agric. Food Chem.*, **43**, 2662–6.

Guarino, P.A. (1982) Isolation of vanillin from vanilla extract for stable isotope ratio analysis: interlaboratory study. *J.A.O.A.C.*, **65**, 835–7.

Hagerdorn, M.L. (1992) Differentiation of natural and synthetic benzaldehydes by 2H nuclear magnetic resonance. *J. Agr. Food Chem.*, **40**, 634–7.

Hanneguelle, S., Thibault, J.-N., Naulet, N. and Martin, G.J. (1992) Authentication of essential oils containing linalool and linalyl acetate by isotopic methods. *J. Agr. Food Chem.*, **40**, 81–7.

Hener, U., Braunsdorf, R., Kreis, P., Dietrich, A., Maas, B., Euler, E., Schlag, B. and Mosandl, A. (1992) Chiral compounds of essential oils. X. The role of linalool in the origin evaluation of essential oils. *Chem. Mikrobiol. Tech. Lebensm.*, **14**, 129–33.

Hener, U., Mosandl, A., Hagenauer-Hener, U. and Dietrich, H. (1995) Isopenanalyse mittels Headspace-GC-IRMS ein Beitrag zur Analytik alkoholhaltiger Getränke. *Wein Wissenschaft*, **50**, 113–17.

Hess, D. (1974) Zum Nachwies von aus Konzentrat hergestellten Säften. *Flüssiges Obst*, **41**, 235–6, 238, 240, 242.

Hillaire-Narcel, C., Carro-Jost, O. and Jacob, C. (1977) Composition isotopique ^{13}C/^{12}C du saccharose et du glucose de diverses origines et contrôle de l'authenticité des sirops et sucres d'érable. *Can. Soc. of Food Sci. and Tech. J.*, **10**, 333–5.

Hoffman, P.G. and Salb, M. (1979) Isolation and stable isotope ratio analysis of vanillin. *J. Agric. Food Chem.*, **27**, 352–5.

Hoffman, P.G. and Salb, M.C. (1980) Radiocarbon (^{14}C) method for authenticating natural cinnamic aldehyde. *J.A.O.A.C.*, **63**, 1181–3.

Hong, V. and Wrolstad, R.E. (1986). Cranberry juice composition. *J.A.O.A.C.*, **69**, 199–207.

Kornexl, B.E., Rossman, A. and Schmidt, H.-L. (1996) Improving fruit juice origin assignment by combined carbon and nitrogen isotope ratio. *Z. Lebensm. Unters. Forsch.*, **202**, 55–9.

Koziet, J., Rossmann, A., Martin, G.J. and Ashurst, P.R. (1993) Determination of carbon-13 content of sugars of fruit and vegetable juices: a European inter-laboratory comparison. *Anal. Chim. Acta*, **271**, 31–8.

Krueger, D.A. (1987) Determination of adulterated natural bitter almond oil by carbon isotopes. *J.A.O.A.C.*, **70**, 175–6.

Krueger, D.A. (1988a) Detection of synthetic flavoring materials using hydrogen stable isotope ratios, abstract. 102nd International Meeting of the AOAC, 29 August–1 September 1988, Palm Beach, Florida.

Krueger, D.A. (1988b) Application of stable isotope ratio analysis to problems of fruit juice adulteration, in *Adulteration of Fruit Juice Beverages* (eds S. Nagy, J.A. Attaway and M.E. Rhodes), Marcel Dekker, pp. 109–24.

Krueger, D.A. (1992) Stable-carbon isotope ratio method for detection of corn-derived acetic acid in apple cider vinegar: collaborative study. *J. A.O.A.C. Int.*, **75**, 725–8.

Krueger, D.A. (1993) Sample preparation bias in carbon stable isotope ratio analysis of fruit juices and sweeteners. *J. AOAC Int.*, **76**, 418–20.

Krueger, D.A. (1993). Authentication of commercial apple juice, in Juice Technology Workshop, October 18–19, 1993 (ed. D.L. Downing), Special Report number 67 of the New York Agricultural Experiment Station, Geneva, NY.

Krueger, D.A. (1995) Detection of added sugar to fruit juices using carbon and hydrogen stable isotope ratio anlysis, in *Methods to Detect Adulteration of Fruit Juice Beverages*, Volume 1 (eds S. Nagy and R. Wade), AgScience, pp. 41–51.

Krueger, D.A. (1996) Adulteration of Malt Extracts and Malt Beverages, abstract, 212th American Chemical Society National Meeting, August 25–29, Orlando, FL.

Krueger, D.A. and Krueger, H.W. (1983) Carbon isotopes in vanillin and the detection of falsified 'natural' vanillin. *J. Agric. Food Chem.*, **31**, 1265–8.

Krueger, D.A. and Krueger, H.W. (1984) Comparison of two methods for determining intramolecular ^{13}C/^{12}C ratios of acetic acid. *Biomed. Mass Spec.*, **11**, 472–4.

Krueger, D.A. and Krueger, H.W. (1985a) Detection of fraudulent vanillin labeled with ^{13}C in the carbonyl carbon. *J. Agric. Food Chem.*, **33**, 323–5.
Krueger, D.A. and Krueger, H.W. (1985b) Isotopic composition of carbon in vinegars. *J.A.O.A.C.*, **68**, 449–52.
Krueger, D.A., Krueger, R.-G. and Krueger, H.W. (1986) Carbon isotope ratios of various fruits. *J.A.O.A.C.*, **69**, 1035–6.
Krueger, D.A., Krueger, R.-G. and Maciel, J. (1992) Composition of pineapple juice. *J.A.O.A.C.*, **75**, 280–82.
Krueger, H.W. and Reesman, R.H. (1982) Carbon isotope analyses in food technology. *Mass Spec. Rev.*, **1**, 205–36.
Lamprecht, G., Pichlmayer, F. and Schmid, E.R. (1994) Determination of the authenticity of vanilla extracts by stable isotope ratio analysis and component analysis by HPLC. *J. Agric. Food Chem.*, **42**, 1722–7.
Lee, H.S. and Wrolstad, R.E. (1988) Apple juice composition: sugar, nonvolatile acid, and phenolic profiles. *J.A.O.A.C.*, **71**, 795–7.
Lesaint, C., Merlivat, M.M., Bricout, J., Fontes, J.-C. and Gautheret, R. (1974) Sur la composition en isotopes de l'eau de la tomate et du maïs. *C.R. Acad. Sc. Paris*, Series D, **278**, 2925–30.
Lindner, P., Bermann, E. and Gamarnik, B. (1996) Characterization of citrus honeys by deuterium NMR. *J. Agric. Food Chem.*, **44**, 139–40.
Lipp, J., Ziegler, H. and Conrady, E. (1988) Detection of high fructose – and other syrups in honey using high-pressure liquid chromatography. *Z. Lebensm. Unters. Forsch.*, **187**, 334–8.
Low, N.H., Brause, A.R. and Wilhelmsen, E. (1994) Normative data for commercial pineapple juice from concentrate. *J.A.O.A.C. Int.*, **77**, 965–75.
Martin, G.E., Alfonso, F.C., Figert, D.M. and Burggraff, J.M. (1981). Stable isotope ratio determination of the origin of vanillin in vanilla extracts and its relationship to vanillin/potassium ratios. *J.A.O.A.C.*, **64**, 1149–53.
Martin, G.G., Hanote, V., Lees, M. and Martin, Y.-L. (1996). Interpretation of combined 2H SNIF/NMR and ^{13}C SIRA/MS analyses of fruit juices to detect added sugar. *J. AOAC Int.*, **79**, 62–72.
Martin, G.J., Danho, D. and Vallet, C. (1991) Natural isotope fractionation in in the discrimination of sugar origins. *J. Sci. Food Agric.*, **56**, 419–34.
Martin, G.J. (1995) Multisite and multicomponent approach for the stable isotope analysis of aromas and essential oils, in *Fruit Flavors: Biogenesis, Characterization and Authentication*, ACS Symposium Series 596 (eds R.L. Rouseff and M.M. Leahy), American Chemical Society, Washington DC, pp. 79–93.
Martin, M.L., Martin G.J. and Guillou, C. (1991) A site-specific and multi-element isotopic approach to origin inference of sugars in foods and beverages. *Mikrochim. Acta*, **II**, 81–91.
McKinney, C., McCrea, J., Epstein, S., Allen, H. and Urey, H. (1950) Improvements in mass spectrometers for the measurement of small differences in isotopic abundance ratios. *Rev. Sci. Instrum.*, **21**, 724.
Meinschein, W.G., Rinaldi, G.G.L., Hayes, J.M. and Schoeller, D.A. (1974) Intramolecular isotopic order in biologically produced acetic acid. *Biomed. Mass Spec.*, **1**, 172.
Mendelsohn, D., Immelman, A.R., Vogel, J.C. and von La Chevallerie, G. (1986) Carbon-13 in natural cholesterol. *Biomed. Environ. Mass Spec.*, **13**, 21–4.
Morselli, M.F. and Baggett, K.L. (1984) Mass spectrometric determination of cane sugar and corn syrup in maple syrup by use of $^{13}C/^{12}C$ ratio: collaborative study. *J.A.O.A.C.*, **67**, 22–4.
Mosandl, A. (1992) Capillary gas chromatography in quality assessment of flavours and fragrances. *J. Chrom.*, **624**, 267–92.
Mosandl, A. (1995a) Enantioselectivität und Isotopendiskriminierung als biogenetisch fixierte Parameter natürlicher Duft- und Aromastoffe. *Lebensmittelchemie*, **49**, 130–33.
Mosandl, A. (1995b) Enantioselective capillary gas chromatography and stable isotope ratio spectrometry in the authenticity control of flavors and essential oils. *Food Reviews International*, **1**, 597–64.
Mosandl, A., Braunsdorf, R., Bruche, G., Dietrich, A., Hener, U., Karl, V., Köpke, T., Kries, P., Lehmann, D. and Maas, B. (1995) New methods to assess authenticity of natural flavors and essential oils, in *Fruit Flavors: Biogenesis, Characterization and Authentication*, ACS Symposium Series 596 (eds R.L. Rouseff and M.M. Leahy), American Chemical Society, Washington DC, pp. 94–113.

Nissenbaum, A. and Field, M. (1980) Rapid preparation of orange juice and other biological fluids for natural ^{18}O and D analysis. *Int. J. Applied Rad. Isotopes*, **31**, 127–8.
Nissenbaum, A., Lifshitz, A. and Stepek, Y. (1974) Detection of citrus juice adulteration using the distribution of natural stable isotopes. *Lebensm.-Wiss. Technol.*, **7**, 152–4.
Nitz, S., Weinrich, B. and Drawert, F. (1992) Multidimensional gas chromatography–isotope ratio mass spectrometry (MDGC-IRMS). Part A: System description and technical requirements. *J. High. Res. Chrom.*, **15**, 387–91.
Paquin, R., Bilodeau, M., Marquis, V. *et al.* (1994) Malic Acid as Internal Standard for Authentication of Maple Syrup by Isotopic Ratio Mass Spectrometry. Abstract, 108th AOAC International Meeting and Exposition, 12–15 September 1994, Portland, Oregon.
Parker, P.L. (1982) The chemical basis for the use of Carbon-13/Carbon-12 ratios to detect the addition of sweeteners to fruit juice concentrates. *Fluess. Obst.*, **49**, 692–4.
Rauschenbach, P., Simon, H., Stichler, W. and Moser, H. (1979) Vergleich der Deuterium- und Kohlenstoffe-13-Gehalte in Fermentations- und Syntheseethanol. *Z. Naturforsch.*, **34C**, 1–4.
Remaud, G., Guillou, C., Vallet, C. and Martin, G.J. (1992) A coupled NMR and MS isotopic method for the authentication of natural vinegars. *Fresnius J. Anal. Chem.*, **342**, 457–61.
Rossell, J.B. (1994) Stable carbon isotope ratios in establishing maize oil purity. *Fat. Sci. Technol.*, **93**, 526–31.
Rossmann, A., Lullman, C. and Schmidt, H.-L. (1992) Massenspektometrische Kohlenstoffe- und Wasserstoffe-Isotopoen-Verhältnismessung zur Authentizität-prüfung bei Honigen. *Z. Lebensm. Unters. Forsch.*, **195**, 307–11.
Rossmann, A., Rieth, W. and Schmidt, H.-L. (1990) Moeglichkeiten und Ergebnisse der Kombination von Messungen der Verhaltnisse stabiler Wasserstoff- und Kohlenstoff-Isotope mit Resultaten konventioneller Analysen (RSK-Werte) zum Nachweis des Zuckerzusatzes zu Fruchtsaften. *Z. Lebensm. Unters. Forsch.*, **191**, 259–64.
Rossmann, A. and Schmidt, H.-L. (1989) Nachweis der Herkunft von Ethanol und der Zuckerung von Wein durch positionelle Wasserstoffe- und Kohlenstoffe-Isotopenverhältnis-Messung. *Z. Lebensm. Unters. Forsch.*, **188**, 434–8.
Schmid, E.R., Fogy, I. and Schwarz, P. (1978) Beitrag zur Unterscheidung von Gärungsessig und Synthetischem Säuressig durch die massenspektrometrische Bestimmund des $^{13}C/^{12}$C-Isotopenverhältnisses. *Z. Lebensm. Unters. Forsch.*, **166**, 89–92.
Schmid, E.R., Grundmann, H., Fogy, I., Papesch, W. and Rank, D. (1981) Intramolecular $^{13}C/^{12}C$ isotope ratios of acetic acid and biological and synthetic origin. *Biomed. Mass. Spec.*, **8**, 496–9.
Schmidt, H.-L. (1986) Food quality control and studies on human nutrition by mass spectrometric and nuclear magnetic resonance isotope ratio discrimination. *Fresnius Z. Anal. Chem.*, **324**, 760–66.
Schmidt, H.-L., Butzenlechner, M., Rossmann, A., Schwarz, S., Kexel, H. and Kempke, K. (1993) Inter- and intramolecular isotope correlations in organic compounds as a criterion for authenticity identification and origin assessment. *Z. Lebensm. Unters. Forsch.*, **196**, 105–10.
Serra-Bonheri, J. and Ventura-Coll, F. (1995) Determination of stable isotope ratio δ^{13}C by mass spectrometry in Spanish honey. *Food Sci. Tech. Int.*, **1**, 25–8.
Simpkins, W.A. and Rigby, D. (1982) Detection of the illicit extension of potable spiritous liquors using $^{13}C/^{12}C$ ratios. *J. Sci. Food Agric.*, **33**, 898–903.
Smith, B.N. (1975) Carbon and hydrogen isotopes of sucrose from various sources. *Naturwissenschaften*, **62**, 390.
Smith, B.N. and Epstein, S. (1971) Two categories of $^{13}C/^{12}C$ ratios for higher plants. *Plant Physiol.*, **47**, 380–84.
Spanos, G.A. and Wrolstad, R.E. (1987) Anthocyanin pigment, nonvolatile acid, and sugar composition of red raspberry juice. *J.A.O.A.C.*, **70**, 1036–46.
White, J.W. (1980) High-fructose corn syrup adulteration of honey: confirmatory testing required with certain isotope ratio values. *J.A.O.A.C.*, **63**, 1168.
White, J.W. (1992) Internal standard stable carbon isotope ratio method for determination of C-4 plant sugars in honey: collaborative study, and evaluation of improved protein preparation procedure. *J. AOAC Int.*, **75**, 543–8.
White, J.W. and Doner, L.W. (1978a) Mass spectrometric detection of high-fructose corn syrup in honey by use of $^{13}C/^{12}C$ ratio: collaborative study. *J.A.O.A.C.*, **51**, 746–50.
White, J.W. and Doner, L.W. (1978b) The $^{13}C/^{12}C$ ratio of honey. *J. Apic. Res.*, **17**, 94–9.

White, J.W. and Robinson, F.A. (1983) $^3C/^{12}C$ ratios of citrus honeys and nectors and their regulatory implications. *J.A.O.A.C.*, **66**, 1–3.

White, J.W. and Winters, K. (1989) Honey protein as internal standard for carbon stable isotope ratio detection of adulteration of honey. *J.A.O.A.C.*, **72**, 907–11.

Winkler, F.J. (1984) Application of natural abundance stable isotope mass spectrometry in food control, in *Chromatography and Mass Spectrometry in Nutrition Science and Food Safety* (eds A. Frigerio and H. Milon), Elsevier, pp. 173–90.

Winkler, F.J. and Schmidt, H.-L. (1980) Scope of the application of ^{13}C isotope mass spectrometry in food analysis. *Z. Lebensm. Unters. Forsch.*, **171**, 85–94.

Yunianta, Zhang, B.-L., Lees, M. and Martin, G.J. (1995) Stable isotope fractionation in fruit juice concentrates: application to the authentication of grape and orange products. *J. Agric. Food Chem.*, **43**, 2411–17.

Ziegler, H., Lüttge, U. and Stichler, W. (1979a) $\delta^{13}C$ and δD values in the organic material of some nectars. *Naturwissenschaften*, **66**, 580–81.

Ziegler, H., Stichler, W., Maurizio, A. and Vorwohl, G. (1977) Use of stable isotopes for the characterization of honeys, their origin and adulteration. *Apidologie*, **8**, 337–47.

Ziegler, H., Maurizio, A. and Stichler, W. (1979b) The characterization of honey samples according to their content of pollen and of stable isotopes. *Apidologie*, **10**, 301–11.

3 Nuclear Magnetic Resonance spectroscopy
I.J. COLQUHOUN and M. LEES

3.1 Introduction

There has been interest in using NMR spectroscopy for the analysis of foods since the early days of the technique. When applied to mixtures, it has the advantage that a very wide range of compounds can be detected in a single experiment. If, for example, the NMR spectrometer is tuned to the hydrogen resonance frequency all the hydrogen-containing compounds in the mixture are, in principle, observable; if tuned to carbon, all the carbon-containing compounds should similarly be observable. Another fundamental property is that the intensity of each NMR signal (measured by the peak area) is dependent on the number of nuclei giving rise to the signal. This means that if the experimental conditions are properly controlled, the relative amounts of different compounds within a mixture can readily be determined, as can the absolute amounts of compounds in different mixtures. Added to these features is the well-known power of NMR as a structure determination method. If the mixture contains unknown compounds, the positions and splitting pattern of the NMR signals may be sufficient to identify the compound on the basis of the NMR evidence alone, i.e. without recourse to standard compounds.

Despite this apparently highly favourable set of circumstances, NMR has not been widely used in food authenticity investigations. Apart from several references to the SNIF-NMR® technique, the volume 'Food Authentication' (Ashurst and Dennis, 1996) makes no mention of the use of conventional (i.e. ^{1}H or ^{13}C) NMR spectroscopy. Nevertheless, potential applications of NMR to many of the commodities discussed in that volume can be foreseen: to fruit juices, wines, vegetable oils, honey, coffee and milk products. The principal reasons for the current neglect are the perceptions that the technique is expensive, that it has inherently low sensitivity, and that there are insuperable difficulties associated with the interpretation and assimilation of very large numbers of complex spectra. As with most techniques capital expense and sensitivity are inextricably linked, in this case because the most obvious way of increasing the sensitivity is to use very high-field (and costly) superconducting magnets. Signal intensity increases with the factor $B_0^{3/2}$, where B_0 is the magnetic field strength (Gunther, 1995). In practice, most NMR food authentication studies to date have been carried out using medium to high-field magnets with a ^{1}H operating frequency of 400, 500 or 600 MHz. While the initial capital investment is undoubtedly high, it can be justified, as shown by the success of SNIF-NMR®, if it is the only means of obtaining the required results. Even when alternative methods are available, the initial expense may still be justified by increased sample throughput, less demanding sample preparation requirements, reduced labour and running costs, increased reliability, and the adaptability of the technique to new problems.

The other question raised above was the possibility of exploiting all the information potentially available from, say, a hundred spectra of superficially similar samples, when each spectrum contains perhaps hundreds of lines. Fortunately the answer is yes. The same problem is faced by many researchers using NMR, in medicine and biochemistry as well as in food science. The solution seen by all these groups of workers lies in the application of chemometrics to NMR data. The advent of modern desk-top computers combined with decreasing costs of mass data storage, has made chemometrics a very powerful and much more widely accessible tool for a diverse range of tasks: computer-aided medical diagnosis, automatic classification of food samples and the quantification of individual compounds in multi-component mixtures.

3.2 Practical considerations

3.2.1 *Background*

Samples for routine high-resolution NMR spectroscopy should be liquids. Best results are obtained if the sample is free of any solid particles although certain fine dispersions may be acceptable. In general any solids present will not contribute to the spectrum, they will only degrade the resolution and sensitivity of signals from components in solution. The term 'high resolution' is used to distinguish the narrow line spectra given by highly mobile molecules in the liquid or solution state from the broad line featureless spectra of typical crystalline or amorphous solids. For the purposes of this chapter, we will mainly be concerned with low molecular weight species, typically compounds such as mono- and oligosaccharides, organic and amino acids, phenolics, mono-, di- and triglycerides. Larger molecules, for example sterols, may also contribute identifiable signals to the spectra but macromolecules such as proteins and most polysaccharides will provide only a broad background signal. In biological samples like blood plasma the protein background gives the spectrum an irregular rolling baseline making it difficult to integrate signals from low molecular weight metabolites. This rolling baseline is not a feature of most of the spectra of food samples discussed here.

3.2.2 *Sample preparation*

Samples can be examined in water or in organic solvents. Volumes required are $c.$ 0.5 ml for 5 mm NMR tubes (the standard for ^{1}H-NMR experiments) or 3 ml for 10 mm tubes. Deuterated solvents are used to provide a deuterium NMR signal for magnetic field stabilization (field-frequency lock) and to allow optimization of the resolution (magnetic field homogeneity) for each sample. Fruit juices can be examined directly followed addition of 5–10% D_2O but in most circumstances it is better to dissolve freeze-dried samples in D_2O rather than H_2O since this minimizes the residual NMR signal from the water. It is also usual to add a measured amount of a reference compound. This serves as both a chemical shift and intensity reference and is also useful as a check that the resolution does not change during a series of measurements. Suitable reference compounds are TMS (tetramethylsilane) for organic solvents and TSP-d_4 (sodium salt of 3-(Trimethylsilyl)propionic-2,2,3,3-d_4 acid) or DSS (sodium 2,2-dimethyl-2-silapentane-5-sulphonate) for aqueous solutions. TMS is not a particularly good intensity reference because of its volatility. In aqueous food samples the chemical shifts of many compounds are pH sensitive. It is therefore advisable to adjust the pH of all samples within a given set to the same value. This is important if it is intended to use some automated form of data analysis. For oils, the samples in a set should all have the same concentration since ^1H and ^{13}C chemical shifts and line-widths can be concentration sensitive.

Fourier transform NMR. Readers who are unfamiliar with modern experimental techniques, or with terms such as rf (radio-frequency) pulse, chemical shift, coupling constant and relaxation time are encouraged to consult one of the standard texts on NMR (e.g. Derome (1987) or Gunther (1995)). We will only consider certain experimental aspects related to sensitivity, quantification and reproducibility. These aspects assume particular importance when it is required to make quantitative or semi-quantitative comparisons between NMR spectra given by large numbers of basically similar samples. All the NMR spectra to be discussed were obtained using Fourier transform (FT) NMR techniques. In the simplest and most widely used version of the experiment the NMR signal (free induction decay, FID) excited by a single rf pulse is sampled at regular time intervals, digitized and stored in a computer. After an optional delay the whole sequence is repeated, the new signal is added to the existing data in memory and the procedure is continued until n scans have been acquired. Signal to noise increases as $\sqrt{n}$, i.e. as the square root of the total time for the experiment. If a signal is to be quantified it should have an intensity that is at least ten times the standard deviation of the noise (Martin, 1995).

3.2.3 *Acquisition time*

The various time periods within a sequence – the pulse length, data acquisition time and delay time – are chosen to optimize the S/N that can be achieved within certain constraints imposed by the quantitative requirements. The S/N in the frequency domain spectrum may be enhanced by multiplying the time domain data by an appropriate 'window' function before the FT is carried out. One commonly applied window function is an exponential with the same decay time constant as the time domain NMR signal. This gives optimum S/N but with a doubling of the natural line-width of the signals (assuming a uniform line-width). Digital resolution is the separation in hertz between consecutive data points in the spectrum. It must be small relative to the signals' line-width if the signals are to be integrated properly. This imposes a minimum value on the acquisition time since digital resolution is simply (acquisition time)$^{-1}$. The acquisition time [number of data points × (spectral width)$^{-1}$] is typically in the range 1–5 s.

3.2.4 Quantitative NMR

For a single scan, the maximum NMR signal is obtained with a 90° rf pulse. The 90° pulse is so called because its duration (a few microseconds) is set to rotate or 'flip' the nuclear magnetization from the z-axis, where it lies parallel to the magnetic field direction, into the x–y plane where the magnetization gives rise to the detected NMR signal. When many pulses are applied, however, a stronger final signal is often obtained if shorter pulses are applied. For conventional quantification experiments with 90° pulses there should be a minimum delay time of $5T_1$ between pulses, where T_1 is the longest relaxation time found in the sample. This delay is required to obtain the proper relative intensities among the different signals in the spectrum. It is likely to be substantially longer (10–30 s might be typical for protons) than the time actually needed to acquire the data and therefore reduces the number of scans that can be acquired in a fixed experiment time. The delay may be shortened if pulses of less than 90° are used or if quantification is to be achieved by calibration methods using similar samples of known concentration and identical experimental conditions. In quantitative work it is also important that the pulse power should be sufficient to give the same flip angle over the whole of the spectral range. In other words signals from equivalent numbers of protons (or carbon-13 nuclei etc.) should have the same intensity whether their signals lie close to the pulse frequency or at the limits of the spectral range. Modern FT NMR spectrometers meet this requirement under standard operating conditions for ^{1}H and ^{13}C nuclei.

3.2.5 Dynamic range and water suppression

The ability of FT NMR spectrometers to faithfully record weak signals in the presence of strong ones is determined by the dynamic range of the instrument. The critical feature is the word length of the analogue to digital converter. Standard 12-bit ADCs (dynamic range ±2047) are being superseded by 16-bit (±32767) units in modern spectrometers and the dynamic range is even further increased by use of the oversampling technique. These recent developments (Hull, 1994) allow the improved sensitivity of high-field spectrometers to be fully realized. The dynamic range problem is most acute in the case of ^{1}H–NMR of aqueous samples where the H_2O signal is many times stronger than all the signals of interest. Many methods of suppressing the water signal have been devised, but the simple presaturation method is still the one most commonly applied. Saturation of the water signal is achieved by applying low power selective rf irradiation at the water frequency during the delay time. The irradiation is switched off immediately before the rf excitation pulse and it remains off during data acquisition. A typical delay time would be 2–3 s for an acquisition time of 1–2 s. This simple procedure is effective in reducing the water signal to below, for example, the height of the sugar signals in fruit juice samples. It is inevitable, however, that signals in the vicinity of the water will be somewhat reduced in intensity. Poor suppression of the water signal can leave an irregular hump or sloping baseline. The quality of spectrum that can be obtained depends very much on the field homogeneity (shim settings) and the type of probe used. Much better results are obtained with proton dedicated or inverse probes than with multinuclear probes which use the ^{1}H decoupler coils for excitation and detection. On modern spectrometers improved water suppression may be achieved with the three pulse NOESY-presat pulse sequence (Hull, 1994), and this is now used routinely for aqueous biological and food samples.

3.2.6 *Detection limits*

Compounds with concentration >10 μM are detectable in 'standard' ^{1}H-NMR experiments on biofluids (Nicholson and Wilson, 1989). A similar, or possibly slightly lower, detection limit should apply to liquid foods since the background is generally less troublesome. 'Standard' conditions implies an operating frequency of about 500 MHz and total acquisition time of 15–20 min. Of course the exact detection limit will depend on the NMR characteristics of the compound, whether the detected resonance is from a CH_3, CH_2 or CH group, whether it is a singlet or multiplet, and whether it overlaps with signals from other compounds. An important feature in the NMR spectra of some foods is the very great difference in concentration between major and minor components. It is difficult to detect a minor compound in a standard 1D (one-dimensional) ^{1}H-NMR spectrum if all its resonances lie in the same range as the signals from one of the major components. Overlap is much less of a problem in ^{13}C-NMR since, with proton decoupling, signals from all carbons appear as singlets and the effective chemical shift range is 20 times greater than for protons. However ^{13}C-NMR is more than 5000 times less sensitive than ^{1}H-NMR so its routine use for authentication is limited to strong samples such as oils, extracts or concentrates.

3.2.7 *Signal assignments*

Signal assignments are made by several methods. Extensive literature compilations of chemical shifts and signal multiplicities are available for commonly occurring metabolites (e.g. Fan, 1996). While these are very useful for samples like fruit juices, much of the data has been gathered for studies on biofluids, and there are still gaps in the information for food components. Thus the existing data need to be supplemented by acquiring spectra of reference compounds that are known, from conventional analytical studies, to be present in the samples. When this is done it is advisable to make the sample conditions (especially pH) as near as possible to those actually used for spectra of the real samples. A complementary and extremely powerful approach is to assign the signals directly using 2D (two-dimensional) NMR experiments (Croasmun and Carlson, 1994). The basic methods are ^{1}H COSY (intramolecular HH correlation via the spin coupling) and 'proton-detected' ^{1}H/^{13}C chemical shift correlation (via the one-bond HC spin coupling). These 2D NMR methods remove most of the ambiguities that arise in assignment of 1D spectra and also reveal many of the signals that are hidden by the spectral overlap mentioned above. At present 2D NMR is generally used for signal assignment and compound identification in only a few representative food samples out of a set. Current developments in instrumentation and computing suggest that its routine use may become much more common.

3.2.8 *Data preparation for multivariate analysis*

Techniques of multivariate data analysis (MVA) are being used increasingly for classification of samples on the basis of their NMR spectra. The spectra are generally transferred from the spectrometer's own computer to a PC or workstation where the MVA is carried out using commercial or 'in-house' software. Researchers have developed several independent approaches for conversion of the 'raw' NMR spectra into a form suitable for MVA, depending somewhat on the make of spectrometer used and the software supplied by the manufacturers. One procedure (Belton *et al.*, 1997b) is to Fourier transform the FIDs using an appropriate line broadening factor, phase and baseline correct the frequency domain spectra, reduce spectra to the real part, and align all the spectra by right or left shifting so that the peak of the reference signal is always in the same data channel. The spectra are stored as ASCII files and then transferred to a PC where the number of variables (data points) may be reduced before MVA by averaging neighbouring data points. A rather similar approach to data reduction is to divide the spectra into segments (called 'buckets') and to sum the intensities of the data points in each segment (Holmes *et al.*, 1994; Spraul *et al.*, 1994). The most popular alternative has been to employ the spectrometers' own 'peak-picking' routines which give peak positions and heights. One can then take all peaks above a certain threshold (Vogels *et al.*, 1993), or again divide the spectra into segments, taking for the intensity in each segment just the highest peak (Howells *et al.*, 1992) or the sum of peak heights (Spraul *et al.*, 1994). Spectra have been normalized by two fundamentally different methods: by setting the intensity of the reference signal to unity in every spectrum (same amount of reference material added to all samples) (Howells *et al.*, 1992) or by setting the total intensity of each spectrum (excluding solvent and reference peaks) to one (Vogels *et al.*, 1993).

A special form of data preparation for NMR spectra, called partial linear fit (PLF) has been proposed (Vogels et al., 1993) and discussed in detail (Vogels et al., 1996b). In contrast to the simple alignment procedure mentioned above which uses a single reference peak and affects all signals equally, PLF aims to eliminate inter-sample chemical shift displacements for individual signals. These displacements, resulting from small pH differences, concentration effects etc., give otherwise identical protons slightly different chemical shifts in different samples, and are an unhelpful source of additional variance. When a set of samples of known origin is available the number of variables (data points or peaks) per spectrum may be reduced by a weighting and selection procedure. Samples are assigned to the different classes and variable weighting factors are calculated according to the ratio (interclass variance/intraclass variance). Only those variables with a high ability to discriminate are retained in the MVA (Vogels et al., 1993; Holmes et al., 1994; Forveille et al., 1996).

The first step of the MVA is usually principal components analysis (PCA). PCA entails the computation of eigenvectors of a matrix $(\mathbf{X}^T\mathbf{X})/(n-1)$, where $\mathbf{X}$ contains n observations (spectra of n samples, with data prepared as above) entered into the matrix row-wise. $\mathbf{X}^T$ is the transpose of $\mathbf{X}$. Different variations of PCA can be performed, by varying the nature of the data in $\mathbf{X}$. If $\mathbf{X}$ is mean-centred (column means subtracted), then $(\mathbf{X}^T\mathbf{X})/(n-1)$ is a covariance matrix: this form of PCA is called the 'covariance method'. If the data in $\mathbf{X}$ are standardized (mean-centred and columns scaled to unit variance), then $(\mathbf{X}^T\mathbf{X})/(n-1)$ is a correlation matrix, and this form of PCA will be termed the 'correlation method'. An advantage of the covariance method is that the eigenvectors (or 'loadings') retain the scale of the original data, and may be interpretable in chemical terms. For this reason the covariance method has been favoured by NMR spectroscopists. Although the loadings obtained by the correlation method are usually very unfamiliar in appearance, an advantage of this approach is that the PCA is influenced by all spectral features equally, whereas in the covariance method, larger signals tend to dominate. Consequently, the correlation method can be useful when variations in minor constituents, with small spectral contributions, are of primary interest. An example of the use of both methods is given below.

3.3 Applications

3.3.1 *Fruit juices and fruits*

Any assessment of the potential of high resolution NMR spectroscopy for fruit juice authentication requires a knowledge of the compounds likely to be detected within a reasonably short acquisition time. 600 MHz ^{1}H-NMR spectra of several types (grape, apple, pineapple, orange, grapefruit) of fruit juice were recorded with a typical acquisition time of about 10 min (Belton et al., 1996). These spectra showed obvious differences between the fruits although there were numerous signals common to all the juices. 2D NMR was used to assign many of the weak and overlapped signals in the 1D spectrum of grape juice, especially those compounds with signals in the region 0.5–2.6 ppm. A successful semi-quantitative comparison was made of compound concentrations estimated from the NMR signal intensities with average concentrations published in the literature.

Features that are common to spectra of most fruit juices are illustrated by the 600 MHz ^{1}H spectra of apple juices (Belton *et al.*, 1997a) shown in Figs 3.1 to 3.3. It is convenient to divide the spectra into three regions, partly because each

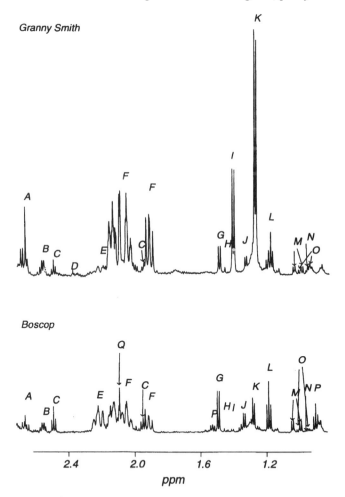

Figure 3.1 600 MHz ^{1}H-NMR spectra of two freshly squeezed apple juices: high-field region. Spectra plotted with a vertical gain of $c.$ 100 with respect to Fig. 3.2. Key to assignments: A, succinic ac.; B, glutamic ac.; C, γ-amino butyric ac.; D, proline (or hydroxy methyl proline); E, chlorogenic ac.; F, quinic ac.; G, alanine; H, citramalic ac.; I, unknown; J, lactic ac.; K, unknown; L, ethanol; M, valine; N, leucine; O, isoleucine; P, propanol; Q, acetic ac. (Belton *et al.*, 1997a).

44 ANALYTICAL METHODS OF FOOD AUTHENTICATION

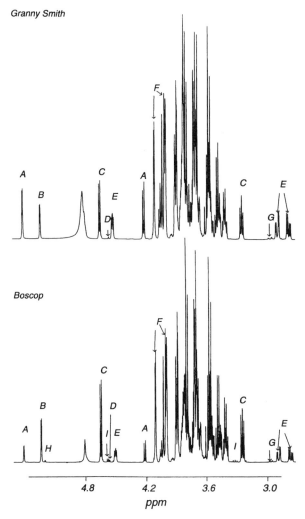

Figure 3.2 600 MHz ^{1}H-NMR spectra of freshly sqeezed apple juices (same samples as Fig. 3.1): mid-field region. Key to assignments: A, sucrose; B, α-glucose; C, β-glucose; D, tartaric ac.; E, malic ac.; F, fructose; G, asparagine; H, α-galactose; I, β-galactose. In the main carbohydrate region only a few well resolved multiplets have been labelled. (Belton *et al.*, 1997a).

region contains readily identifiable signals from different types of compound, and partly to display all the spectral information given the very large variation in signal intensities between regions. The reader should note the great increase in vertical gain used to plot the regions in Figs 3.1 and 3.3 compared with Fig. 3.2. Specific signal assignments are given in the captions to each figure.

In the high-field region (Fig. 3.1) the main contributors are quinic and chlorogenic acids, amino acids and alcohols (ethanol and propanol). In the mid-field region (Fig. 3.2) the major signals are from sucrose, glucose and fructose

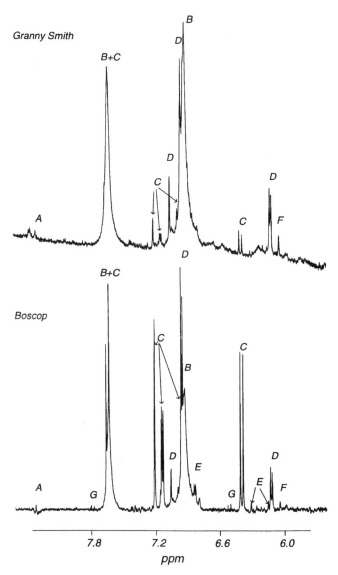

Figure 3.3 600 MHz ^{1}H-NMR spectra of freshly squeezed apple juices (same samples as Fig. 3.1): low-field region. Spectra plotted with a vertical gain of c. 350 with respect to Fig. 3.4. Key to assignments: A, formic ac.; B, polyphenols; C, chlorogenic ac.; D, epicatechin; E, phloridzin; F, phloretin; G, p-coumaric ac. (Belton et al., 1997a).

with overlapping signals covering the range from 5.4 to 3.2 ppm. Only a few of the non-overlapping sugar signals have been listed in the caption. Some minor sugars (e.g. galactose) with resolved multiplets can also be identified from the 1D spectrum. It is difficult, however, to find any signals from sorbitol, a compound with a typical concentration in apple juice which should be far above the detection limit. This is because all signals from sorbitol lie in the range 3.6–3.9 ppm, a region dominated by signals of the three main sugars. Nevertheless, the presence of sorbitol will be reflected in the overall intensity pattern and its influence should be detectable by multivariate analysis of a group of spectra. The second important feature of the mid-field region is a multiplet at ~2.8 ppm arising from the CH_2 group of malic acid. In apple juice this multiplet overlaps with another one from the most significant amino acid, asparagine. In spectra of citrus juices, raspberry and strawberry purées, the characteristic signals (two doublets) of citric acid appear, again at ~2.8 ppm. Citrate and malate provide the most prominent signals with chemical shifts affected by the pH.

The low-field region spectra (Fig. 3.3) contain signals from aromatic protons. The spectra of most fruits have two broad signals (at 6.9 and 7.6 ppm) assigned to condensed polyphenol species, as well as various sharper lines which can be assigned to phenolics (e.g. chlorogenic acid, epicatechin), aromatic amino acids and other acids (formic, shikimic). The phenolics in particular are reactive species which start to be transformed in enzymic oxidation reactions as soon as the fruits are pressed. The resulting chemical changes are of course reflected in the spectra. One should also be aware of other changes that can take place during storage of the samples with effects on the appearance of spectra. These include acid hydrolysis of sucrose and microbiological activity resulting in production of ethanol, lactate or acetate.

Differences in the spectra of the two apple varieties shown, Granny Smith and Boscop, suggested that NMR might be capable of distinguishing individual fruit varieties as well as different types of fruit. To test this a multivariate analysis was carried out on a set of 500 MHz ^1H spectra of apple juices from three different varieties: Spartan, Bramley and Russet (Belton et al., 1997). Apples grown in the same year but in different English locations were collected directly from the producers. The analysis was carried out on the regions of Fig. 3.1 and Fig. 3.2 taken together ('full spectrum') and on the two regions taken separately. The low-field region containing the aromatic peaks and the data points in the vicinity of the water signal were not used. The first step, a PCA, was needed to reduce the dimensionality of the data since the number of variables (860 data points for each full spectrum) was vastly in excess of the number of samples (26). 'Covariance' and 'correlation' PCA methods were compared in view of the large range of signal intensities encountered in the spectra. The PCA is a linear transformation that provides a new description of the data in terms of the PC 'scores' of each sample referred to a set of PC axes. The result of the transformation is that the essential variation between samples is now expressed in terms of a much smaller number of uncorrelated variables. The relationship between original and new variables is seen in the PC loadings. When the covariance PCA method is used, an examination of the loadings (which resemble the original NMR spectra) can show which chemical species are responsible

NUCLEAR MAGNETIC RESONANCE SPECTROSCOPY 47

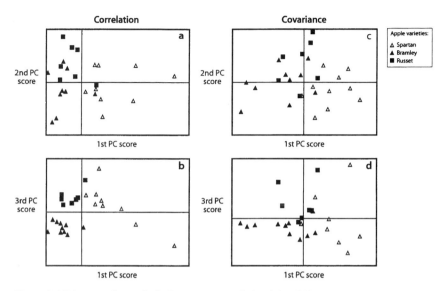

Figure 3.4 PC scores from principal component analysis of the 'full' spectrum (4.55–0.8 ppm) – apple juices: (a), (b) correlation method; (c), (d) covariance method. (Belton et al., 1997.)

for the spread of samples in PC space. Figure 3.4 shows plots of the PC scores for the most important PCs and compares results of covariance and correlation PCA methods applied to the full spectrum. Figure 3.4 provides clear indications of a tendency for samples of a given variety to cluster in the PC space, but the plots can only show that part of the total information contained in the first three PCs. In fact the first five PCs explained over 95% of the total variance for the covariance method, but 15 PCs were needed to reach this figure with the correlation method. This reflects the greater emphasis given to variations in the concentration of minor components when the latter method is used.

When used in conjunction with spectroscopic data an important purpose of the PCA is to reduce the number of variables to less than the number of samples. This allows further multivariate methods such as linear discriminant analysis (LDA) and canonical variates analysis (CVA) to be applied (Belton et al., 1997b). While the two-dimensional plots of Fig. 3.4 give a useful visual impression, the likely success rate of a multi-dimensional PC-based model in correctly assigning future unknown samples is more effectively gauged by LDA. In LDA the samples of known origin are first assigned to groups (here, according to cultivar). The positions of the group means, in the multi-dimensional PC space are then calculated from the scores using a chosen number of PCs. Using the squared Mahalonobis distance as a criterion, the nearest group mean to each sample is then identified, the samples are reclassified into groups according to this criterion and the number of correct assignments is determined. This procedure allows an objective comparison of the success rate achieved in classification, not only between covariance and correlation methods, but also between data matrices based on the full spectrum and on the two regions taken separately. Figure 3.5 shows that a high success rate is achieved when five PCs are used in the model, whichever region or method is taken. The PC loadings confirmed that variations in the amounts of the major components (sugars, malic acid) were important for the discrimination. It is emphasized, however, that both the PCA and the LDA on the high-field region alone showed that discrimination was possible on the basis of variations in the minor components (amino acids, etc.). CVA is a data rotation method which involves finding a linear combination of a subset of the PC scores which maximizes the ratio of between-groups to within-groups variance. It simultaneously maximizes the distances between the groups and minimizes the scatter within each group, providing an optimal (two-dimensional) impression of the capacity of the data to discriminate between groups. Figure 3.6 shows its application to the apple juice data. To avoid over-fitting in both LDA and CVA the number of PCs should be substantially fewer

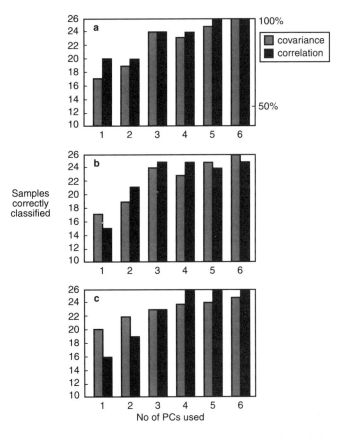

Figure 3.5 Success rate of discriminant analysis procedure for apple juice varieties employing different spectral regions and PCA methods: (a) 'full spectrum' (4.55–0.8 ppm); (b) region 1 (4.55–2.5 ppm); (c) region 2 (2.5–0.8 ppm).

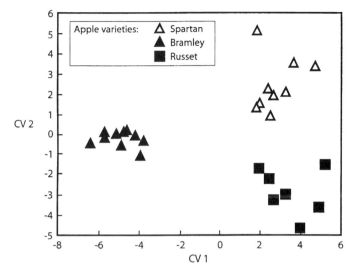

Figure 3.6 CV scores from canonical variates analysis (full spectrum, correlation method, first 5 PCs used) – apple juices. (Belton et al., 1997b.)

than the total number of samples and preferably smaller than the size of the smallest group. Both training and test sets should be available to test the validity of any proposed model. The example of the apple juices has been chosen to illustrate the possible approaches, but certainly the number of samples and varieties needs to be increased to demonstrate the true potential of the methods.

Similar pattern recognition techniques have been used in conjunction with ^{1}H NMR to detect adulteration of orange juices (Vogels et al., 1996a). The experimental approach was slightly different in that samples were freeze-dried and then dissolved in D_2O. Two groups of samples were investigated. In the first case the investigators added known amounts of sucrose, beet medium invert sugar or sodium benzoate to two authentic juices. 1–10% (weight percentage with respect to the original juice) of the chosen adulterant was added. The second set consisted of 13 different authentic juices and another 13 samples which had been found to be suspect by independent analytical methods (pulp wash, low amino acid content, presence of D-malic acid, low proline content, addition of grapefruit, low potassium content). 400 MHz ^{1}H spectra were measured in triplicate and the authors applied a partial linear fit procedure to remove small inter-sample chemical shift variations attributable to pH and other experimental uncertainties. Following PCA it was found that scores on PC1 and PC2 could separate all samples with added sugar from pure juices, although the first two PCs only explained 34% of the variance. The sample with added sodium benzoate was not discriminated by the first two PCs although a subsequent DA procedure on the PC scores did separate it. Some explanation of the chemical factors responsible for the discrimination was given from examination of the loadings. For the second set of samples, the plot of PC1 vs PC2 (29% of variance) revealed three clusters consisting of the authentic juices, the pulp washes and the other suspect samples (as a single group). The K-nearest neighbour classification technique showed that just the first two PCs were needed, despite the low proportion of variance explained, and that addition of further PCs to the model did not improve the result. It should be noted that neither the presence of D-malic acid nor a low potassium content would be picked up directly by ^{1}H NMR so that accompanying factors must have been responsible for the discrimination. Again, a discussion of factors responsible was given, but this area of interpretation is one where both a more systematic mathematical approach and a greater knowledge of signal assignments are desirable.

A somewhat different use of the NMR-multivariate analysis approach is illustrated by a study that aimed to differentiate between grapevine cultivars and their clones (Forveille et al., 1996). There is no established method of determining the identity of different clones of a particular cultivar. The study involved three cultivars (Cabernet Franc, Cabernet Sauvignon and Merlot Noir) with several clones of each from two geographical sites. The samples examined were extracts of polyphenols from the grape seeds and leaves. Inverse ^{1}H/^{13}C 2D NMR experiments were employed in order to obtain both ^{1}H and ^{13}C chemical shifts. Experiments which gave long range (through 2- or 3-bonds) H—C correlations were carried out as well as conventional experiments to establish 1-bond correlations. The chemical structure of the polyphenols, with many quaternary carbons, makes the long-range experiment particularly useful. Each experiment, however, takes 2–3 h to perform at present, in contrast to the much shorter times needed for simple 1D ^{1}H-NMR experiments. The data points for analysis are cross-peaks in the 2D spectrum, characterized by both a ^{1}H and ^{13}C chemical shift, and with intensities given by the cross-peak volume integrals. About 50–100 cross-peaks were measured in a typical experiment but a reduced number of variables (about 10 for the 18 sample grape seed set) was used in the multivariate analysis. These variables were selected to be the most discriminating ones by calculating an F factor (ratio of 'inter-group' to 'intra-group' variance) for all the variables from prior knowledge of the samples' identities. Exploratory data analysis showing clear evidence of grouping was carried out with PCA and HCA (hierarchical cluster analysis) on the reduced set of variables. FDA (factorial discriminant analysis), which involves assignment of samples to their known groups, was then applied to give a model for testing of future unknown samples. The model was tested by cross-validation and jack-knifing procedures. Excellent discrimination between cultivars was achieved using the first two axes (FD1 and FD2). Within a given cultivar FD1 and FD2 also provided differentiation between clones (repeat measurements) and, in the one cultivar where it was tested, clustering of different clones by geographical origin was evident. For the present, this approach remains experimentally demanding, but given future technical advances, it may prove to be very powerful for a wide range of fruits.

Apart from the example given above, ^{13}C-NMR has not been used directly for authentication of fruits and fruit juices. However ^{13}C spectra of a number of juices and concentrates have been measured (Rapp et al., 1991) and the amino acid signals have been assigned. Quantitative measurements of amino acid composition have been made for orange juice concentrates, with results that compare well with conventional amino acid analyses. The drawback was the long acquisition time needed for each sample (about 4 h on a 250 MHz spectrometer) but use of higher magnetic fields, improved probes, larger sample volumes and, possibly inverse 2D NMR techniques would all improve the situation. ^{13}C-NMR also revealed the presence in orange juice of relatively high concentrations of N,N-dimethylproline, a compound not previously identified in juices. ^{13}C-NMR has been used to determine the structure of fructose and glucose containing oligosaccharides isolated from beet medium invert syrup (Swallow and Low, 1993). These oligosaccharides are formed in addition to fructose and glucose during acid hydrolysis of sucrose syrups. Their presence (detected chromatographically) in citrus juices is a sign that the natural juice has been adulterated with BMIS.

3.3.2 *Wines*

^{1}H (400 MHz) and ^{13}C (100 MHz) NMR spectra were used in conjunction with multivariate analysis to classify wines from three German regions (Rheingau, Rheinhessen and Mosel-Saar-Ruwer) according to geographical origin (Vogels et al., 1993). The 53 wines were from 6 different vintages. Wines (2 ml) were freeze-dried and then dissolved in D$_2$O (0.7 ml) for the NMR experiments. The experiments took 0.5 h per run for ^{1}H (3 runs were carried out for each sample) and 4.5 h per run for ^{13}C (1 run for each sample). The classification procedure consisted of PCA followed by discriminant analysis on the PC scores. The number of PCs used in the DA model was quite high (39 PCs for ^{1}H, 13 PCs for ^{13}C) but was restricted to less than one-quarter the total number of spectra (including repeats for ^{1}H). Results were visualized by plotting the scores on the first two discriminant axes. Taking the raw data, a better classification was obtained from ^{1}H than from ^{13}C-NMR. K-nearest neighbours classification was used to determine the effects of different data pre-processing steps. A partial linear fit procedure (alignment of spectral lines) was beneficial in all cases, and selection of the most discriminating variables according to the variance ratio method also improved the classification when carried out after PLF. PLF followed by variable selection reduced the number of variables (spectral lines) from ~2000 to 520 (^{1}H) or 190 (^{13}C). Both steps were required to obtain a good classification from the ^{13}C data. Relatively higher concentrations of certain compounds (e.g. proline) in the wines of a particular region were suggested by examination of the factor loadings but, as with fruit juices, this is an area where more work is required.

Characteristic signals of many compounds have been detected and assigned in the ^{13}C NMR spectra of wines and wine concentrates (Rapp and Markowetz, 1993). The most prominent signals in wines themselves are from ethanol and glycerol. Illegal additions of methanol and diethylene glycol (Rapp et al., 1986) have been detected using ^{13}C NMR. A much wider range of compounds can be detected and quantified following a 5–10-fold concentration of the wine. The method of quantification is based on comparison of the signal intensity of the analyte with that of a known amount of internal standard (1,3-propanediol). Samples are acidified to a standard pH of 2 before measurement. This pH gives narrow lines for all signals and ensures that chemical shifts are the same as those recorded for reference compounds in the literature (Rapp et al., 1991; Rapp and Markowetz, 1993). ^{13}C signals identified include those of organic acids (tartaric, malic, lactic, succinic, citric, acetic); sugars (glucose, fructose, galactose, arabinose, xylose, rhamnose, trehalose and ribose); sugar alcohols and acids (mannitol, erythritol, sorbitol, myo-inositol, galacturonic, glucuronic and gluconic acids); amino acids; and preservatives (sorbic acid). Use of illegal carbohydrate sweeteners can be detected through the presence of distinctive signals of oligosaccharides (maltose, maltotriose, gentiobiose). Particular attention has been given to the quantification of amino acids in wine concentrates (Rapp et al., 1991) and NMR results have been compared with those from amino acid analysers, generally with good agreement. The methods of Rapp et al. have been applied to the analysis of amino acids in red and white wines from six European countries (Holland et al., 1995) and the wide variations in the concentrations of proline, arginine, alanine and glutamic acid have been discussed. There were not enough samples of each type to carry out a multivariate analysis. Nevertheless, the method appears promising if the acquisition time per sample can be reduced and if the database can be greatly enlarged.

3.3.3 Vegetable oils

Two of the most important methods in oil authentication are the determination of fatty acid composition and the measurement of triglyceride profiles (Firestone and Reina, 1996). Although it does not give such a detailed picture of oil composition as GC or HPLC, NMR does provide a direct and potentially very rapid means of obtaining much of the required information. Oils are usually examined as 10–20% solutions in deuterochloroform. On modern spectrometers (e.g. 400 MHz for ^{1}H) both ^{1}H and ^{13}C spectra of the triglycerides can be obtained with excellent signal to noise in a few minutes at these concentrations. If the main interest is in minor components (diglycerides, free fatty acids, etc.) acquisition times for ^{13}C spectra need to be considerably longer.

One simple application of ^{1}H-NMR is determination of the degree of unsaturation. The number of olefinic protons may be determined relative to the number of protons in the glyceride unit by integrating the multiplets at 5.29 ppm (olefinic protons plus glyceride methine) and 4.2 ppm (glyceride methylene protons). A good correlation was obtained for olive oil and a number of other oils between values determined in this way and the iodine value (Sacchi *et al.*, 1989). Typical variations in the ^{1}H spectra of different types of vegetable oil are shown in Fig. 3.7 (the spectra shown here do not include the olefinic and glyceride resonances referred to above). The three oils represented are olive (high oleic acid content), sunflower (high linoleic) and rapeseed (high oleic, appreciable linolenic). The effects of these compositional differences are obvious for the methyl triplets, A and B, and for the methylene multiplets, E and G, which come from CH_2 groups adjacent to the double bonds. Differences are also evident in the main chain $(CH_2)_n$ signals, C, but are not so readily interpretable. The spectra suggest that NMR should be useful for discriminating between oils of different origin and for detecting addition of, for example, rapeseed or sunflower to olive oil. Formulae for calculating the composition of oils in terms of mole % saturated, monoene (mainly 18:1) and 18:2 and 18:3 acids from integrated NMR spectra have been given (Wollenberg, 1991). The long chain saturated acids (16:0, 18:0) are not differentiated by this NMR analysis.

^{13}C spectra of the same three oils are shown in Fig. 3.8 (omitting the triglyceride carbon signals at 68.8 ppm (CH) and 62.0 ppm (CH_2). This region is very similar in all the oils). Differences are evident in all the regions shown,

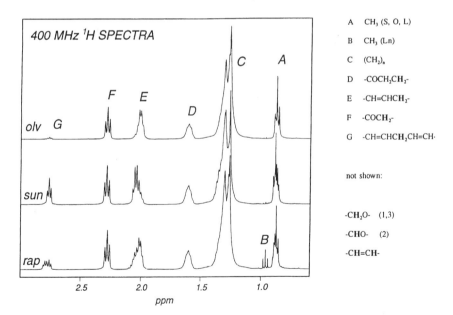

Figure 3.7 400 MHz ^{1}H-NMR spectra and signal assignments of vegetable oils: olive (olv); sunflower (sun); rapeseed (rap); saturated (S); oleic (O); linoleic (L); linolenic (Ln). The glyceride and olefinic signals are not shown.

most obviously for the olefinic and methyl carbons. The spectra in Fig. 3.8 were obtained under 'routine' acquisition conditions (about 1.3 s per scan). With a 20 s acquisition time, the differences between oil types become more obvious in the carbonyl region as well, with the carbonyl signals for the different unsaturated chains becoming resolvable (Wollenberg, 1990). The ^{13}C spectrum allows the distribution of chains between (1,3) and (2) glyceride positions to be determined, as illustrated by the carbonyl region of Fig. 3.8. For the three oils shown it can be seen that the saturated chains are confined to the (1,3) position as expected for a natural oil. The sensitivity of chemical shifts to the position of substitution extends further down the chain and is responsible for the splittings

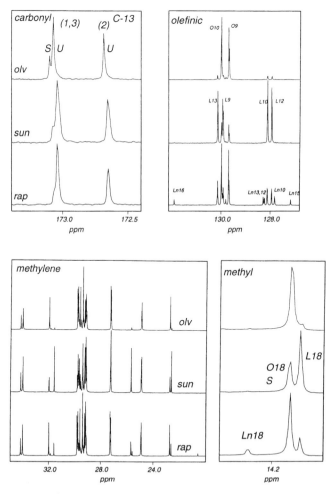

Figure 3.8 100 MHz ^{13}C-NMR spectra of vegetable oils: olive (olv); sunflower (sun); rapeseed (rap); saturated (S); unsaturated (U); oleic (O); linoleic (L); linolenic (Ln). O10 is carbon 10 of the oleic acid chain etc. Glyceride signals not shown.

of O9, O10 signals, etc. in the olefinic region. Assignments and listings of ^{13}C chemical shifts in triglycerides are available from several literature sources (Gunstone, 1993; Wollenberg, 1990). Compositional data may be obtained from quantitative ^{13}C as well as from ^{1}H spectra although the experimental requirements needed to ensure that all signals have the correct relative intensities are somewhat more demanding. These requirements have been discussed in detail (Shoolery, 1977; Wollenberg, 1990).

Chemometrics methods have been used for oil classification in several studies (Firestone and Reina, 1996). The fatty acid (Forina *et al.*, 1983) or triglyceride (Tsimidou *et al.*, 1987a, b) compositions, obtained by classical analytical methods, provided the basic data for these studies. In view of the relationship between triglyceride composition and ^{1}H and ^{13}C-NMR spectra, attempts have been made to discriminate between oils of different origin by applying PCA and discriminant analysis methods directly to the NMR data (Colquhoun and Puaud, unpublished). Figure 3.9 shows results of PCA on sets of ^{1}H and ^{13}C spectra for five types of oil (rapeseed, groundnut, sunflower, corn and soya). The oils were from both retail and production sources – each symbol represents a different oil (no duplicate runs are plotted). The results from ^{1}H and ^{13}C spectra are quite comparable as can be seen from the plots of PC1 vs PC2. PC1 clearly separates rapeseed and groundnut from the other three oil types while PC2 separates rapeseed from groundnut. One of the corn oil samples lies well outside the main cluster. The ^{1}H spectra provide a slightly better separation than ^{13}C for the

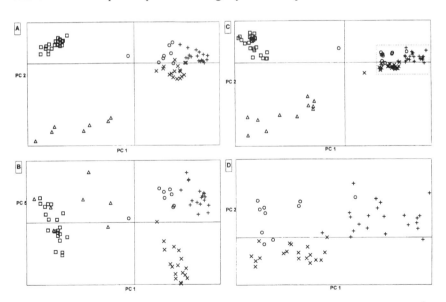

Figure 3.9 PC scores from principal component analysis of oils NMR spectra. (A) based on ^{1}H spectra, PC1 vs PC2; (B) ^{1}H spectra, PC1 vs PC5; (C) ^{13}C spectra, PC1 vs PC2; (D) expansion of boxed region in (C). Oils: □ rapeseed; △ groundnut; ○ corn; + sunflower; × soya. (Colquhoun and Puaud, unpublished.)

remaining three oil groups. This, however, is only a two-dimensional plot of the two most significant PCs. The higher PCs also contain significant information as shown by the separation of the soya oil group from corn and sunflower oils on PC5 (^{1}H spectra). Although the separation of corn, sunflower and soya groups does not appear to be very good in the PC1 vs PC2 plots, both LDA and CVA give 100% correct assignment of samples when six PCs are used (89 samples for ^{13}C, 71 for ^{1}H). The separation on PC1 is clearly related to the oleic/linoleic acid ratio in the oils: the loadings plot for this PC shows negative loadings at frequencies associated with oleic acid, positive loadings at linoleic acid frequencies; negative peaks associated with linolenic acid are prominent in the PC5 loading. This is in agreement with the known composition of the oils: rapeseed and groundnut have a high oleic acid content, the others are high in linoleic acid. Of the three groups with high linoleic acid content, soya oil is the only one with an appreciable content of linolenic acid. The combined NMR-chemometrics approach clearly offers a rapid and convenient means of measuring and classifying oil samples on the basis of their fatty acid compositions. What still needs to be determined is the reproducibility of the method (i.e. what proportion of the dispersion within the groups in Fig. 3.9 is due to genuine compositional differences, what proportion is attributable to measurement error) and the concentration limits for detection of adulteration. PCA and PLS (partial least squares) methods have been used to discriminate between virgin olive oils on the basis of their ^{13}C-NMR spectra (Shaw *et al.*, 1996). The oils, representing a number of Italian varieties and growing regions, were prepared under controlled laboratory conditions. A systematic approach was used to select the most discriminating variables (NMR peaks). Classification of the oils by variety was more successful than classification by region of origin.

As well as the triglycerides, many minor oil components give rise to detectable NMR signals. There has been special interest in using ^{13}C-NMR to quantify these components in the different grades of olive oil. The main aim is to distinguish virgin olive oils, produced only by mechanical pressing of sound olives, from refined (neutralized, deodorized and bleached) oils, residue (solvent extracted) oils, or mixtures of the types. More recently there have been attempts to use the same method (and very high field ^{1}H-NMR) to discriminate between authentic virgin oils from different regions and cultivars. Evidence that oils have been heat treated in the presence of decolorizing earth has been obtained by detection of minor ^{13}C signals in the region of 32 ppm. The signals have been assigned to methylene groups adjacent to *trans* double bonds of mono- and polyunsaturated chains that result from isomerization during heating (Sacchi *et al.*, 1989). Free fatty acid and diglyceride content are two important parameters by which olive oil quality may be judged. Free oleic acid may be detected by ^{13}C-NMR since there is a large difference between its carbonyl chemical shift (176.8 ppm) and the corresponding shifts in triglycerides (~173 ppm). Similarly the ^{13}C glyceride signals of (1,2) and (1,3)-diglycerides are quite characteristic and are well resolved from each other and from the much stronger signals of the triglycerides. All these signals occur in the range 60–72 ppm. Quantitative analyses have been made, for a range of different oil grades, of: (a) mole % of free fatty acid present; (b) fraction of diglycerides present as a percentage of total di- and triglycerides; (c) ratio of (1,2) to (1,3) diglycerides. In principle, monoglycerides can also be quantified but are usually not detected. The amount of free acid was plotted against the amount of diglyceride for different oil types. A dividing line could be drawn between oils produced only by pressing and the lower grade oils. Although the free acid content was generally less than 1% in the lower grade oils, the diglyceride content was relatively high (7–15% compared with under 3% in extra virgin oils). The ratio of (1,2) to (1,3) diglycerides was found to decrease on storage (Sacchi et al., 1990) but it has also been shown that the ratio is quite variable in freshly pressed oils and is probably dependent on cultivar (Vlahov, 1996a). It was mentioned above that the positional distribution of fatty acid chains on the glycerol moiety could be determined because of the shift between carbonyl signals of chains attached to the (1,3) and (2) positions and because the signals of saturated and unsaturated chains are resolved. (Figure 3.8 shows an authentic olive oil with no saturated chains at the (2) position.) This feature has been used to identify synthetic (esterified) olive oils since chemical esterification is a process that results in a random distribution of the different chains between glyceride positions, and any saturated chains in the (2) position are readily seen (Sacchi et al., 1992). The same technique has been used to study the positional distribution of fatty acid chains in oils of a single cultivar from different Italian regions (Vlahov, 1996b).

A disadvantage of all these ^{13}C-NMR methods is the length of the acquisition time required to quantify the minor components. The sensitivity, spectral dispersion and dynamic range now offered by high field (600 MHz) spectrometers promises to make ^{1}H-NMR a competitive technique for minor components, especially if statistical methods are used to handle the added spectral complexity. It has been shown (Sacchi et al., 1996) that a wide range of compounds can be detected in the 600 MHz ^{1}H-NMR spectra of olive oils (the solvent was d-chloroform with a trace of DMSO to solubilize all components). Signals assigned included those of sterols, diglycerides, phenolics and volatiles (alcohols and aldehydes). A method of quantification was suggested for linolenic acid and sterols which involves integrating the methyl signals of these compounds relative to the neighbouring ^{13}C satellite signals of the main triglyceride methyl resonance. Accurate quantification of such signals would be important for the detection of certain types of adulteration. A multivariate analysis, carried out for signals in the range 4.5–5 ppm and 8–10 ppm, showed some clustering of oil samples. By selecting these regions emphasis is placed on the signals of volatile compounds which have an influence on perceived quality. In all the work discussed above spectra of the oils were taken directly without any form of sample preparation other than dilution in a solvent. In another approach to classification of olive oils, the unsaponifiable matter (from treatment of the oil with potassium hydroxide) was isolated and examined by ^{13}C-NMR (Zamora *et al.*, 1994). There were obvious differences between the spectra of the unsaponifiable material obtained from virgin, refined and refined pomace oils. Signals associated with squalene, sterols, triterpenic alcohols and long chain alcohols were identified. A successful classification was achieved with the help of variable selection (most discriminating peaks), CVA and cluster analysis. Further work is needed to determine whether the method is sensitive enough to quantify proportions of virgin and refined oils in mixtures.

3.3.4 *Fish and meat*

Few NMR studies of these materials have been explicitly concerned with authentication. One area where NMR might prove useful is in discriminating between fresh and frozen–thawed meat or fish. Measurements of the water transverse relaxation time, T_2, can be made directly on the samples using relatively low cost benchtop NMR instruments. The relaxation behaviour of the water in muscle is very different from that in pure water. This is because a variety of exchange and diffusion phenomena control the observed relaxation behaviour: the relaxation time is shortened and may be made up of several components. Details are determined by the sizes of the compartments that the water occupies, and by the state of the proteins which have sites that can exchange hydrogen atoms with water. Physical treatments that affect these properties may result in changes of the observed relaxation behaviour. Examples of underlying changes are damage to tissues caused by growth of ice crystals, and protein denaturation or aggregation. Two T_2 relaxation components were measured for the water in fresh cod (Lambelet *et al.*, 1995). A third component with a longer relaxation time was detected in frozen–thawed cod after 24 h freezing. The relaxation time of this component increased with increasing storage temperature, and also increased with storage time at $-10\,°C$ and $-20\,°C$ but not at $-40\,°C$. An automatic method (using NMR imaging equipment) has been devised which allows relaxation times for a large number of meat samples to be obtained simultaneously (Evans *et al.*, 1996). T_1, T_2 and magnetization transfer (water-protein cross relaxation) rates were measured for pork, beef and lamb samples and a statistical analysis of the results was carried out. It was found that the freeze–thaw treatment had no effect on the T_1 and T_2 times but the magnetization transfer rate was increased. Magnetization transfer measurements are made by perturbing the protein protons and observing the effect on the water signal. They provide an indirect method of monitoring the dynamics of protein motions: in this case it appears that the freeze–thaw treatment leads to aggregation and loss of mobility of the myofibrillar proteins.

High resolution ^{1}H-NMR spectra have been obtained for perchloric acid extracts of bovine muscle, and many of the signals have been assigned (Lundberg *et al.*, 1986). There have been no systematic attempts, however, to compare fresh and frozen meat or meats of different species by this method. Breakdown of trimethylamine oxide in frozen and fresh fish has been studied by high resolution ^{1}H-NMR of aqueous extracts. 'Fresh' cod from retail sources showed considerable (bacterial?) breakdown of TMAO to trimethylamine, a compound not detected in fish that was frozen when truly fresh (Howell et al., 1996). Instead, a limited breakdown to dimethylamine was observed in fish stored at $-20\,°C$ and no changes for TMAO or other metabolites in fish stored at $-30\,°C$. NMR can be used for analysis of fish and meat lipids in the same way as described for vegetable oils. Following a simple extraction step ^{13}C-NMR spectra of lipids from tuna were examined to determine free fatty acid content and fatty acid composition and distribution. Such measurements can potentially be used to determine the freshness of fish, the extent of lipolytic reactions, and effects of processing and storage (Sacchi et al., 1993). ^{13}C-NMR spectra of fatty livers ('foie gras') from ducks and geese could be obtained directly from the livers without any lipid extraction step (Dufour et al., 1996). The fatty acid composition was closely related to the feeding regime and could not be used to determine the species or geographical origin.

3.4 Site-specific natural isotope fractionation by NMR

3.4.1 Introduction

One of the most notable contributions of high-resolution nuclear magnetic resonance to food authentication is its use to measure the isotopic content, at natural abundance level, of specific molecular sites in a given species. The technique, known as SNIF-NMR®*for site-specific natural isotope fractionation studied by NMR, was developed in the early 1980s by Professors Gérard and Maryvonne Martin at the University of Nantes. The reader is referred to two extremely comprehensive reviews by the pioneers of this technique (Martin and Martin, 1990, 1995b).

The earliest use of SNIF-NMR® was to detect the 'chaptalisation' or enrichment of wines – the practice of adding sugar to grape must before fermentation to increase the end alcoholic content of the wine – for which it was adopted as the official method by the Office International du Vin et de la Vigne (OIV) and subsequently by the European Community (Commission Regulation EEC No 2676/90). It has since been applied to a number of other areas: origin identification of grain and fruit alcohols, detection of added sugar in fruit juices and concentrates, determination of the natural or synthetic status of flavourings such as vanilla, anethole, benzaldehyde. In general, the information associated with the site-specific isotope ratios of chemical species extracted from a food product or ingredient can provide an insight into the botanical and/or geographical origin of the species, making SNIF-NMR® a powerful tool for food and beverage authentication.

* SNIF-NMR® is a registered trade mark of EUROFINS Laboratories

3.4.2 Natural isotope fractionation

A full discussion of the basic principles of isotope analysis is given in Chapter 2. The major elements in a natural product, carbon, hydrogen, oxygen, nitrogen, exist in their naturally occurring isotopic forms, ^{12}C, ^{13}C; ^{1}H, ^{2}H; ^{14}N, ^{15}N; ^{16}O, ^{18}O. The slight difference in mass between the isotopes of the same element, and the influence of physical, chemical and biochemical processes during its elaboration, can lead to the same chemical compound having a quite different isotopic make-up depending on its source. These subtle differences are most commonly determined by stable isotope ratio mass spectrometry (SIRMS) which provide a measure of the overall or average isotope content for an entire compound or product. Variations in isotope content also exist between different molecular sites within the same species and in a number of cases, SNIF-NMR® can provide direct access to this information. In theory, most elements are potential targets for site-specific isotope content determination. In practice, however, and from the point of view of NMR spectroscopy, the nucleus studied must fulfil a number of requirements. ^{18}O, for instance, has a nuclear spin quantum number, I, of zero, is therefore NMR inactive and thus cannot be used to measure $^{18}O/^{16}O$ ratios by the technique. Other elements such as deuterium (I = 1), ^{13}C and ^{15}N (I = $\frac{1}{2}$) are accessible to NMR spectroscopy. In general, the major applications of the SNIF-NMR® technique have so far been developed for deuterium, for which there are a wide range of interesting examples of food authentication.

3.4.3 Determining site-specific ratios by NMR

Isotopic distribution in products is defined in terms of their isotopic abundances, A, or isotopic ratios, R.

$$A = \frac{heavy}{heavy + light}, \quad R = \frac{heavy}{light} \quad (3.1)$$

R and A are expressed in per cent (%) or parts per million (ppm) depending on the nature of the atomic constituent being considered. Similarly, a site-specific isotopic ratio, R_i, can be defined as the ratio of heavy to light atoms of the same element in a specific site i. For deuterium, for example, a site-specific ratio would represent the number of deuterium (^{2}H or D) atoms to the number of protons (^{1}H) in a specific site i, given as:

$$R_i = \left(\frac{D}{H}\right)_i \quad (3.2)$$

In this case, the number of D atoms in site i can be considered as equivalent to the number of monodeuterated isotopomers of type i – an isotopomer being defined as equivalent chemical species of different isotopic content. The overall isotope ratio $\bar{R}$ is then the weighted average of the isotope ratios R_i of its n

monodeuterated isotopomers i:

$$\bar{R} = \sum_{i=1}^{n} F_i R_i \qquad (3.3)$$

Where F_i is the statistical molar fraction corresponding to a statistical distribution of deuterium in the n isotopomers with p_i equivalent positions, and f_i the true or experimental molar fraction, which can be obtained directly from the corresponding signal intensity in the deuterium NMR spectrum.

$$F_i = \frac{p_i}{\sum_{i=1}^{n} p_i} \qquad (3.4)$$

$$R_i = \frac{f_i}{F_i} \bar{R} \qquad (3.5)$$

Site-specific ratios R_i can be determined directly from equation (3.5) above, provided that the overall isotope ratio is available from SIRMS measurements. This is the **external referencing method**. Mass spectrometry measurements are not required if an internal standard is used in the NMR determination. In this **internal referencing procedure**, a precisely known quantity of a working standard, WS, is added to a weighed amount of sample X. The isotope ratios of each site i can then be calculated from the following equation:

$$R_i^x = \frac{P^{WS}}{P_i^x} \times \frac{M^x}{M^{WS}} \times \frac{m^{WS}}{m^x} \times \frac{T_i}{t^x} R^{WS} \qquad (3.6)$$

where P^{WS} and P_i^x are the stoichiometric numbers of protons in WS and site i of X, M^{WS} and M^x, and m^{WS} and m^x are, respectively, the molecular weights and masses of WS and X, t^x is the purity (w/w) of the sample X and T_i is the ratio of the areas of the ^{2}H-NMR signal of site i of X and of the working standard WS.

3.4.4 ^{2}H-NMR for quantitative determinations of site-specific ratios

Because of its very low natural abundance – there are approximately 150 deuterium atoms for every one million protons – ^{2}H-NMR is considerably less sensitive than the more commonly used proton NMR. Deliberate deuterium labelling has been used to study chemical or biochemical reaction mechanisms and since deuterium chemical shifts are almost identical to the corresponding proton chemical shift, the investigation of specifically labelled sites in the ^{2}H spectrum can provide valuable information to help elucidate a complex ^{1}H spectrum of the same compound. An advantage of SNIF-NMR®, however, is that it measures D/H ratios at the natural abundance level without recourse to lengthy labelling experiments.

Obtaining ^{2}H-spectra in the quantitative conditions required for the determination of isotopic ratios is crucial and requires careful consideration of the acquisition and data treatment parameters. A field-frequency locking device, usually a fluorine lock, is required to avoid field drift and subsequent broadening of the signals occurring. Although this may not be necessary at lower field strengths, it is particularly desirable for a deuterium frequency of 61.4 MHz or higher, which corresponds to a nominal ^{1}H frequency of 500 MHz (9.4 T), or if an automatic sampler is implemented on the spectrometer. Broad-band decoupling is applied to remove line splitting due to ^{2}H ~ ^{1}H interactions – the probability of bideuterated species at the natural abundance level is extremely low, and therefore ^{2}H ~ ^{2}H coupling can be ignored. As for other nuclei with I ≥ 1, deuterium has a quadrupole moment, and its relaxation is governed by an efficient quadrupolar mechanism. This and the low relative efficiency of the dipolar mechanism, means that nuclear Overhauser effects can be neglected in proton-decoupled spectra which is particularly advantageous in quantitative deuterium determinations. Other typical conditions described by Martin and Martin (1995b) include an acquisition time six times the maximum longitudinal relaxation time, a pulse angle of 90°, a spectral width equal to eight times the resonance range, and a sufficient number of scans to obtain a signal-to-noise ratio greater than 150.

Well-resolved signals for the different isotopomers in the ^{2}H-NMR spectrum will have a direct bearing on the accuracy of the isotope ratio determinations. Spectral data processing (phase corrections, curve fitting procedures and so on) are also a source of differences in results between laboratories. A considerable improvement in analytical precision can be obtained using specifically developed software, EUROSPEC-LISS, which is based on a complex least-squares curve fitting technique (Martin, 1994).

In an intercomparison exercise designed to assess the influence of spectrometer performance on the precision of the quantitative data obtained (Guillou et al., 1988), three ethanol samples from different origins were distributed to 15 laboratories operating very different spectrometer systems, with basic magnetic fields varying from 4.7 to 9.4 T. The study gave a repeatability of 0.8% and a confidence interval (CI 95%) of 0.25%. It also showed that the reproducibility of 2–3% obtained on the different NMR systems could be reduced to 1% when dedicated and highly automated procedures were used.

Such considerations of precision and accuracy are important, in particular for a technique for which the results have official implications. Certified reference materials – a set of three ethanol samples of different origins in sealed NMR tubes, CRM 123 – are available from the Community Bureau of Reference of the European Commission for instrument calibration (Martin and Trierweiler, 1994; Martin et al., 1994) and since the early studies, the use of dedicated analytical systems for SNIF-NMR®, as discussed below, have become more widespread.

3.4.5 Choice of isotopic probe for SNIF-NMR® analysis

A further source of systematic variations in results can be traced back to sample preparation – extraction, purification, derivatization – of the isotopic probe to be used. Each experimental stage must be carefully executed and assessed for potential isotopic fractionation effects. The ease with which the molecule to be investigated can be isolated and the absence of discriminating fractionation effects will determine its usefulness as an isotopic probe. In addition, the compound must be suitable for ^{2}H-NMR spectroscopy in terms of sensitivity, integration of NMR signals and availability of an internal reference.

Major applications of the ^{2}H-SNIF-NMR® method, applied to ethanol as isotopic probe, are discussed below. Examples of isotope ratio determinations of other molecular species, acetic acid, vanillin, and so on are outlined.

3.4.6 Ethanol as isotopic probe

Ethanol, CH_3CH_2OH, has three monodeuterated isotopomers at natural abundance level as shown in Fig. 3.10. The deuterium can be located either on the methyl site (I), the methylene site (II) or the hydroxyl site (III) of the ethanol molecule. The site-specific deuterium to hydrogen ratio is therefore the proportion of each monodeuterated species to the fully hydrogenated molecule, $(D/H)_I$, $(D/H)_{II}$ or $(D/H)_{III}$. Because of the low natural abundance of deuterium relative to ^{1}H, the probability of having a bideuterated species is very low and can be ignored. The ^{2}H-NMR proton-decoupled spectrum at the natural abundance level is composed of a single line for each species as shown in the example in Fig. 3.11. The four peaks correspond to each of the three monodeuterated species and to the internal reference used.

Site-specific isotope ratios are calculated using the information from the spectrum in equation (3.6) above. The working standard, WS, is N,N-tetramethylurea, TMU, previously calibrated with respect to a primary international standard V.SMOW (Vienna Standard Mean Ocean Water). In the case of ethanol, signal heights are used instead of signal areas to improve accuracy of the results.

A relative parameter, r, can be obtained directly from the NMR spectrum using the intensities I_I and I_{II} of the CH_2D and CHD signals. (Note: In other publications relating to SNIF-NMR®, the reader may encounter the use of the letter R to designate this intramolecular ratio. For the purposes of this chapter

CH_2DCH_2OH
(I)

CH_3CHDOH
(II)

CH_3CH_2OD
(III)

Figure 3.10 Monodeuterated isotopomers of ethanol.

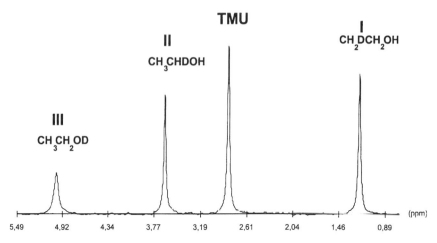

Figure 3.11 2H-NMR spectrum of ethanol.

and to avoid confusion with R = isotope ratio, the lower case letter r has been used.)

$$r = 2 \frac{(D/H)_{II}}{(D/H)_{I}} = 3 \frac{I_{II}}{I_{I}} \quad (3.7)$$

r describes the internal distribution of deuterium and represents the deuterium content of the methylene site, with the methyl site chosen as reference and given the statistical value of 3. If ^{2}H was randomly distributed in the ethanol molecule, r would be equal to 2. Experimental values of r are generally greater than 2 and differ according to the source of the alcohol, reinforcing the premise of a non-statistical distribution of deuterium depending on origin.

3.4.7 Wine authentication

In the original application of ^{2}H-SNIF-NMR® to detect the chaptalization or enrichment of wines, it was shown that the deuterium/hydrogen ratios measured on the different molecular sites of wine ethanol differ significantly according to the origin of the sugars that produced the alcohol (Martin and Martin, 1981). Figure 3.12 shows how beet sugar can be quantified in a 1981 Sauvignon wine from the Gironde by an increase in the internal parameter r (Martin et al., 1986a). The technique has become an important tool for the European Community to enforce legislation controlling wine enrichment. Since the latter is calculated from reference values obtained on authentic wine samples, a European Data Bank has been set up at the European Commission Joint Research Centre in Ispra, Italy (Guillou et al., 1995) which collects isotopic data from all the wine-producing member states on an annual basis (EC Regulations Nos. 2347/91 and 2348/91).

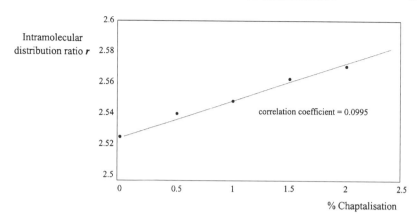

Figure 3.12 Increase in internal parameter, r, with increasing beet sugar addition (chaptalization) is a 1981 Sauvignon wine from the Gironde.

Further studies have shown that site-specific isotopic parameters can be used to distinguish between wines of different geographical origins, particularly from regions with quite different climates (Martin et al., 1988). Improved assessment of geographical origin has been obtained by combining SNIF-NMR® data with results of trace metal analyses by atomic absorption on the same authentic grape musts (Day et al., 1993, 1994).

3.4.8 Fruit juice authentication

The SNIF-NMR® technique has been extended to fruit juices (Martin and Martin, 1995a) to detect the undeclared addition of sugar either as beet sucrose or modified sugar syrups to the juice or concentrate. This application requires the fermentation of the fruit juice sample and determination of $(D/H)_i$ ratios on the resulting alcohol after its removal by distillation. Preparation of the ethanol sample for isotope analysis requires careful monitoring to ensure complete conversion of the sugars and quantitative recovery of the alcohol after distillation.

The authenticity of an unknown juice can be assessed by comparing the measured $(D/H)_i$ ratios with data obtained on authentic natural products contained in a reference database. Since deuterium distribution is influenced by both climatic and geographical conditions, authentic products from the same year and region should be considered.

Studies have shown that the deuterium content of the methyl site on the ethanol is directly related to the deuterium content on sites 1, 6 and 6′ of the starting sugars (Martin et al., 1986b). Successful measurement of the $(D/H)_I$ ratio therefore provides an indication of the origin of the ethanol.

During the fermentation of the fruit juice, there is also some exchange between the hydrogen of the water of the fermentation medium, and the methyl, and to a greater extent, the methylene sites of the ethanol (Martin et al., 1986b).

The isotope content of the fermentation water, $(D/H)_W^S$, can vary and this must be taken into consideration in the determination of the site-specific parameters. The latter are therefore corrected or standardized using equations (3.8) to (3.10) to a common fermentation water which has a deuterium content equivalent to that of the international standard V.SMOW (Martin et al., 1996a). Data from juices fermented in different starting water, in the same or in different laboratories, can then be safely compared.

$$\left(\frac{D}{H}\right)_I^{Norm.V.SMOW} = \left(\frac{D}{H}\right)_I - 0.19 \times \left[\left(\frac{D}{H}\right)_W^S - 155.76\right] \quad (3.8)$$

$$\left(\frac{D}{H}\right)_{II}^{Norm.V.SMOW} = \left(\frac{D}{H}\right)_{II} - 0.78 \times \left[\left(\frac{D}{H}\right)_W^S - 155.76\right] \quad (3.9)$$

$$R^{Norm.V.SMOW} = R - 0.011 \times \left[\left(\frac{D}{H}\right)_W^S - 155.76\right] \quad (3.10)$$

$$\left(\frac{D}{H}\right)^{V.SMOW} = 155.76 \text{ ppm (Craig, 1961)}$$

An interpretation strategy to assess orange, grapefruit and apple authenticity has been proposed by Martin et al. (1996a). This is based on an adulteration triangle (Fig. 3.13) constructed from isotopic parameters measured for authentic reference alcohols from beet, cane and fruit sugars. V.SMOW-standardized $(D/H)_I$ values are represented along the horizontal axis, and the ^{13}C content of the alcohols, measured using isotope ratio mass spectrometry and denoted $\delta^{13}C$ with respect to international standard PDB (Pee Dee Belemite) along the vertical

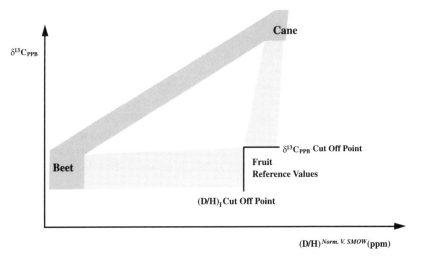

Figure 3.13 Adulteration triangle used to interpret isotopic data. The samples are represented in the plane of the 13 content and of the normalized hydrogen isotope ratio $(D/H)_I^{Norm.V.SMOW}$ of the ethanol fermented from the juice.

Table 3.1 Mean values and cut-off points for different reference groups of fruit (Source: Martin et al., J. A.O.A.C. Int. (1996))

Fruit type	Geographical origin	Mean values $(D/H)_I^{Norm.V.SMOW}$ ppm	Cut-off points $(D/H)_I^{Norm.V.SMOW}$ ppm
Orange	Florida	104.8	103.5
Orange	Brazil	104.3	103.0
Orange	Israel	106.9	105.0
Orange	Unknown	105.0	103.0
Apple	Unknown	99.5	97.0
Grapefruit	Unknown	104.0	102.0

axis. Possible adulteration of a sample with added cane or beet sugar is indicated by its position on the triangle. Cut-off points (COP) are defined for each fruit type from an authentic database and beet sugar adulteration can be quantified using equation (3.11).

$$\text{Adulteration}_{min}\% = \text{Beet}_{min}\% = 100 \times \frac{(D/H)_I^{Sample} - (D/H)_I^{COP}}{(D/H)_I^{Beet} - (D/H)_I^{COP}} \quad (3.11)$$

Similar equations are described to detect cane sugar, and mixtures of beet and cane. Mean values and cut-off points for deuterium/hydrogen ratios according to fruit type and geographical origin are given in Table 3.1.

The SNIF-NMR® method for the determination of beet sugar in fruit juices is now an official AOAC (Association of Official Analytical Chemists – Method 995.17) for which it has been collaboratively studied. Its performance was assessed as part of the AOAC collaborative study and an excellent correlation was observed between the percentage of added beet sugar and the measured isotopic parameter (Martin et al., 1996b). Repeatability and reproducibility results, expressed as s_r and s_R according to ISO standard 5725, ranged from 0.19 to 0.25 ppm, and 0.21 to 0.37 ppm, respectively, depending on the fruit type studied.

Guide values for $(D/H)_i$ ratios have been included in the 1997 version of the AIJN (Association of the Industry of Juices and Nectars from Fruits and Vegetables) Code of Practice for a number of fruit types. Minimum $(D/H)_I$ (ethanol) values are proposed for orange juice (103–107 ppm), for grapefruit juice (102–106 ppm), for grape juice (99–106 ppm) and for pineapple (107–111.5 ppm) – for all geographical origins. These guidelines provide an indication of product authenticity, although expert interpretation is required to take into account specific geographical origin and, if necessary, seasonal variations.

3.4.9 Sample treatment: the SNIF-NMR® concept

As mentioned earlier, a high degree of automation of the analytical chain and the isotopic determinations considerably improve the precision of the results obtained. A complete SNIF-NMR® analysis of a fruit juice consists of a number of successive steps which have to be followed with considerable care. For example sample mishandling and risks of evaporation can all lead to isotopic fractionation and erroneous results for the (D/H) ratios measured. The SNIF-NMR Concept has been developed (see Fig. 3.14) as a completely integrated push-button system that covers each step of the analytical chain.

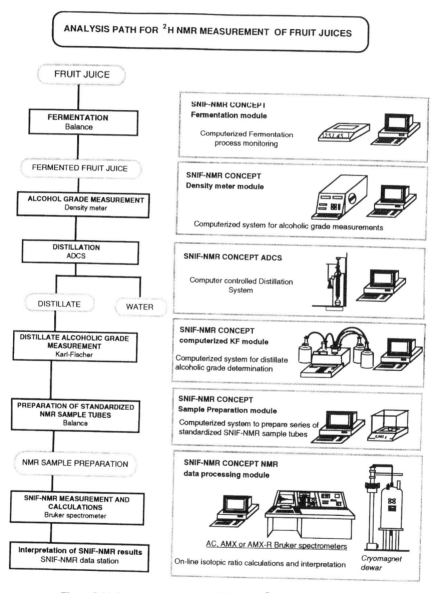

Figure 3.14 Steps in the automated SNIF-NMR® analysis of fruit juices.

A fruit juice concentrate, for example, is first diluted to a standard °Brix value using water of known isotopic content (parameter $(D/H)_W^S$ in equations (3.8) to (3.10) above). Fermentation of the resulting solution is then carried out with a standard yeast strain (Martin *et al.*, 1983b) and takes between three and five days depending on the type of fruit being analysed. A commercial enzyme kit is used at the end of this step to ensure complete transformation of all the sugar. The ethanol is then distilled off using a computer-controlled Cadiot column fitted with a Teflon spinning band (the automatic distillation control system, ADCS). The recovery yield of ethanol is calculated from the alcohol content determined before and after distillation. This must be greater than 95% by mass to avoid any occurrence of isotopic fractionation.

A standard preparation system monitors all mass changes during filling and handling of the tightly sealed NMR tubes for the SNIF-NMR® measurement, warning the operator of any significant weight change due to evaporation of the alcohol. Finally, a data processing module provides on-line calculations of isotopic parameters and interpretation of the results using a reference knowledge base.

3.4.10 *Authentication of other sugar-rich products*

A similar analytical procedure is used to detect added beet and cane sugar in high sugar products such as honey (Lindner *et al.*, 1996) and maple syrup (Martin *et al.*, 1996c). For the latter, the high costs associated with maple syrup production, its labour-intensive collection and low yields, and the fact that its major component, sucrose, is also readily obtainable from a variety of other much cheaper sources, make it a likely target for economic adulteration. Studies have shown that an 'adulteration triangle' constructed from ethanols from authentic maple syrup samples as in the case of the fruit juices above, can be used to detect added beet sugar, when such a sugar represents between 5 and 20% of total solids.

3.4.11 *Origin determination of alcoholic beverages*

^{2}H-SNIF-NMR® has been proposed as a non-destructive method for identifying the origin of ethanols in various spirits and beverages (Martin *et al.*, 1983a). The internal distribution parameter r is highest for vodka, 2.55 (0.05), generally obtained from C3 type cereals or a mixture of C3 cereals and potatoes, and lowest for gin, $r = 2.23$ (0.02), and rum 2.24 (0.03) from corn and cane sources (C4) respectively. Between these two, bourbon whiskies (2.31; 0.03) can be distinguished from Scotch malt whiskies (2.43; 0.03).

The same relative deuterium distribution parameter, r, in addition to site-specific parameters $(D/H)_i$, can also provide information on beers brewed in different European countries (Martin *et al.*, 1985a). In this case the variations can be linked not just to the different geographical origin of the beer, but also to the specific brewing procedure used.

3.4.12 Acetic acid as isotopic probe

Vinegars are commonly obtained by bacterial or chemical oxidation of ethanol which itself has been produced by the fermentation of sugars of various origins. Determination of isotopic parameters on the acetic acid, CH_3COOH, produced in this way has been used by Remaud *et al.* (1992) as a means of authenticating vinegar. For the analysis, the vinegar is extracted with ethyl ether in a Soxhlet apparatus and the acetic acid distilled off using a spinning band Cadiot column similar to that used for the distillation of alcohol for the isotopic analysis of fruit juice. The $(D/H)_i$ content of the methyl site of the acetic acid is measured from the ^{2}H-NMR spectrum using *N,N*-tetramethylurea (TMU) as working standard. The results are comparable to those obtained from alcohols from the same origin and, in conjunction with the overall ^{13}C content obtained by SIRMS, can be used to confirm the origin of the vinegar (wine, cider, malt or beet) and detect as low as 15% addition of synthetic acetic in a natural vinegar of unknown origin.

3.4.13 SNIF-NMR® applied to other isotopic probes

As NMR and related technology has progressed so has the potential of SNIF-NMR® as an authentication method increased, with the possibility of determining isotopic ratios of more complex molecules than the examples given above. The best known example is authenticity of natural flavourings, of which vanilla is probably the most important.

Vanillin (Fig. 3.15) contains six monodeuterated isotopomers, although since two of the aromatic positions are equivalent, only five signals are observed in the ^{2}H-NMR spectrum (Fig. 3.16). All five site-specific ratios $(D/H)_i$ (i = 1 to 5) can be measured providing a wealth of information for discriminating between natural vanillin extracted from vanilla beans (*Vanilla planifolia*) and semi-synthetic vanillin from lignin or synthetic vanillin from guaiacol (Maubert *et al.*, 1988; Martin *et al.*, 1993b).

In recent years there have been some significant improvements in SNIF-NMR® analysis from enhanced analytical precision as a result of new software for automatic signal treatment, refined procedures for sample extraction and

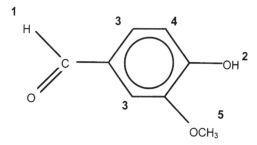

Figure 3.15 Vanillin molecule. The numbering of the positions corresponds to the signals in the ^{2}H-NMR spectrum in Fig. 3.16.

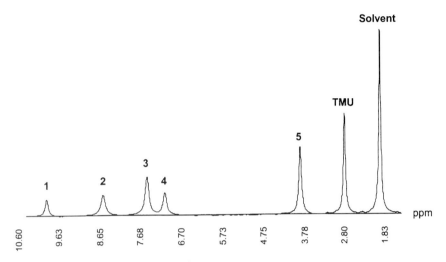

Figure 3.16 ^{2}H-NMR spectrum of vanillin.

purification, to the considerable progress in the statistical methods available for data treatment. A combination of these, described by Remaud *et al.* (1997), have now made it possible to assess the authenticity of vanilla flavour in finished food products such as ice-cream, yogurts and desserts.

Similar methodology to that described for vanillin above has been applied to a large number of compounds of interest as flavourings. These include molecules of the phenyl propanoid group, such as anetholes and estragols (Martin *et al.*, 1993b), of the terpenoid group, such as geraniol, linalool, limonene (Hanneguelle *et al.*, 1992; Martin *et al.*, 1993b) and aliphatic esters such as ethyl acetate, ethyl butyrate, isoamylacetate (Martin *et al.*, 1993b).

3.4.14 *The future*

As the example of flavour compounds above illustrates, increasingly complex molecules are becoming more accessible to ^{2}H-SNIF-NMR. Work is currently being financed by the European Commission Directorate General for Science, Research and Development (DG XII) to investigate the potential of this technique for the authentication of virgin olive oil (AIR project, 1994–1997). This is based on earlier studies of deuterium fractionation in the biosynthesis of plant fatty acids (Quemerais *et al.*, 1995) and involves the determination of site-specific parameters from the ^{2}H-spectrum of the fatty acid mixture obtained after saponification of the oil.

In principle, any compound which can be extracted from a food matrix without incurring isotopic fractionation and in sufficient quantity that a sufficiently well-resolved spectrum can be obtained, is a potential target for SNIF-NMR®. Site-specific isotope ratios measured using this methodology are a valuable source of information on the origin of the compound, and by inference, on that of the food matrix from which it was extracted, making SNIF-NMR® a powerful addition to any analyst's armoury in the field of food authentication and detection of economic adulteration.

References

AIR Project No AIR3-CT94-1224 (1994–1997) Molecular and isotopic characterisation of virgin olive oil.
Ashurst, P.R. and Dennis, M.J. (eds) (1996) *Food Authentication*, Blackie, London.
Belton, P.S., Delgadillo, I., Holmes, E., Nicholls, A., Nicholson, J.K. and Spraul, M. (1996) Use of high-field NMR spectroscopy for the analysis of liquid foods. *J. Agric. Food Chem.*, **44**, 1483–7.
Belton, P.S., Delgadillo, I., Gil, A.M. et al. (1997a) High-field proton NMR studies of apple juices. *Magn. Reson. in Chem.* (submitted).
Belton, P.S., Colquhoun, I.J., Kemsley, E.K.K. et al. (1997b) Application of chemometrics to the ^{1}H NMR spectra of apple juices: discrimination between apple varieties. *Food Chem.* (in press).
Craig, H. (1961) Standard for reporting concentrations of deuterium and oxygen-18 in natural waters. *Science* **133**, 1833–4.
Croasmun, W.R. and Carlson, R.M.K. (eds) (1994) *Two-Dimensional NMR Spectroscopy*, VCH, New York.
Day, M.P., Zhang, B.L. and Martin, G.J. (1993) Determination of the geographical origin of wine using joint analysis of elemental and isotopic composition. II. Differentiation of the principal production zones in France for the 1990 vintage. *J. Sci. Food Agric.*, **67**, 113–23.
Day, M.P., Zhang, B.L. and Martin, G.J. (1994) The use of trace element analysis to complement stable isotope methods in the characterization of grape musts. *Am. J. Enol. Vitic.*, **45**(1), 79–85.
Derome, A.E. (1987) *Modern NMR Techniques for Chemistry Research*, Pergamon Press, Oxford.
Dufour, S., Durand, T., Martin, D. et al. (1996) Carbon 13 NMR of liver from different avian species. Characterization of fatty livers from duck and goose. *J. Magn. Reson. Anal.*, **2**, 95–102.
E.C. Regulation No. 2676/90 Official Journal of the European Communities No. L 202/33, 3/10/90.
E.C. Regulation No. 2347/91 Official Journal of the European Communities No. L 214/92, 2/08/91.
E.C. Regulation No. 2348/91 Official Journal of the European Communities No. L 214/92, 2/08/91.
Evans, S.D., Nott, K.P. and Hall, L.D. (1996) The effect of freezing and thawing on the magnetic resonance imaging parameters of beef, lamb and pork meat. *J. Magn. Reson. Anal*, **2**, 179 (abstract).
Fan, T.W.-M. (1996) Metabolite profiling by one- and two-dimensional NMR analysis of complex mixtures. *Prog. NMR Spectrosc.*, **28**, 161–219.
Firestone, D. Reina, R.J. (1996) Authenticity of vegetable oils, in *Food Authentication*, (eds P.R. Ashurst and M.J. Dennis) Blackie, London, pp. 198–258.
Forina, M., Armanino, C., Lanteri, S. and Tiscornia, E. (1983) Classification of olive oils from their fatty acid composition, in *Food Research and Data Analysis* (eds H. Martins and H. Russwurm), Applied Science, London, pp. 189–214.
Forveille, L., Vercauteren, J. and Rutledge, D.N. (1996) Multivariate statistical analysis of two-dimensional NMR data to differentiate grapevine cultivars and clones. *Food Chem.*, **57**, 441–50.
Guillou, C., Reniero, F. and Serrini, G. (1995) *A review of the European Wine Data Bank*. Proceedings of the Third European Symposium on Food Authenticity, October 12–13, Nantes, France.
Guillou, C., Trierweiler, M. and Martin, G.J. (1988) Repeatability and reproducibility of site-specific isotope ratios in quantitative ^{2}H-NMR. *Magnetic Resonance in Chemistry*, **26**, 491–6.

Gunstone, F.D. (1993) High resolution ^{13}C NMR spectroscopy of lipids, in *Advances in Lipid Methodology*, 2 (ed. W.W. Christie), The Oily Press, Dundee, pp. 1–68.
Günther, H. (1995) *NMR Spectroscopy* (2nd edn), Wiley, Chichester.
Hanneguelle, S., Thibault, J.N., Naulet, N. and Martin, G.J. (1992) *J. Agric. Food Chem.*, **40**, 81–97.
Holland, M.V., Bernreuther, A. and Reniero, F. (1995) The use of amino acids as a fingerprint for the monitoring of European wines, in *Magnetic Resonance in Food Science* (eds P.S. Belton, I. Delgadillo, A.M. Gil and G.A. Webb), Royal Society of Chemistry, Cambridge, UK, pp. 136–45.
Holmes, E., Foxall, P.J.D., Nicholson, J.K. *et al.* (1994) Automatic data reduction and pattern recognition methods for analysis of ^{1}H nuclear magnetic resonance spectra of human urine from normal and pathological states. *Anal. Biochem*, **220**, 284–96.
Howell, N., Shavila, Y., Grootveld, M. and Williams, S. (1996) High-resolution NMR and magnetic resonance imaging (MRI) studies on fresh and frozen cod (*Gadus morhua*) and haddock (*Melanogrammus aeglefinus*). *J. Sci. Food Agric.*, **72**, 49–56.
Howells, S.L., Maxwell, R.J., Peet, A.C. and Griffiths, J.R. (1992) An investigation of tumor ^{1}H nuclear magnetic resonance spectra by the application of chemometric techniques. *Magn. Reson. Med.*, **28**, 214–36.
Hull, W.E.. (1994) Experimental aspects of two-dimensional NMR, in *Two-dimensional NMR Spectroscopy* (2nd edn) (eds W.R. Croasmun and R.M.K. Carlson), VCH, New York., pp. 67–456.
Lindner, P., Bermann, E. and Gamarnik, B. (1996) Characterisation of citrus honey by deuterium NMR. *J. Agric. Food Chem.*, **44**(1), 139–40.
Lambelet, P., Renevey, F., Kaabi, C. and Raemy, A. (1995) Low-field nuclear magnetic resonance study of stored or processed cod. *J. Agric. Food Chem.*, **43**, 1462–6.
Lundberg, P., Vogel, H.J. and Ruderus, H. (1986) Carbon-13 and proton NMR studies of post-mortem metabolism in bovine muscles. *Meat Science*, **18**, 144–60.
Martin, G.G., Hanote, V., Lees, M. and Martin, Y.L. (1996a) Interpretation of combined ^{2}H SNIF/NMR and ^{13}C SIRA/MS analyses of fruit juices to detect added sugar. *J.A.O.A.C. Int.*, **79**(1), 62–72.
Martin, G.G., Martin, Y.L., Naulet, N. and McManus, H.J.D. (1996c) Application of ^{2}H SNIF/NMR and ^{13}C SIRA/MS analyses to maple syrup. Detection of added sugars. *J. Agric. Food Chem.*, **44**, 3206–13.
Martin, G.G., Remaud, G. and Martin, G.J. (1993a) Isotopic methods for control of natural flavours authenticity. *Flavour and Fragrance Journal*, **4**, 97–107.
Martin, G.G., Wood, R., Martin, G.J. (1996b) Detection of added beet sugar in concentrated and single strength fruit juices by deuterium nuclear magnetic resonance (SNIF-NMR method): collaborative study. *J.A.O.A.C. Int.*, **79**(4), 917–28.
Martin, G.J. (1995) Analytical performance of high resolution NMR, in *Magnetic Resonance in Food Science* (eds P.S. Belton, I. Delgadillo, A.M. Gil and G.A. Webb), Royal Society of Chemistry, Cambridge, UK, pp. 105–19.
Martin, G.J., Benbernou, M. and Lantier, F. (1985a) Application of site-specific natural isotope fractionation (SNIF-NMR) of hydrogen to the characterization of European beers. *J. Inst. Brew.*, **91**, 242–9.
Martin, G.J., Guillou, C., Naulet, N. *et al.* (1986a) Control of origin and enrichment of wine by specific isotope analysis. Study of different methods of wine enrichment. *Sciences des Aliments*, **6**, 385–405.
Martin, G.J., Guillou, C., Martin, M.L. *et al.* (1988) Natural factors of isotope fractionation and the characterisation of wines. *J. Agric. Food Chem.*, **36**, 316–22.
Martin, G.J., Hanneguelle, S. and Remaud, G. (1993b) Application of site-specific natural isotope fractionation studied by nuclear magnetic resonance (SNIF-NMR) to the detection of flavour and fragrance adulteration. *Ital. J. Food Sci.*, **3**, 191–213.
Martin, G.J. and Martin, M.L. (1981) Deuterium labelling at the natural abundance level as studied by high field quantitative ^{2}H NMR, *Tetrahedron Lett.* **22**, 3525–8.
Martin, G.J., Martin, M.L., Mabon, F. and Michon, M.J. (1983a) A new method for the identification of the origin of ethanols in grain and fruit spirits: high-field quantitative deuterium nuclear magnetic resonance at the natural abundance level. *J. Agric. Food Chem.*, **31**, 311–15.
Martin, G.J. and Martin, G.G. (1995a) in *Methods to Detect Adulteration of Fruit Juice Beverages*, (eds S. Nagy and R.L. Wade), Vol. 1, AgScience, Inc., Auburndale, FL, pp. 1–27.

Martin, G.J. and Martin M.L. (1995b) Stable isotope analysis of food and beverages by Nuclear Magnetic Resonance in *Annual Reports on NMR Spectroscopy*, Vol. 31, Academic Press, pp. 81–104.

Martin, G.J., Sun, X.Y., Guillou, C. and Martin, M.L. (1985b) NMR determination of absolute site-specific natural isotope ratios of hydrogen in organic molecules. Analytical and mechanistic applications. *Tetrahedron*, **41**, 3265–96.

Martin, G.J. and Trierweiler, M. (1994) Ethanol reference materials for ^{2}H NMR determinations (SNIF-NMR). *BCR Information* EUR 14395 EN.

Martin, G.J., Trierweiler, M., Ristow, R., Hermann, A. and Belliardo, J.J. (1994) The certification of the three reference ethanols by SNIF-NMR: BCR Certified Reference Material CRM 123, *BCR Information*, EUR 15347 EN.

Martin, G.J., Zhang, B.L., Martin, M.L. and Dupuy, P. (1983b) Application of quantitative deuterium NMR to the study of isotope fractionation in the conversion of saccharides to ethanols. *Biophy. Res. Commun.*, **111**, 890–6.

Martin, G.J., Zhang, B.L., Naulet, N. and Martin, M.L. (1986b) Deuterium transfer in the bioconversion of glucose to ethanol studied by specific isotope labelling at the natural abundance level. *J. Am. Chem. Soc.*, **108**, 183–93.

Martin, M.L. and Martin, G.J. (1990) Deuterium NMR in the study of site-specific natural isotope fractionation, in *NMR Basic Principles and Progress*, Vol. 23 (ed. H. Günther), Springer-Verlag, Heidelberg, pp. 1–60.

Martin, Y.L. (1994) A global approach to accurate and automatic quantitative analysis of NMR spectra by complex least-squares curve fitting. *J. Magn. Res.* Series AIII, 1–10.

Maubert, C., Guerin, C., Mabon, F. and Martin, G.J. (1988) Détermination de l'origine de la vanilline par analyse multidimensionnelle du fractionnement isotopique naturel spécifique de l'hydrogène. *Analysis*, **16**, 434–9.

Nicholson, J.K. and Wilson, I.D. (1989) High resolution proton magnetic resonance spectroscopy of biological fluids. *Prog. NMR Spectrosc.*, **21**, 449–501.

Quemerais, B., Mabon, F., Naulet, N. and Martin, G.J. (1995) Site-specific isotope fractionation of hydrogen in the biosynthesis of plant fatty acids. *Plant, Cell Environ.*, **19**, 989–98.

Rapp, A., Spraul, M. and Humpfer, E. (1986) ^{13}C NMR spectroscopic determination of ethylene glycol in wine. *Z. Lebensm. Unters. Forsch*, **182**, 419.

Rapp, A., Markowetz, A. and Niebergall, H. (1991) Application of ^{13}C-NMR spectroscopy for detection and quantitative determination of amino acids in wine and fruit juices. *Z. Lebensm. Unters. Forsch.*, **192**, 1–6.

Rapp, A. and Markowetz, A. (1993) NMR spectroscopy in wine analysis. *Chemie in unserer Zeit*, **27**, 149–55.

Remaud, G., Guillou, C., Vallet, C. and Martin G.J. (1992) A coupled NMR and MS isotopic method for the authentication of natural vinegars. *Fresenius J. Anal. Chem.*, **342**, 457–61.

Remaud, G., Martin, Y.L., Martin G.G. and Martin G.J. (1997) Detection of natural vanilla flavors and extracts. Application of the SNIF-NMR® method to vanillin and parahydroxybenzaldehyde. *J. Agric. Food Chem.* **45**(3), 859–66.

Sacchi, R., Addeo, F., Giudicianni, I. and Paolillo, L. (1989) Nuclear magnetic resonance spectroscopy in the analysis of olive oils. *Riv. Ital. Sostanz. Grazze*, **66**, 171–8.

Sacchi, R., Addeo, F., Giudicianni, I. and Paolillo, L. (1990) Application of ^{13}C-NMR spectroscopy to the determination of monoglycerides and free fatty acids in virgin and refined olive oil. *Riv. Ital. Sostanz. Grasse*, **67**, 245–52.

Sacchi, R., Addeo, F., Giudicianni, I. and Paolillo, L. (1992) Analysis of the positional distribution of fatty acids in olive oil triacylglycerols by high resolution ^{13}C-NMR of the carbonyl region. *Ital. J. Food Sci.*, **4**, 117–23.

Sacchi, R., Medina, I., Aubourg, S.P. et al. (1993) Quantitative high-resolution ^{13}C NMR analysis of lipids extracted from the white muscle of Atlantic tuna (*Thunnus alalunga*). *J. Agric. Food Chem.*, **41**, 1247–53.

Sacchi, R., Patumi, M., Fontanazza, G. et al. (1996) A high-field ^{1}H nuclear magnetic resonance study of the minor components in virgin olive oils. *J. Am. Oil Chem. Soc.*, **73**, 747–58.

Shaw, A.D., di Camillo, A., Vlahov, G. et al. (1996) *Discrimination of different olive oils using ^{13}C NMR and variable reduction*. Proceedings of Food Authenticity 96 Conference, Norwich, UK.

Shoolery, J.N. (1977) Some quantitative applications of ^{13}C NMR spectroscopy. *Prog. NMR Spectrosc.*, **11**, 79–93.

Spraul, M., Neidig, P., Klauck, U. *et al.* (1994) Automatic reduction of NMR spectroscopic data for statistical and pattern recognition classification of samples. *J. Pharm. Biomed. Anal.*, **12**, 1215–25.

Swallow, K.W. and Low, N.H. (1993) Isolation and identification of oligosaccharides in a commercial beet medium invert syrup. *J. Agric. Food Chem.*, **41**, 1587–92.

Tsimidou, M., Macrae, R. and Wilson, I. (1987a) Authentication of virgin olive oils using principal component analysis of triglyceride and fatty acid profiles: Part 1 – classification of Greek olive oils. *Food Chem.*, **25**, 227–39.

Tsimidou, M., Macrae, R. and Wilson, I. (1987b) Authentication of virgin olive oils using principal component analysis of triglyceride and fatty acid profiles: Part 2 – detection of adulteration with other vegetable oils. *Food Chem.*, **25**, 251–8.

Vlahov, G. (1996a) Improved quantitative ^{13}C nuclear magnetic resonance criteria for determination of grades of virgin olive oils. The normal ranges for diglycerides in olive oil. *J. Am. Oil Chem. Soc.*, **73**, 1201–03.

Vlahov, G. (1996b) The structure of triglycerides of monovarietal olive oils: a ^{13}C-NMR comparative study. *Fett/Lipid*, **98**, 203–205.

Vogels, J.T.W.E., Tas, A.C., van den Berg, F. and van der Greef, J. (1993) A new method for classification of wines based on proton and carbon-13 spectroscopy in combination with pattern recognition techniques. *Chem. Intell. Lab. Syst.*, **21**, 249–58.

Vogels, J.T.W.E., Terwel, L., Tas, A.C. *et al.* (1996a) Detection of adulteration in orange juices by a new screening method using proton NMR spectroscopy in combination with pattern recognition techniques. *J. Agric. Food Chem.*, **44**, 175–80.

Vogels, J.T.W.E., Tas, A.C., Venekamp, J. and van der Greef, J. (1996b) Partial Linear Fit: a new NMR spectroscopy pre-processing tool for pattern recognition applications. *J. Chemom.*, **10**, 425–38.

Wollenberg, K.F. (1990) Quantitative high resolution ^{13}C nuclear magnetic resonance of the olefinic and carbonyl carbons of edible vegetable oils. *J. Am. Oil Chem. Soc.*, **67**, 487–94.

Wollenberg, K.F. (1991) Quantitative triacylglycerol analysis of whole vegetable seeds by ^{1}H and ^{13}C magic angle spinning NMR spectroscopy. *J. Am. Oil Chem. Soc.*, **68**, 391–400.

Zamora, R., Navarro, J.L. and Hidalgo, F.J. (1994) Identification and classification of olive oils by high-resolution ^{13}C nuclear magnetic resonance. *J. Am. Oil Chem. Soc.*, **71**, 361–4.

4 Infrared spectroscopy

C.N.G. SCOTTER and R. WILSON

4.1 Theory

Infrared spectroscopy (Banwell, 1994) is divided into three quite separate forms called 'near-infrared', 'mid-infrared' and 'far-infrared'. These distinctions are largely historical and based on the exact nature of the absorptions giving rise to the corresponding spectra, as well as differences in instrumental design and experimental approach. All are forms, however, of vibrational spectroscopy and arise from transitions between vibrational energy levels. In the simplest approach to vibrational spectroscopy, one can consider the bond between two atoms of masses m_1 and m_2 as behaving as a tiny spring of 'strength', or force constant, k. This system will vibrate at some natural resonance frequency, f, given by Hooke's law:

$$f = \frac{1}{2\pi}\sqrt{\frac{k}{\mu}} \qquad (4.1)$$

μ is called the 'reduced mass' and is defined as:

$$\mu = \frac{m_1 m_2}{m_1 + m_2}$$

For a vibration to give rise to an infrared spectrum there must be a large electronegativity defference between atoms m_1 and m_2, creating a dipole moment. Thus absorption features arise in infrared spectra from covalent bonds like $O-H$, $C-H$, $N-H$, $C=O$. Band broadening and shifting phenomena may also occur as a result of weak interactions between hydrogen bonds, when, for example, water, carbohydrates or proteins undergo structural change.

This simple approach can be used to explain the observed difference in absorption frequencies between different functional groups, on the basis of differing force constants or reduced masses but quantum theory also has to be considered. The energy, E, of a photon of wavelength gamma, is:

$$E = hf = h\frac{c}{\lambda} \qquad (4.2)$$

In this equation h is Planck's constant and c is the velocity of light. Hooke's law, for a simple harmonic oscillator model predicts a potential energy curve as a function of internuclear distance that is parabolic, with the potential energy minimized at the equilibrium nuclear distance. Increasing internuclear distance leads to increased potential energy in a continuous manner. In a quantum mechanical approach, however, only certain energy levels are permitted. These energy levels are given by:

$$E(n) = (n + \tfrac{1}{2})hf \qquad (4.3)$$

where $n = 0, 1, 2,$ are the vibrational quantum numbers. These energy levels are illustrated in Fig. 4.1. Transition between energy levels can only occur in discrete steps when sufficient energy, E, is provided: $\Delta E = hf$.

Permitted transitions require that $\Delta n = \pm 1$. The energy of the photon will be absorbed by the molecule if it matches that required for the transition between energy levels. However, there is an additional consideration that requires a change in the dipole moment associated with the vibration. The transitions occurring in which $\Delta n = \pm 1$ are called fundamental vibrations and it is these that are observed in mid-infrared spectroscopy. The energy required to stimulate these transitions occurs at wavelengths between 2.5 and 25 µm (mid-infrared spectroscopists normally use frequency units called wavenumbers or cm^{-1}, in which case the mid-infrared region extends from 4000–400 cm^{-1}).

In addition, some NIR instruments incorporate near-visible (850–1100 nm) and visible (400–850 nm) wavelengths. In these regions the spectra reflect both weak vibrational overtones and electronic transitions. As the term implies, electronic transitions provide absorptions in the spectra as a result of electron

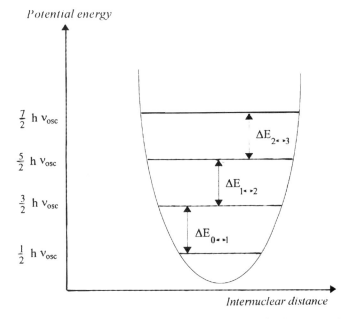

Figure 4.1 Potential energy diagram for harmonic oscillator. Energy levels are spaced at constant separation = ΔE.

energy transitions rather than atoms or nuclei as described above in the vibrational spectroscopy model. Typically, electron transition absorptions in spectra are related to colour.

Vibrating bonds, however, are anharmonic oscillators (Fig. 4.2). When the internuclear distance is small, atomic repulsion provides a large potential energy difference between energy levels (0–1, 1–2, etc. in Fig. 4.1). As the internuclear separation increases the bond stretches but the energy differences between the overtone energy levels are reduced compared with the transitions between, say, the fundamental vibrational energy level and the 1st or 2nd overtones. This anharmonic behaviour can be incorporated into the Schrödinger equation and leads to a new expression for permitted energy levels;

$$E(n) = (n + \tfrac{1}{2})hf - (n + \tfrac{1}{2})^2 x_e hf \tag{4.4}$$

x_e is a small and positive anharmonicity constant. As a result, anharmonicity energy levels become closer as n increases and transitions of the type $n\Delta = \pm 2$, $\Delta n = \pm 3$, or overtones are allowed (Fig. 4.2). Other transitions can also occur in which $E = hf_1 + hf_2$ etc. These are called combination bands. Combinations

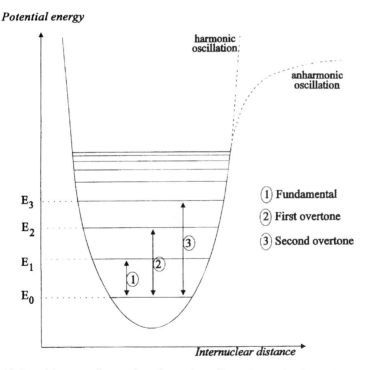

Figure 4.2 Potential energy diagram for anharmonic oscillator. Energy levels spacing becomes reduced with increasing potential energy. Molecule dissociates at higher internuclear distance and overtones occur. $E_0, E_1, \ldots$ etc. are individual energy levels.

and overtones are seen at higher energy in the near infrared region (1–2.5 μm, 10 000–4000 cm^{-1}). However, not all fundamental vibrations will give rise to overtones and, in practice, the greatest intensity occurs for vibrations of the type C—H, O—H and N—H. The possible number of overtones and combinations resulting from a few fundamental vibrations is one reason that near-infrared spectra are difficult to interpret and assign. On the other hand, because these vibrations are formally forbidden in quantum mechanics, they are relatively weak compared to the fundamentals. This is a practical advantage for near-infrared spectroscopy since it allows quite long optical paths to be used which simplifies sample presentation and ensures representative sampling. Mid-infrared spectra may be easier to interpret but the high absorption coefficients mean that short optical paths must be used and this can give rise to problems in heterogeneous systems. Near-infrared spectroscopy has been used for many years for quantitative analysis of foods and mid-infrared has become so used in recent years as new sample presentation methods have been developed. Both techniques are relatively new to the field of food authentication.

4.2 Mid-infrared spectroscopy: equipment and methods

Most mid-infrared spectroscopy today is carried out using Fourier transform instrumentation, based on the Michelson interferometer. A typical arrangement is shown in Fig. 4.3. The Michelson interferometer comprises a beamsplitter and two mirrors, one of which is moveable. Light from the infrared source is sent into the interferometer and each wavelength is modulated (encoded). This

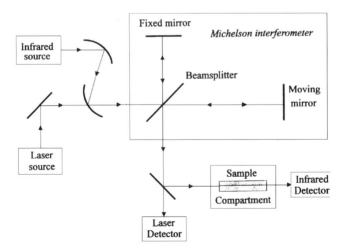

Figure 4.3 Schematic of typical Fourier transform infrared spectrometer. Function of components is described in text.

occurs as follows: light impinges upon the beamsplitter where it is divided into two beams that travel to the two mirrors. After reflection at these mirrors the light returns to the beamsplitter where it can recombine. When the mirrors are equidistant from the beamsplitter light of all wavelengths arrives back at the beamsplitter having travelled the same distance and will be in-phase. This leads to constructive interference at the beamsplitter and maximum intensity (brightness) of light emerging from the interferometer. If the moveable mirror is moved some distance, introducing an optical path difference or optical retardation (δ) into one of the beams then, depending upon the path difference, light will no longer be in-phase on return to the beamsplitter. For any given wavelength (λ) there will come a time when the path difference is such that destructive interference occurs and no light of that wavelength emerges from the interferometer. If the mirror continues to move sufficient path difference will be induced such that constructive interference again occurs, at $\delta = \lambda$, 2λ, 3λ etc. (with destructive interference at $\delta = 0.5\lambda$, 1.5λ). Each wavelength will be converted into a sinusoid of characteristic frequency. The summation of all the sinusoids is called an interferogram. This is related to the spectrum via the process of Fourier transformation. The light from the interferometer is passed through the sample and on to the infrared detector. A visible laser system is incorporated in order to determine the position of the moving mirror and to provide evenly spaced digitization of data. The use of a Michelson interferometer allows much more energy to reach the sample, provides reproducible wavelength reproducibility and allows spectra to be collected in a very short time. All these principles are described in detail elsewhere (Griffiths and DeHaseth 1986). However, these advantages when combined, allow sample presentation techniques to be used which are ordinarily inefficient optically but which, with Fourier transform infrared spectroscopy, become viable techniques overcoming some of the traditional sample presentation problems associated with mid-infrared spectroscopy.

The most important sample presentation methods are diffuse reflectance (called DRIFT) and attenuated total reflectance (ATR) which are described in detail by Wilson and Goodfellow (1994). Both methods require the minimum of sample preparation. Diffuse reflectance is more or less the same as used in near-infrared but, because mid-infrared wavelengths are similar to the size of particles in many powders, diffuse reflectance spectra can suffer from distortion due to scattering, surface reflection and intense absorption (Belton and Wilson, 1990).

In ATR a crystal of high refractive index is used (Fig. 4.4) which is cut into a parallelogram or trapezoidal shape. Infrared light is sent into the crystal at such an angle that it is totally internally reflected. Depending upon the exact geometry and length, the light will undergo multiple reflections and emerge from the crystal. At each reflection an evanescent wave is established that decays exponentially into the medium in contact with the crystal. If this medium is absorbing then there will be transfer of energy from inside the crystal to the surrounding

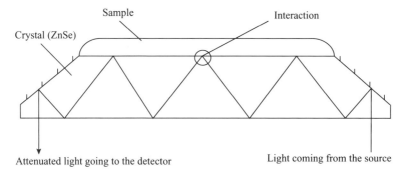

Figure 4.4 Schematic of an attenuated total reflectance cell (refer to text for details).

medium and the emerging beam will be attenuated. ATR does not rely on the sample, which constitutes the surrounding medium, being transparent or transmitting in the conventional sense. As a result, ATR allows opaque or highly scattering samples to be used; the only proviso is that the sample must make intimate contact with the ATR crystal. The effective penetration (optical path) at any reflection is very short, typically a few μm, so that ATR can be used to overcome the strong absorption of materials such as water that necessitate the use of short optical paths. In ATR these optical paths are very reproducible from one sample to another. ATR then allows easy sample measurement and is one reason why there has been an upsurge in interest in the mid-infrared region. For authenticity studies this provides the rapid sample throughput required of screening techniques. The major problem with ATR, and one of the mid-infrared region generally, is that heterogeneous samples give rise to problems in that the volume sampled by the infrared beam may not be representative of bulk composition.

Fourier transform infrared spectrometers are single-beam devices. The collection of the spectrum of a sample requires the initial collection of a 'background' single-beam spectrum. This is usually collected with the sample accessory in place without a sample, or for aqueous systems, with water in the cell. In diffuse reflectance finely ground potassium bromide is often used. The background spectrum is a black-body profile of the source, modified by the detector response and showing evidence of atmospheric absorptions (carbon dioxide and water vapour). Another single-beam spectrum is collected with the sample in place and this is ratioed to the background spectrum to eliminate the atmospheric absorptions and the black-body profile. The resulting transmittance spectrum is then converted to an absorbance spectrum for quantitative purposes (Wilson and Goodfellow 1994). Each single beam spectrum is produced by the Fourier transformation of a number of co-added interferograms. The number co-added is typically 64–256 and in authenticity studies it is advisable to use a new background for each sample in order to avoid introducing any group structure resulting from commonality of backgrounds.

4.3 Near-infrared instrumentation and methods

4.3.1 Instrumentation

Prior to the late 1960s the near-infrared region was largely rejected as a means of organic compositional analysis. The NIR band in the electromagnetic spectrum is narrow (750–2500 nm) compared with the mid-infrared (2800–50000 nm) and the absorption peaks were observed to be overlapping and difficult to resolve with their interpretation being most uncertain. Small extinction coefficients and low intensities are associated with overtones and combinations compared with the mid-IR spectral features, and posed a severe challenge to instrument designs to provide analytically functional signal to noise levels and instrumental stability (Burns and Ciurczak, 1992).

It was not until the late 1960s that Karl Norris at the US Department of Agriculture demonstrated the potential for NIR instrumentation to deliver useful quantitative data for components in agricultural products. The first commercial instrument employing a tungsten halogen source, research grade interference filters and lead sulphide detectors in a 45° array was launched in 1971. In 1975 the Canadian Grain Commission adopted NIR analysis as an official method. Five years later the American Federal Grain Inspection Service followed the same path (Osborne *et al.*, 1993). Between 1975 and the present time a range of instrumental principles and designs have been produced by manufacturers. Both qualitative and quantitative methods of analysis have been developed in the fields of agriculture, food, chemical and textile production and pharmaceuticals. Scotter (1995) has listed 56 quantitative and qualtitative NIR applications to 55 food raw materials or products. The majority are off-line analyses, but on-line measurements have also been noted.

Near-infrared spectrometers range from relatively simple filter instruments with a few wavelengths for a dedicated purpose to full wavelength scanning instruments with a range extending from the low visible to the top of the NIR range (400–2500 nm). Five main types of wavelength generation are available:

1. interference filter
2. holographic grating
3. Fourier transform
4. acousto-optically tunable filter (AOTF), and
5. light emitting diode (LED).

In addition, a wide range of detector types and arrangements are available. Noble (1995) and Osborne *et al.* (1993) provide detailed information about types and manufacturers of these instruments.

4.3.2 NIR methods

The reasons for the application of NIR instrumentation to the measurements of a wide variety of constituents and qualities in industries from agriculture to chemicals, lies in the relative ease and rapidity with which the analyses can be performed. Once calibrated, instruments can provide multicomponent analysis of the required accuracy with one sample scan taking only a few seconds. Many of the instruments are also suitable for operating in harsh industrial environments. Materials for analysis can be presented in almost any form, whole or without any preparation, as solids, powders, pastes, slurries and liquids. Both transmission and diffuse reflectance measurement geometries are available. Osborne *et al.* (1993) give a detailed account of methods of presentation.

4.4 Applications of mid-infrared spectroscopy

4.4.1 *Fruit products*

The use of mid-infrared spectroscopy for the authentication of fruit-based products dates back to 1993 (Wilson *et al.*, 1993) and a paper describing the use of FTIR for the determination of the fruit content of jam. Although the quantitative estimation of fruit content using the potassium bromide pellet method in combination with partial least squares (PLS) analysis was only partially successful, the authors showed some diffuse reflectance spectra which indicated the potential for authentication. Sample preparation involved collecting the washed solids of jam on a filter paper which was then cut into small discs for examination. The authors showed that each fruit gave rise to what appeared to be characteristic 'fingerprints', although the spectra were affected by spectral distortions which are sometimes known to afflict diffuse reflectance spectra in the mid-infrared region. The author interpretation was that the solid components of jam, which were largely materials of the fruit cell walls, had unique compositional and optical properties. Later workers (Belton *et al.*, 1995) performed more controlled experiments in which the infrared spectra of extracted plant cell walls were collected using infrared microspectroscopy (McCann *et al.*, 1992). These workers were able to show that is was possible to discriminate between apple plant cell walls and the cell walls from other fruits. This reinforced the earlier conclusions that plant cell wall composition could be an indicator of plant species.

More diffuse reflectance results were reported (Defernez and Wilson 1995), but this time a different optical geometry was used within the diffuse reflectance cell that eliminated the spurious optical effects seen earlier. Better quality data were produced and discrimination between strawberry and other jams was achieved. In a search for more rapid data presentation methods these authors used ATR for the analysis of homogenized whole jams but found that sugar content was the dominant effect. However, ATR was found to be useful for the authentication of fruit purées used in jam production (Defernez *et al.*, 1995). Using discriminant analysis with 149 spectra these authors showed that it was possible to identify fruit species (100% classification success). Other factors could be determined, such as the addition of sulphur dioxide, the ripeness of fruit and whether the fruit was fresh or freeze-thawed. Typical ATR spectra are shown in Fig. 4.5. The authors used linear discriminant analysis and made no attempt to determine the ability to detect adulteration. Recently, the same workers have reported the use of PLS regression on to a dummy variable (Fig. 4.6) with nearly nine hundred spectra for detecting adulterated samples (Kemsley *et al.*, in press). The levels detectable vary with the adulterant used but the authors conclude their method to be suitable for screening purposes. Chemometric methods for adulteration detection are significantly different to those used for classification. A number of methods have been recently compared using infrared spectra as model systems (Kemsley 1996). Some interpretation of the molecular basis for discrimination has been possible and whereas cell-wall components such as pectin are important, in purées solutes such as sugars, oligosaccharides and organic acids are probably more relevant.

4.4.2 *Coffee*

In coffee more spectral interpretation has been possible. One of the main authenticity issues addressed is that of discrimination between *Coffea arabica* and *Coffea canephora* variant *Robusta* beans in ground and instant coffee powders. Some workers have looked at the use of infrared spectroscopy for the classification of ground green beans (Suchanek *et al.*, 1996; Dupuy *et al.*, 1995).

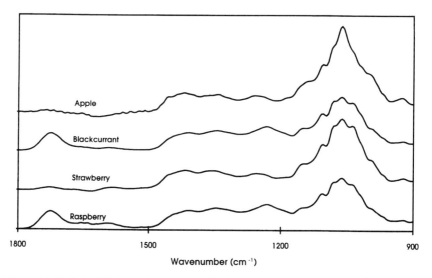

Figure 4.5 Typical mid-infrared ATR spectra of fruit purées between 1800–900 cm^{-1} ('fingerprint region'). Structure in region 1200–900 cm^{-1} reflects differing composition of sugars. Pectin, a key esterified polysaccharide component and organic acids contribute to spectrum in region 1800–1400 cm^{-1}.

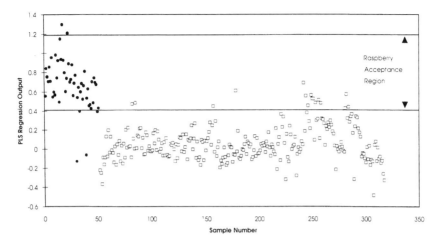

Figure 4.6 Presentation of mid-infrared chemometric analysis of fruit purées. Method used was PLS regression on to a dummy variable with authentic raspberry purées assigned nominal value of 1 and all other purées assigned as 0. Threshold acceptance value set as shown. ●, Raspberry; □, non-raspberry.

Whereas one set of workers developed a system of analysis using potassium bromide pellets (Suchanek *et al.*, 1996), the others (Dupuy *et al.*, 1995) used the dry-extract approach in which a ground sample was suspended in water and dried on to a silicon window coated with a calcium fluoride layer. However, as the number of samples used in these studies was very small (4 and 15), no statistically robust conclusion could be drawn and the sample presentation methods were quite time consuming. Furthermore, green beans can be classified by visual examination. Ground roast beans are more difficult to identify, but a study (Kemsley *et al.*, 1995) showed that this was possible. These workers used diffuse reflectance, thus overcoming sample preparation difficulties but a common problem with mid-infrared diffuse reflectance is that there can be great variation with repeatability of replicate loadings. To overcome these problems, a data pre-treatment protocol was established involving baseline correction and area normalization of spectra before use in a discriminant analysis. Kemsley *et al.* were able to obtain a discrimination using 28 samples which, although still a relatively small dataset, was sufficient to allow some validation to be performed. The authors were able to offer some interpretation on the molecular basis of discrimination, showing that lipid content and polysaccharide profiles were important factors.

The problem of authentication becomes more difficult in instant coffee powders. Briandet *et al.* (1996a) have carried out an infrared study using laboratory-prepared samples and again used diffuse reflectance sampling. Using a total of 52 samples and independent validation, a classification success of

100% was obtained. Discrimination was also carried out using solutions of powdered coffee applied to an ATR plate and, again, 100% classification success was achieved. Interpretation of principal components loadings of the ATR spectra showed that the levels of caffeine and chlorogenic acid were important in the discrimination. This result was consistent with the known compositional differences between *arabica* and *robusta* and agreed with near-infrared results on similar samples (Downey and Boussion, 1996). Using partial least squares analysis and the ATR spectra of *robusta/arabica* blends, Briandet *et al.* were able to show the potential of the mid-infrared technique for the quantitative estimation of *robusta* content. The authors obtained a correlation of 0.99 with a standard error of prediction, SEP, of 1.3% w/w. Instant coffee can be adulterated by a wide range of materials, including parchment or added sugars and starch as well as extraneous vegetable materials. Briandet *et al.* (1996b) have carried out a mid-infrared study, again using diffuse reflectance and ATR, aimed at determining the ability to detect, and quantify, a range of potential adulterants. The adulterants chosen included xylose, glucose and fructose as well as starch and chicory and a number of chemometric methods

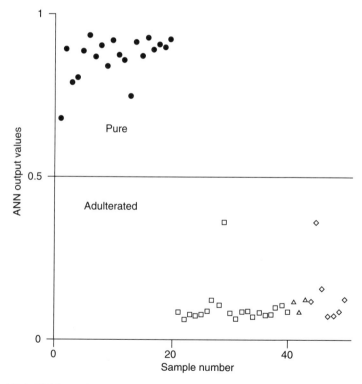

Figure 4.7 Artificial neural network assignments (independent test set). Clear separation between pure coffees and laboratory-adulterated samples obtained. Reproduced with permission from Briandet *et al.*, 1996b. ●, Pure coffee; □, coffee + glucose; △, coffee + starch; ◇, coffee + chicory.

such as linear discriminant analysis (LDA), PLS and artificial neural networks (Fig. 4.7) were compared for their ability to identify adulterated samples. Artificial neural networks appeared to be particularly good at identifying adulterated samples and quantitative estimation of added sugars was also possible (Fig. 4.8).

4.4.3 *Edible oils*

Edible oils, particularly olive oils, have also attracted the attention of midinfrared spectroscopists. Lai *et al.* (1994) have used ATR to classify various vegetable oils. Over 40 samples were used and good classification ability was found but the number of samples was not high and there was some doubt as to sample authenticity. These workers were able to discriminate between extra virgin and other olive oils using a total of 13 training and 5 validation samples. In a later paper (Lai *et al.*, 1995) the same authors demonstrated the possibility of using infrared spectroscopy to quantify mixtures of extra virgin olive oils

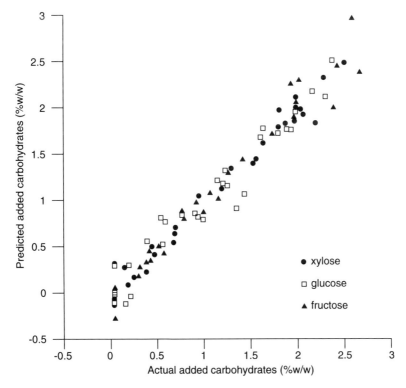

Figure 4.8 Calibration for added carbohydrates in coffee. Predicted (by PLS1) versus actual added xylose, fructose and glucose. Reproduced with permission from Briandet *et al.*, 1996b.

with refined olive oils. Similar work has been reported using Raman spectroscopy (Baeten *et al.*, 1996).

4.5 Applications of NIR spectroscopy

Practical applications of NIR to feed and food analysis were developed in the late 60s by Karl Norris at the USDA (United States Department of Agriculture) laboratory at Beltsville, Maryland, USA. Since that time quantitative and qualitative applications of the technique have been developed widely in the industrial fields of chemical and textile production, pharmaceuticals and, of course, agriculture, animal feeds and food, as noted by Burns and Ciurczak (1992).

4.5.1 *Cereals*

Wheat hardness. Before the mid-1980s few references can be found to authenticity (Food Science and Technology abstracts). One of the earliest applications of NIR spectroscopy to a cereals authenticity issue arose from the difficulty of defining and discriminating hard and soft wheats. Norris *et al.* (1989) described an NIR reflectance technique for defining wheat hardness, obtaining a coefficient of determination of 7% (the Standard Error expressed as a percentage of the range). Since 1989 Delwiche and Norris (1993) and Brown *et al.* (1993) have further developed the measurement.

Wheat variety. Bertrand *et al.* (1987) studied the use of NIR as a tool to discriminate between wheat varieties. Using 206 samples of wheat, comprising 6 typical French varieties using multi-dimensional analysis based on the principal components derived from the sample spectra, Bertrand and his co-workers succeeded in constructing a discriminant calibration model which predicted 95% of the new test samples correctly.

Wheatflour – type for purpose. Sirieix and Downey (1993) conducted an NIR investigation to determine if the NIR method in reflectance could be used to determine differences between six different types of white wheatflour, namely biscuit, self-raising, household, cake, baker's and soda-bread. Using a factorial discriminent analysis model the workers were able to predict 96 out of 99 samples correctly. It was noted that the discrimination could be attributed to particle size, protein content and inorganic additives.

Durum wheat-pasta. Durum wheat (*Triticum durum*) is traditionally used to make dried pasta because the final product is of superior quality compared with that produced from cheaper common wheat (*Triticum aestivum*). The MAFF Working Party on Food Authenticity first considered the misdescription of wheat type used to make pasta, a high priority authenticity issue in 1993 (MAFF 1995).

Bertrand et al. (1996) scanned 24 samples of ground durum and whole wheat grains from 6 variety sources as well as 16 end-product semolinas and flours formulated in an experimental semolina mill. Principal component followed by stepwise discrimant analysis was employed on the reflectance NIR spectra and an 80% correct classification was obtained for the grain, flour and semolina samples.

McAig et al. (1992) have described a technique for measuring durum pigment using an NIR instrument operating in the visible range. Using a pigment peak observed at 498 nm, tested NIR calibrations provided an SEP of 0.77 ppm of pigment measured by solvent extraction.

Rice – authenticity of basmati US long grain. Whitworth et al. (1995) have used NIR spectroscopy, rapid viscoanalysis and image analysis to discriminate between basmati, US long grain and other rice types. Whitworth noted that further NIR studies were required to ensure that the discrimination was not based purely on water wavelengths. If, however, there was genuine NIR discriminatory ability then the combination of image analysis and NIR should provide a powerful discriminatory tool in this context.

4.5.2 *Fruit products*

Fruit juices – authenticity. Several groups of workers have applied NIR to fruit juice authentication. Scotter et al. (1995) have described an NIR reflectance technique, combined with Canonical Variates Discriminant Analysis, as the basis for providing a screening facility for the authentication of six species origin fruit juices. A blind test of the system, using 46 samples, provided 85% correct recognition of whether they were authentic or not. Meurens et al. (1995) have described a quantitative NIR approach to fruit analysis which could provide the basis for a rapid authenticity screening technique. Twomey et al. (1995) used vacuum-dried orange juice samples from which were obtained NIR reflectance spectra. Some of the samples had been adulterated, separately and in combination, with: 1. sugars, 2. acids, 3. pulp wash, and 4. grapefruit juice. With factorial discriminant analysis based on the Principal Component data derived from the raw spectra, Twomey and his team could obtain up to 94% accuracy of classification of the orange juices.

Fruit purées – authenticity. Wesley and McGill (1996) examined the use of NIR in reflectance mode in the wavelength range 400–2400 nm to authenticate the fruit origin of fruit pulps. SIMCA (soft independent modelling of class analogies) and discriminant analysis both using PCs (Principal Components) were used to provide discriminant models. In this preliminary study, Wesley and McGill demonstrated, using SIMCA, that strawberry pulp could be distinguished from other fruit pulps or mixtures of them, in 13 out of 15 cases (15 independent test samples applied to a cross-validated calibration model).

Jam – accuracy of the fruit content. Day and Scotter (1994) showed that, using NIR transmission and reflectance spectroscopy, that acceptable calibrations could be generated for jam component analyses which could provide a better measure of fruit content than is currently available.

Based upon a set of 32 strawberry jams formulated at Leatherhead Food Research Association, UK, an NIR calibration for fructose could be obtained using transmission data (680–1235 nm) from a well-mixed sample which gave a Multiple Correlation coefficient (MR) of 0.915 and a Standard Error of Calibration (SEC) of 2.05 which was an 8.7% error over the range of the constituent. Two wavelength terms were used in the calibrations. Using reflectance data from freeze-dried samples, two-wavelength-term calibration was also obtained for calculated fruit content giving an MR of 0.839 and an SEC of 1.33 which is 5.8% of the fruit content range used.

Day and Scotter (1991) also carried out a preliminary calibration for fruit content based on the labelled fruit content on a set of 26 commercial jams. A two-term calibration was obtained with an MR of 0.829 and an SEC of 3.95 which is an 11.3% error over the declared fruit content range.

4.5.3 Food-grade oils – substituting in part or wholly inferior oils for more expensive types

Several groups of workers have demonstrated clearly an ability for NIR transmission and reflectance spectra to be able to be used to discriminate between different species sources of food oils.

Chen and Chen (1995) in their transmission work showed, for example, that a soybean oil containing 20% peanut oil could be well distinguished from a pure soya oil on the basis of an NIR calibration for oleic and linoleic fatty acid ratios. Bewig *et al.* (1995) and his co-workers using an NIR reflectance approach and a Mahalanobis distance discriminant analysis were able to classify a test set of 11 out of 13 oil samples correctly. Cottonseed, canola, soybean and peanut samples were included in the calibration and test sets.

Scotter and Voyiagis (1994) and Downey (1995) carried out two types of discriminant analysis on the same spectra. Employing canonical variates discriminant analysis, Scotter and Voyiagis showed that the following four groups could be separated 100% correctly in a CV calibration analysis based on 42 samples of:

1. virgin and extra-virgin olive oils
2. other olive oil grades
3. soya oils
4. other oils including corn, grapeseed, rapeseed, sunflower and walnut.

Downey, using factorial discriminant analysis on the same dataset, demonstrated with a tested model that 84.2% correct identifications could be made. Downey, however, split the group of Virgin and extra-virgin olive oils and it was between these two groups that the misclassifications occurred. Wesley *et al.* (1995) also showed in preliminary calibrations work that other seed oil adulterants in olive oil could be detected with NIR reflectance spectroscopy at a level of 1%.

4.5.4 *Vegetable fat adulteration of dairy products*

Sato *et al.* (1989) produced NIR transflectance spectra to assess the ability of NIR to detect non-dairy fats or oils in dairy products. They calibrated for mixtures of butterfat and margarine and for addition of various vegetable oils to butterfat and concluded that overall levels of approximately 3% non-dairy oil/fat could be detected by this type of analysis.

4.5.5 *Animal fat substitutes in sausages*

Ellekjaer *et al.* (1992) performed NIR reflectance and transmission experiments, with accompanying sensory data, to determine the relative ability of the techniques to distinguish between sausages containing various amounts of fat, and sausages containing fat substitutes like iota-carrageenan, xanthan and guar gums. A total of 57 sausagemeat samples were submitted to discriminant analysis using a Bayesian technique. The NIR reflectance technique was most successful, the analysis of the data providing a 90% correct classification, and in a number of cases provided a discrimination where the sensory data did not.

4.5.6 *Meat species discrimination*

Spray *et al.* (1990) used NIR transmission and reflectance spectra to assess how well raw and cooked meats from different species could be distinguished. Principal components were generated from the spectral data and examination of Principal Component biplots for the first 4 PCs showed that the transmission spectra in particular demonstrated good separation of the species types both when the meat was raw and cooked.

4.5.7 *Classification of Iberian pig carcass quality*

In southern Spain a premium quality dry cured pork product is manufactured from pork of a particular quality. Oak acorn fed pigs provide the required quality meat for curing. There is thus a significant illegal profit to be made from declaring the cured meat as premium acorn fed pork when, in fact, the product has been derived from inferior quality pork. Hervas *et al.* (1994) obtained NIR transmission spectra of 118 fat samples from premium and inferior quality pig carcasses. Artificial neural net approaches were then used to define a calibration model which was tested with an independent sample set.

For inferior quality compound fed pig meat the discrimination was 100% correct, for mixed acorn and compound feeding 93.3% of the samples were identified correctly, and for the premium acorn fed animal samples 96.0% were correctly identified.

4.5.8 *Discrimination of fresh from frozen-thawed meat*

Downey and Beauchêne (1997a, b, in press) record in two separate publications two NIR approaches to distinguishing between fresh and frozen-thawed meat. Firstly, in a preliminary study combined visible and near-infrared wavelengths in reflectance were used in combination with factorial discriminant analysis on 32 samples of fresh and frozen-thawed *longissimus dorsi*. The discriminant model identified 95.3% of the samples correctly. In the second study, Downey and Beauchêne used the dried meat drip juice scanned in NIR reflectance as the sample material. In this case a separate test sample set was presented to the calibration factorial discriminant model and 93.44% of 61 samples were correctly allocated to a fresh or frozen-thawed class. Using PLSI and dummy variables (0 = fresh, 1 = frozen-thawed), 100% correct identification was achieved.

4.5.9 *Tea – inappropriately described type*

In the tea industry human tea tasters are the arbiters of tea quality and the availability of instrumental, preferably rapid methods, to determine the quality of black leaf tea would be advantageous. Hall *et al.* (1988) used NIR reflectance spectra of 134 black tea samples of wide ranging type, origin and quality to show that useful theaflavin (a prime quality parameter of black tea) and quality calibrations estimated by tea tasters could be produced. Theaflavin content could be measured with a calibration error of 3.3 µmoles/g, the percentage error over the range being 13.2%. The overall quality scored by five tasters over a scale of 25–74 was predicted with an accuracy of ±15.6 units on the scale. This result compares favourably with the result of one tea taster compared with the mean of the other four.

Osborne and Fearn (1988) carried out a preliminary reflectance study on 55 samples of black tea relating to the detection of a tea adulteration in one part of the set that was absent in the other part of the sample set. Using a Mahalanobis distance discrimination function they were able to distinguish between the adulterated and genuine tea samples to an accuracy of 91%.

4.5.10 *Coffee – type and coffee content*

Downey *et al.* (1994) used IR reflectance spectra to discriminate between *arabica* and *robusta* coffee bean varieties. Using a factorial discriminant procedure they succeeded in correctly classifying 96.2% of whole bean unblended samples. Inclusion of blended samples reduced this figure to between 82.9% and 87.6%. In the case of ground samples, including blends, 83.2% samples were correctly classified. In all cases the figures are for the ability of a calibration model to predict a test set of new samples.

Acknowledgements

R. Wilson thanks the BBSRC for funding and Marianne Defernez, James Holland and Kate Kemsley for providing diagrams.

C. Scotter would like to thank Margaret Voyiagis for expert technical assistance relating to work carried out at CCFRA, and David Evans for his expert assistance with statistical considerations.

Appendix 4A Some chemometrics methods in NIR analysis

Regression

1. Simple linear regression (1)
2. Forward stepwise multiple linear regression (1)
3. Multiple linear regression by backward elimination (1)
4. Locally weighted regression (2)
5. Ridge regression (1)
6. Partial least squares regression (1,2)
7. Principal component regression (1,2)
8. Artificial neural nets (1,2)

Spectral transformation

1. Smoothing algorithms (5)
2. 1st–4th derivative functions (5)
3. Standard normal variate and detrend (7)
4. Multiplicative scatter correction (5)
5. Kubelka–Munk function (6)
6. Fourier transform (5)

Discriminant procedures

1. Principal component analysis (1,3)
2. Fisher discriminant analysis (1)
3. Discriminance by Mahalanobis distance (1)
4. Factorial discriminant analysis (5)
5. PLS-based discriminant analysis (see Reg Wilson)
6. Canonical variates discriminant analysis (3)
7. SIMCA (3,4)
8. Cluster analysis (3)

9. Discriminant analysis using a Bayesian classification (2,3)
10. Artificial neural nets (1,3)

Parametric statistics

1. Standard error of calibration (4)
2. Standard error of prediction (4)
3. Root mean square error of prediction (4)
4. Root mean square error of cross validation (4)
5. PRESS statistic (4)

Note: Reference (4) gives a more comprehensive list of calibration and prediction statistics.

References for NIR chemometrics methods

1. Osborne, B.G., Fearn, T. and Hindle, P. (1993) *Practical NIR Spectroscopy.*
2. Naes, T. (1992) Progress in multivariate calibration, in *Near Infrared Spectroscopy*; Bridging the Gap between Data Analysis and NIR Application (Eds K.I. Hildrum, T. Isaksson, T. Naes and A. Taudberg), pp. 51–60.
3. Evans, D.G. (1993) Statistical methods for food analysis. *CCFRA Technical Bulletin, No.* 96.
4. Workman, J.J. (1992) NIR calibration basics, in *Handbook of Near-Infrared Analysis* (Eds D.A. Burns and E.W. Ciurczak), pp. 247–280.
5. Bertrand, D. (1993) Data pre-treatment and signal analysis in spectroscopy. Advanced Comcet Chemometrics School Libramont (Belgium), April 26–28, 1993.
6. Olinger, J.M. and Griffiths, P.R. (1993) Theory of diffuse reflectance in the NIR region, in *Handbook of Near-Infrared Analysis* (Eds D.A. Burns and E.W. Ciurczak), pp. 13–35.
7. Barnes, R.J., Dhanoa, M.S and Lister, S.J. (1989) Standard normal variate transformation and detrending of near-infrared diffuse reflectance spectra. *Applied Spectroscopy*, **43**(5), 772–777.

References

Mid-infrared

Baeten, V., Meurens, M., Morales, M.T. and Aparicio, R. (1996) Detection of virgin olive oil adulteration by Fourier transform Raman spectroscopy. *J. Agric. Food Chem.*, **44**(8) 2225–30

Banwell, C.N. (1994) *Fundamentals of Molecular Spectroscopy*, McGraw-Hill, London.

Belton, P.S., Kemsley, E.K., McCann, M.C. *et al.* (1995) The identification of vegetable matter using Fourier transform infrared spectroscopy. *Food Chem.*, **54**(4), 437–41.

Belton, P.S. and Wilson, R.H. (1990) Infrared sampling methods, in *Perspectives in Modern Chemical Spectroscopy* (ed. D.L. Andrews), Springer-Verlag, New York, pp. 67–86.

Briandet, R., Kemsley, E.K. and Wilson, R.H. (1996a) Discrimination of arabica and robusta in instant coffee by Fourier transform infrared spectroscopy and chemometrics. *J. Agric. Food Chem.* **44**, 170–74.

Briandet, R., Kemsley, E.K. and Wilson, R.H. (1996b) Approaches to adulteration detection in instant coffees using infrared spectroscopy and chemometrics. *J. Sci. Food Agric.*, **71**, 359–66

Defernez, M., Kemsley, E.K. and Wilson, R.H. (1995) Use of infrared spectroscopy and chemometrics for the authentication of fruit purees. *J. Agric. Food Chem.*, **43**, 109–13.

Defernez, M. and Wilson, R.H. (1995) Mid-infrared spectroscopy and chemometrics for determining the type of fruit used in jam. *J. Sci. Food Agric.*, **67**, 461–7.

Downey, G. and Boussion, J. (1996) Authentication of coffee bean variety by near-infrared reflectance spectroscopy of dried extract. *J. Sci. Food Agric.*, **71**(1), 41–40.

Dupuy, N., Huvenne, J.P., Duponchel, L. and Legrand, P. (1995) Classification of green coffees by FT-IR analysis of dry extract. *Applied Spectroscopy*, **49**(5), 580–85.

Griffiths, P.R. and De Haseth, J.A. (1986) *Fourier Transform Infrared Spectrometry*, Wiley-Interscience, New York.
Kemsley, E.K. (1996) Discriminant analysis of high-dimensional data: a comparison of principal components analysis and partial least squares data reduction methods. *Chemometrics and Intelligent Laboratory Systems*, **33**, 47–61.
Kemsley, E.K., Holland, J.K., Defernez, M. and Wilson, R.H. (in press) Detection of raspberry purees using infrared spectroscopy and chemometrics. *J. Agric. Food Chem.*
Kemsley, E.K., Ruault, S. and Wilson, R.H. (1995) Discrimination between *Coffea arabica* and *Coffea canephora* variant *robusta* beans using infrared spectroscopy. *Food Chemistry*, **54**, 321–6
Lai, Y.W., Kemsley, E.K. and Wilson, R.H. (1994) Potential of Fourier transform infrared spectroscopy for the authentication of vegetable oils. *J. Agric. Food Chem.*, **42**, 1154–9.
Lai, Y.W., Kemsley, E.K. and Wilson, R.H. (1995) Quantitative analysis of potential adulterants of extra virgin olive oil using infrared spectroscopy. *Food Chem.*, **53**, 95–98.
McCann, M.C., Hammouri, M.K., Wilson, R.H. *et al.* (1992) Fourier transform infrared spectroscopy: a new way to look at plant cell walls. *Plant Physiol.*, **100**, 1940–7.
Suchanek, M., Filipova, H., Volka, K. and Delgadillo, I. (1996) Qualitative analysis of green coffee by infrared spectrometry. *Fresenius J. Anal. Chem.*, **354**, 327–332.
Wilson, R.H. and Goodfellow. B.J. (1994) Mid-infrared spectroscopy, in *Spectroscopic Techniques for Food Analysis* (ed. R.H. Wilson), VCH, New York, pp. 59–86.
Wilson, R.H., Slack, P.T., Appleton, G.P. *et al.* (1993) Determination of the fruit content of jam using Fourier transform infrared spectroscopy. *Food Chem.*, **47**, 303–308.

Near-infrared references

Bertrand, D., Devaux, M.F. and Robert, P. (1987) Application of pattern recognition techniques in NIR spectroscopy, in *Near-infrared Diffuse Reflectance/Transmittance Spectroscopy* (eds J. Hollo, K.J. Kaffka and J.O. Gonczy), Academiai Kiado, Budapest, pp. 31–42.
Bertrand, D., Novales, B., Devaux, M.F. and Robert, P. (1996) Discrimination of durum wheat products for quality control, in *Near-infrared Spectroscopy. The Future Waves* (eds A.M.C Davies and P. Williams), NIR Publications, pp. 430–35.
Bewig, K.M., Clarke, A.D., Roberts, C. and Unklesbay, N. (1995) Discriminant analysis of vegetable oils using near-infrared spectroscopy, in *Leaping Ahead with Near-infrared Spectroscopy* (eds G.D. Batten, P.C. Flinn, L.A. Welsh and A.B. Blakeney), Royal Spectroscopy Group. Royal Australian Chemical Institute, pp. 324–328.
Brown, G.L., Curtis, P.S. and Osborne, B.G. (1993) Factors affecting the measurement of hardness by NIR spectroscopy of ground wheat. *J. Near-infrared Spectroscopy*, **1**(3), 147–52.
Burns, D.A. and Ciurczak, E.W. (1992) *Handbook of Near-infrared Analysis*, Marcel Dekker, pp. 1–6.
Chen, Y.S. and Chen A.O. (1995) Quality analysis and purity examination of edible vegetable oils by near-infrared transmittance spectroscopy, in *Leaping Ahead with Near-infrared Spectroscopy* (eds G.D. Batten, P.C. Flinn, L.A. Welsh and A.B. Blakeney), Royal Spectroscopy Group. Royal Australian Chemical Institute, pp. 316–23.
Day, L. and Scotter C.N.G. (1991) The use of NIRS in the analysis of strawberry jam (II). *CFDRA Technical Memorandum No. 631.*
Delwiche, S.R. and Norris, K.H. (1993) Classification of hard red wheat by near-infrared diffuse reflectance spectroscopy. *Cereal Chem.* **70**(1), 29–35.
Downey, G. (1995) Oil discrimination by Canonical variate analysis of near-infrared reflectance spectra, in *Food Authentication by Spectroscopic Tecniques (FAST)*. A CEC FLAIR/QUEST Action Group/Guideline document, EC 24–28.
Downey, G. and Beauchêne, D. (1997a) Discrimination between fresh and frozen-then-thawed beef longissimus dorsi by combined visible near-infrared reflectance spectroscopy. A feasibility study. *Meat Science* (in press).
Downey, G. and Beauchêne, D. (1997b) Authentication of fresh versus frozen-then-thawed beef by near-infrared reflectance spectroscopy of dried drip juice. *Lebensmittel–Wissenschaft und Technologie* (in press).
Downey, G., Boussion, J. and Beauchêne, D (1994) Authentification of whole and ground coffee beans by near-infrared reflectance spectroscopy. *J. Near-infrared Spectroscopy*, **2**(2), 85–92.

Ellekjaer, M.R., Naes, T., Isaksson, T. and Solheim, R. (1992) Identification of sausages with fat-substitutes using near-infrared spectroscopy, in *Near-infrared Spectroscopy. Bridging the Gap between Data Analysis and NIR Applications* (eds K.I. Hildrum, T. Isaksson, T. Naes and A. Tandberg), Ellis Horwood, pp. 321–6.

Fábián, Z., Izvekov, V., Salgó, A. and Orsi, F. (1994) Near-infrared reflectance and Fourier transform infrared analysis of instant coffee. *Analytical Proceedings including Analytical Communications*, **31**, 261–3.

Hall, M.N., Robertson, A. and Scotter, C.N.G. (1988) Near-infrared reflectance prediction of quality, theaflavin and moisture content of black tea. *Food Chem.*, **27**, 61–75.

Hervas, C., Garrido, A., Lucena, B. et al. (1994) Near-infrared spectroscopy for classification of Iberian pig carcasses using an artificial neural network. *J. Near-infrared Spectroscopy*, **2**(4), 177–84.

McAig, T.N., Mcleod, J.G., Clarke, J.M. and Depauw, R.M. (1992) Measurement of durum pigment with a near-infrared instrument operating in the visible range. *Cereal Chem.*, **69**(6), 671–2.

MAFF (1995) Authenticity of dried durum wheat pasta. *Food Surveillance Paper, No. 47.* HMSO.

Meurens, M., Weigic, L., Foulou, M and Acha, V. (1995) Rapid analysis of fruit juice by infrared spectrometry. *Cerevisia*, **20**(3), 33–6.

Noble, D. (1995) Illuminating IR. *Anal. Chem.*, 735A-740A.

Norris, K.H., Hruschka, W.R., Bean, M.M. and Slaughter, D.C. (1989) A definition of wheat hardness using near-infrared reflectance spectroscopy. *Cereal Foods World*, **34**(9), 696, 698, 701, 703–5.

Osborne, B.G. and Fearn, T. (1988) Discriminant analysis of black tea by near-infrared reflectance spectroscopy. *Food Chem.*, **29**, 233–8.

Osborne, B.G., Fearn, T. and Hindle, P.H. (1993) *Practical NIR Spectroscopy*, Longman Scientific and Technical.

Sato, T., Yoshimo, M., Suzuki, I. et al. (1989) Use of Near-infrared Spectroscopy to Detect Foreign Fat Adulteration of Dairy Products. Proceedings of the 2nd International ICNIRS Conference, Japan (eds M. Iwamoto and S. Kawano), Korin Publishing, Japan, pp. 148–56.

Scotter, C.N.G. (1995) Near-infrared spectroscopy, in *Encyclopedia of Analytical Science* (ed. A. Townshend), Academic Press, pp. 2189–200.

Scotter, C.N.G., Legrand, A. and Voyiagis, M. (1995) NIR for juice authenticity screening, in *Leaping Ahead with Near-infrared Spectroscopy* (eds G.D. Batten, P.C. Flinn, L.A. Welsh and A.B. Blakeney), Royal Spectroscopy Group, Royal Australian Chemical Institute, pp. 307–11.

Scotter, C.N.G. and Voyiagis, M.N. (1994) Oil discrimination by Canonical variate analysis of near-infrared reflectance spectra, in *Food Authentication by Spectroscopic Techniques* (*FAST*). A CEC FLAIR/QUEST Action Group guideline document, EC 21–3.

Sirieix, A. and Downey, G. (1993) Commercial wheatflour authenticiation by discriminant analysis of near-infrared reflectance spectra. *J. Near-infrared Spectroscopy*, **1**(4), 187–97.

Spray, M.J., Scotter, C.N.G. and Hall, M.N. (1990) Meat speciation by near-infrared spectroscopy. *CFDRA Technical Memorandum No. 585.*

Twomey, M., Downey, G. and McNulty, B. (1995) The potential of NIR spectroscopy for the detection of the adulteration of orange juice. *J. Science, Food and Agriculture*, **67**(1), 77–84.

Wesley, I.J., Barnes, R.J. and McGill, A.E.J. (1995) Measurement of the adulteration of olive oils by near-infrared spectroscopy. *J. American Oil Chemists Society*, **72**(3), 289–92.

Wesley I.J. and Mcgill A.E.J. (1996) The application of near-infrared spectroscopy to the identification of fruit pulps, in *Near-infrared Spectroscopy. The Future Waves* (eds A.M.C. Davies and P. Williams), NIR Publications, 426–9.

Whitworth, M., Greenwell, P., Fearn, T. and Osborne, B.G. (1995) Physical techniques for establishing the authenticity of rice. *CCFRA R&D Report No. 16.*

Zhuo, L. and Scotter, C.N.G. (1994) The calibration for fruit content in strawberry jam by NIR, in *Spectroscopic Calibration Challenges*. A CEC FLAIR/QUEST Action Group Guideline Document, EC 15–22.

5 Oligosaccharide analysis
N. LOW

5.1 Introduction

Adulteration of foods is not a new problem. In the early 1800s unscrupulous tea traders would treat used tea leaves with black lead to 'restore colour' in order to resell the leaves as 'fresh'. Similar situations occurred with foods which were of poor quality or spoiled. Recently, adulteration of foods rich in carbohydrate, such as citrus and apple juices, honey and maple syrup, has become a significant problem in the food industry. Of particular importance to reputable food producers and processors is the undeclared addition of inexpensive sweeteners or syrup to high carbohydrate foods.

In most food products the major soluble solid present is carbohydrate. In the case of most fruit juices carbohydrates account for >90% of the total soluble solids; in foods such as honey and maple syrup this level is even higher (Table 5.1). Because of the high levels of carbohydrates in these foods, their adulteration normally involves replacing the natural carbohydrates present with natural carbohydrates from a less expensive source.

The economics behind the adulteration of food products is best illustrated using orange juice as an example. In 1995–1996 world production of citrus fruit was approximately 3.6×10^7 tonnes. The estimated on-tree value of oranges during that production year was more than (US) $4 billion (United States Department of Agriculture, 1996). In 1995–1996 the price of frozen concentrated orange juice (FCOJ; 42 Brix) was ~$3.15/kg solids (Florida Department of Agriculture, 1996). In this same year the price of inexpensive sweeteners/syrups ranged from $0.53–$0.66/kg solids (Alberta Sugar Company, personal communication). Therefore, the profits which can be realized from adulteration are considerable.

Table 5.1 Carbohydrate content of selected foods

Food	Carbohydrate content[a] (%)
Apple juice[b]	95
Orange juice[b]	91
Honey[c]	>97
Maple syrup[d]	>98

[a] Dry basis.
[b] Low, 1996.
[c] Low and South, 1995.
[d] Stuckel and Low, 1996

Table 5.2 Glucose (G), fructose (F) and sucrose (S) concentration and ratio in selected foods

Food	G (%)	F (%)	S (%)	Ratio
Apple juice[a]	2.43	6.59	1.97	2.7 : 1 (F : G)
Orange juice[b]	2.10	2.26	3.75	1 : 1.1 : 1.8 (G : F : S)
Honey[c]	32.3	38.5	2.0	1.2 : 1 (F : G)
Maple syrup[d]	0.43	0.30	68.0	

[a] Low, 1996.
[b] Wudrich, 1993.
[c] Low and South, 1995.
[d] Stuckel and Low, 1996.

5.1.1 *Adulterants*

The major carbohydrates present in foods are glucose, fructose and sucrose, and the concentration and ratio of these major carbohydrates in foods is variable. However, distinct and analytically measurable concentrations and ratios are known for foods rich in carbohydrate (Table 5.2). Unfortunately for reputable food processors and packers, a number of commercially produced inexpensive sweeteners/syrups not only contain the same major carbohydrates present in these carbohydrate rich foods but also possess the correct ratio of these major carbohydrates. These commercially produced inexpensive sweeteners include:

1. beet/cane sugar: produced by the chemical refining of either sugar beet or sugar cane. The final product is ~100% sucrose (a disaccharide which is comprised of one molecule of glucose linked to one molecular of fructose);
2. beet/cane medium invert sugar (MIS): produced from refined beet or cane sucrose by either acid or enzymatic hydrolysis and results in a finished product containing a 1 : 1 : 2 ratio of glucose, fructose and sucrose;
3. beet/cane total invert sugar (IS): produced from refined beet or cane sucrose by either acid or enzymatic hydrolysis and results in a finished product containing a 1 : 1 ratio of glucose to fructose and low levels of sucrose (~0.8–10%);
4. high fructose starch syrup (HFS): produced by the enzymatic hydrolysis of starch (polymer of glucose). The resulting glucose syrup is partially converted to fructose enzymatically. The final product consists of 42% fructose (high fructose syrup 42) or 55% fructose (high fructose syrup 55), 90% fructose (high fructose syrup 90) or 95% fructose (high fructose 95). Raw materials for this syrup include starches from corn, potato, cassava and palm;
5. high fructose syrup from inulin (HIS): produced by acid treatment of inulin (polymer of fructose with terminal glucose units). The resulting syrup is rich in fructose (fructose to glucose ratios range from 8 : 1 to 3 : 1 in commercial syrups). Raw materials for this syrup include Jerusalem artichoke, chicory root and dahlia tubers.

Although a number of analytical methods have been established to detect the undeclared addition of inexpensive sweeteners and syrups to foods rich in carbohydrate, this chapter will focus on the oligosaccharide method developed by our research group. This method shows that the above adulterants contain unique oligosaccharides which can be used as 'fingerprints' to detect their addition to carbohydrate rich foods such as fruit juices, honey and maple syrup.

5.1.2 Carbohydrate classification

Carbohydrates are classified into three major groups: monosaccharides, also called simple sugars (such as glucose and fructose), oligosaccharides, and polysaccharides.

In general, mono- and oligosaccharides are water-soluble crystalline compounds. These compounds are generally aliphatic aldehydes, or ketones, which contain multiple hydroxyl groups. These compounds can be further classified as an aldose (aldehyde-containing carbohydrate) or ketose (ketone carbohydrate).

Oligosaccharides. Oligosaccharides (Greek *oligos*, meaning a few; Helferich et al., 1930) are relatively low molecular weight (340–1600 daltons) polymers that are covalently bonded through glycosidic linkages which upon hydrolysis yield monosaccharides. Arrows such as $1 \rightarrow 4$ indicate that the #1 carbon of one monomeric unit is covalently attached to the #4 carbon atom of the other monomeric unit via a glycosidic linkage.

The standard convention for the number of monomer units making up an oligosaccharide is 10 (Hassid and Ballou, 1957). Oligosaccharides comprised of two monomer units are called disaccharides, three monomer units are trisaccharides, and so on. Oligosaccharides occur naturally in plants, animals and microorganisms. The most important food oligosaccharide is sucrose, a non-reducing disaccharide.

5.2 Oligosaccharide formation in foods and inexpensive sweeteners

5.2.1 Carbohydrate metabolism in plants

The growth of plants is regulated by the supply of water and nutrients and physical factors such as the intensity and duration of sunlight. One of the essential nutrients required for plant growth is carbon dioxide. Atmospheric carbon dioxide is taken in by leaves and other green tissues of plants and is converted to carbohydrate in chloroplasts during the process known as photosynthesis. Phosphorylated monosaccharides are the final product of photosynthesis.

5.2.2 Oligosaccharide synthesis in plants

Sucrose. Sucrose is the most abundant oligosaccharide found in plants. It is a non-reducing disaccharide, α-D-glucopyranosyl β-D-fructofuranoside, which yields D-fructose and D-glucose on acidic hydrolysis.

During photosynthesis dihdroxyacetone phosphate is produced which, in the presence of cytoplasmic triose phosphate isomerase, equilibrates with glyceraldehyde-3-phosphate. The combination of these two substrates via an aldol condensation reaction results in the formation of D-fructose-(1 → 6)-biphosphate. This compound is then converted to D-glucose-6-phosphate via three enzymatically mediated reactions. The enzyme UDP-glucose pyrophosphorylase then catalyses the reaction between UTP and D-glucose-6-phosphate to yield UDP-D-glucose. A glucose unit is then readily transferred from this complex to D-fructose-6-phosphate to yield sucrose phosphate. Free sucrose is released by the action of sucrose phosphatase on this complex.

Raffinose. After sucrose, raffinose is the most abundant oligosaccharide found in plants. It is a non-reducing trisaccharide, O-α-D-galactopyranosyl-(1 → 6)-O-α-D-glucopyranosyl-(1 → 2) β-D-fructofuranoside, which yields D-fructose and 6-O-α-D-galactopyranosyl-D-glucose (melibiose) on mild acidic hydrolysis.

Myo-inositol, a sugar alcohol found in all plants is the precursor for raffinose synthesis. Myo-inositol is enzymatically formed from D-glucose-6-phosphate and is converted to UDP-D-galacturonic acid which in turn is enzymatically converted to UDP-D-galactose. Galactinol (1-O-α-D-galactopyranosyl-myo-inositol) is formed via a transferase reaction between myo-ionsitol and UDP-D-galactose. A second transferase reaction adds a galactosyl residue to sucrose to yield raffinose.

Others. A number of other oligosaccharides, such as maltose (O-α-D-glucospyranosyl-(1 → 4)-D-glucopyranose) and 1-kestose (O-α-D-glucopyranosyl-(1 → 2)-β-D-fructofuranosyl-(1 → 2) β-D-fructofuranoside) have also been found in trace amounts in some plant varieties. The formation of these minor oligosaccharides can be traced to the transferase and/or transglycosylation action of hydrolase enzymes in plants.

5.2.3 Oligosaccharide formation in inexpensive sweeteners

Research (Low and Wudrich, 1993) from our laboratory has shown the presence of trace levels (0.001–1%) of oligosaccharides in all commercial inexpensive sweeteners analysed to date. The formation of oligosaccharides in these products can be attributed to the enzymatic and chemical methods used for production.

High fructose syrup production from starch. Commercial starch consists mainly of carbohydrate (98.9–99.6%) with small amounts of protein, lipid and ash (Swinkels, 1985). Amylose and amylopectin are the main carbohydrate components of starch with amylose being a chain-like polymer of D-glucose units linked primarily by α-1 → 4 glycosidic bonds with some α-1 → 6 branch sites. The amylopectin molecule is a 'bush-like' polymer, where the D-glucose units are linked by α-1 → 4 glycosidic bonds for the linear portion and α-1 → 6 glycosidic bonds at the branched points.

The production of high fructose syrup from starch normally involves an initial treatment of the starch with hydrochloric acid followed by enzymatic conversion with a thermostable α-amylase (liquefaction enzyme; EC 3.2.1.1), glucoamylase (saccharification enzyme; EC 3.2.1.3), and finally xylose isomerase (also called glucose isomerase; EC 5.3.1.18). Initial treatment of starch with acid results in the production of a reduced viscosity, low dextrose equivalent (DE; the amount of reducing sugar produced) syrup. Alpha-amylase is an endoenzyme which hydrolyzes the α-1 → 4 glycosidic bonds of this low DE syrup and yields a syrup of much higher DE and lower viscosity. The resulting glucose oligomers/polymers can be readily attacked by glucoamylase to yield a syrup rich in D-glucose. The D-glucose in this syrup is then enzymatically isomerized to D-fructose and the degree of isomerization determines the final product description (i.e. high fructose syrup 42, 55, 90 or 95).

During the transformation of starch to a high fructose syrup a number of oligosaccharides are formed called fingerprint oligosaccharides. The most important of these in fruit juice authenticity determination is isomaltose (O-α-D-glucopyranosyl-(1 → 6)-D-glucopyranose). As no debranching enzymes (those capable of hydrolyzing the α-1 → 6 glycosidic bond of starch) are employed in the production of high fructose syrup from starch, appreciable (up to 1.5%; Reeve, 1992) concentrations of isomaltose may be present in the final syrup. In addition, isomaltose synthesis can occur via the transglycosylation activity of glucoamylase. For example, maltose hydrolysis by glucoamylase results in α-D-glucose being 'trapped' in the active site of the enzyme either as a covalent or ion pair intermediate (Fig. 5.1).

The second step of the reaction involves the attack of a nucleophile, typically water which results in two D-glucose units being produced from isomaltose hydrolysis. It is important to note that the stereochemistry at the anomeric position is retained as the attacking nucleophile can only react with the alpha face of the molecule. The conditions employed for high fructose syrup production, that is low moisture and high carbohydrate concentration, affords the opportunity for a competing nucleophile such as D-glucose to attack the alpha face which can result in isomaltose formation (and other disaccharides comprised of two glucose units). That isomaltose production would be favoured over other glucose disaccharides may be explained by the reduced steric interaction encountered with primary versus secondary hydroxyl groups.

Results from our laboratory on oligosaccharide analysis of commercial high fructose syrups by high performance anion exchange chromatography with pulsed amperometric detection (HPAE-PAD) indicate that these syrups contain appreciable levels of glucose polymers up to maltoheptaose. Mean results from the HPAE-PAD analysis of ten commercial high fructose syrups showed the presence of isomaltose (1.61%), maltose (1.01%), isomaltotriose (0.12%), maltotriose (0.07%), maltotetraose (0.03%), maltopentaose (0.07%), maltohexaose (0.15%), and maltoheptaose (0.15%) (Stuckel, 1995).

Invert sugar production from beet/cane sucrose. Commercial production of invert sugar involves the treatment of either beet or cane sucrose with hydrochloric acid. The term 'invert sugar' refers to the change in the optical rotation of plane polarized light which occurs when sucrose (+66.5°) is hydrolysed to an equimolar mixture of D-fructose and D-glucose (the equimolar mixture of these two monosaccharides has an optical rotation of −19.9°).

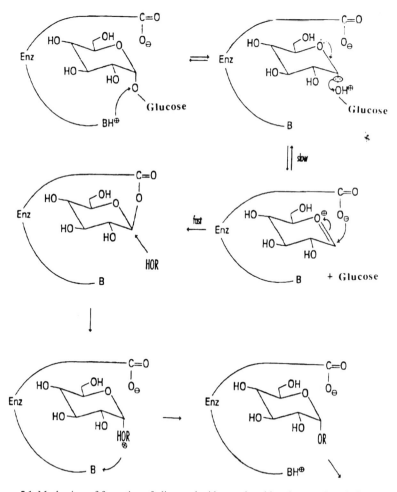

Figure 5.1 Mechanism of formation of oligosaccharides catalysed by glucoamylase during maltose hydrolysis (transglycosylation). HOR is any carbohydrate in solution.

During the commercial production of invert sugar (either total or medium invert) oligosaccharides may also be synthesized by the mechanism known as reversion (Szejtli *et al.*, 1970; Krol, 1978). This mechanisms of oligosaccharide formation is shown in Fig. 5.2 and involves initial protonation of the glycosidic linkage of sucrose. This glycosidic bond is broken, resulting in the formation of either D-glucose and a charged fructofuranose moiety or D-fructose and a charged glucopyranose moiety (Swallow and Low, 1993). The glucose or fructose intermediate carrying the positive charge (oxonium ion) is partially planar due to the partial sp nature of the bond. The 5-membered furanose ring is more planar that the 6-membered pyranose ring. We therefore hypothesized that the formation of the cyclopentyl (i.e. fructofuranosyl) action occurs more readily than does that of the cyclohexyl action (Swallow and Low, 1993). Since the fructofuranosyl oxonium ion intermediate is more readily formed that the

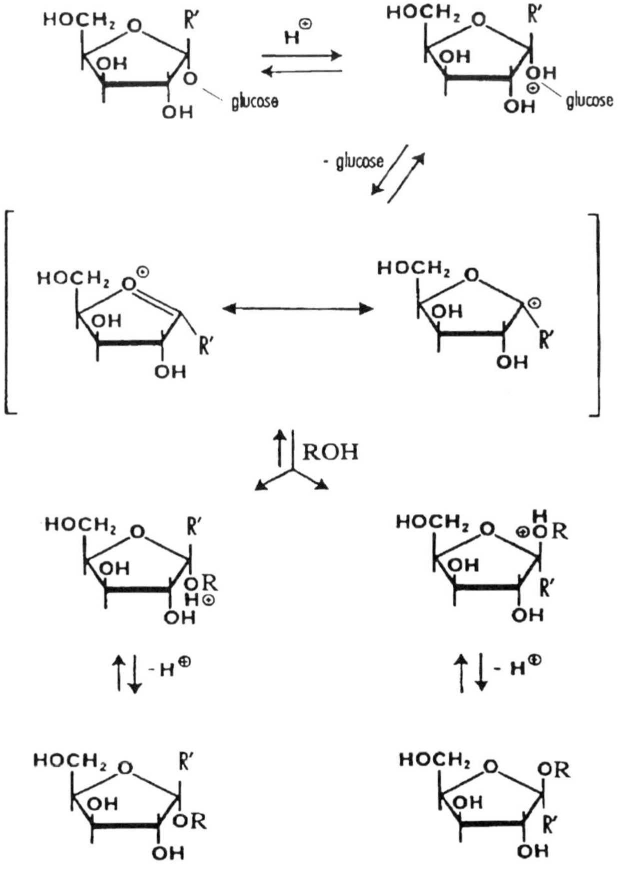

Figure 5.2 Mechanism of formation of oligosaccharides by acid during the hydrolysis of sucrose (reversion reaction). ROH is any carbohydrate in solution.

glucopyranosyl oxonium ion, the probability is much higher that this molecule will react with other carbohydrates in solution. This hypothesis was shown to be true as the four fingerprint oligosaccharides which were isolated and structurally identified in beet medium invert sugar were shown to be trisaccharides formed by sucrose addition to fructofuranosyl oxonium ion. In addition to the planar nature of the fructofuranosyl cation, this intermediate is also in resonance with a tertiary carbonium ion, whereas the glucopyranosyl oxonium is in resonance with a secondary carbonium ion.

Invert sugar may also be produced by sucrose treatment with hydrolase enzymes, specifically α-glucosidase or β-fructosidase. To the author's knowledge enzymatic production of invert sugar is not done commercially. However, the use of hydrolase enzymes for invert sugar production also results in trans-glycosylation products (i.e. fingerprint oligosaccharides) which can be used to detect adulteration with invert sugar (Low and Wudrich, 1993).

High fructose syrup production from inulin. Inulin is a non-structural plant polysaccharide comprised of fructose units which are β-($1 \rightarrow 2$) linked and have terminal glucose units (McCance and Lawrence, 1929). It has served as a staple food for a number of groups including Australian aborigines (Gott, 1983) and some South American populations (Asami *et al.*, 1989).

Inulin is hydrolysed commercially by acid (typically hydrochloric) to yield a syrup which is rich in fructose. This hydrolysed syrup is sold under a number of commercial names and is purported to improve or change the nutritional value of processed foods. These improvements include increasing the amount of dietary fibre, reducing the caloric content, and increasing the Bifidus-promoting capacities of the food.

During acid hydrolysis of inulin two oligosaccharide fingerprints are formed (most likely via the reversion reaction) which are unique to this material. We have isolated these oligosaccharides from hydrolysed inulin but have not yet completed their structural determination. Preliminary results show that these compounds are disaccharides and are made up of D-fructose moieties.

5.3 Principle of the fingerprint oligosaccharide method

This method is based on identifying natural oligosaccharide patterns in pure foods and inexpensive sweetners/syrups. These oligosaccharides are produced via direct enzymatic synthesis, by the transglycosylation action of hydrolase enzymes in the food itself (such as α,β-glucosidase which are present in the honey sac of bees) or by the use of hydrolase enzymes in sweetener/syrup production (where α-amylase and glucomylase are used for high fructose syrup production). In addition, oligosaccharides can be formed chemically ('reversion' reactions) by the action of acid on sucrose, D-fructose and D-glucose.

Oligosaccharide analysis in foods is accomplished in our laboratory by high performance anion exchange liquid chromatography with pulsed amperometric detection (HPAE-PAD), capillary gas chromatography with flame ionization detection (CGC-FID) or by capillary zone electrophoresis (CZE). The remainder of this chapter will focus on a description of the fingerprint oligosaccharide methodology based on these three instruments; and will discuss how the analytical results derived are used and interpreted to determine if a particular food has been adulterated with any of the aforementioned inexpensive sweeteners.

5.4 Sample preparation

Two general sample preparation methods have been presented in literature (Low, 1995). The purpose of sample preparation method 1 is to isolate the carbohydrates from the other chemical components in the sample. This results in a 'simplified' chromatogram and a superior signal to noise ratio. Sample preparation method 2 is more facile and can be performed rapidly with no statistically significant variation in results when compared to method 1. With the exception of our initial publication on oligosaccharide fingerprinting (Swallow et al., 1991), these two sample preparation methods represent those which have been used for examining fruit juices using our developed methodology (Low and Swallow, 1991; White and Cancalon, 1992; Wisenberger et al., 1992; Low and Wudrich, 1993; Wudrich et al., 1993; Hammond, 1993; Low and South, 1995; Low, 1995, 1996; Stuckel and Low, 1995; Low and Hammond, 1996).

5.4.1 Sample preparation method 1 (SPM1)

1. A sufficient amount of juice, honey, or maple syrup is diluted to 5.5 °Brix with HPLC grade water to yield a final volume of 40 ml.
2. The sample is then centrifuged at 3000 × g for 15 min at 4 °C (this step can be omitted for apple juice, honey and maple syrup).
3. During centrifugation two separate ion exchange columns are prepared (we use 10 ml plastic disposable syringe barrels) by the addition of 5–6 g of analytical grade cation exchange resin (Bio Rad AG 50W X8; 100–200 mesh; Richmond, CA); the addition of 5–6 g of anion exchange resin (Bio Rad AG1 X4; 100–200 mesh; formate form). These columns are well washed with HPLC grade water.
4. Following centrifugation, the supernatant is applied to the cation exchange column. The initial 5–6 ml passing through the column are discarded and the next (approximately) 20 ml are retained. This material is then passed through the anion exchange column (again discarding the first 5–6 ml) and retaining the remainder of the sample (approximate final volume is ~10 ml).
5. An octadecyl silyl column (Waters C_{18} Sep Pak cartridge; Waters, Milford, MA) is activated by passage of two volumes (2 × 5 ml) of either ethanol or

methanol, followed by washing with five volumes of HPLC grade water. This C_{18} column is then attached to a syringe and the eluant from 4. is passed through the column, discarding the first 1 ml, and retaining the remainder of the solution.

6. the solution from 5. is then passed through a syringe filter (0.2 μm nylon; Chromatographic Specialities Inc., Brockville, ON).

5.4.2 Sample preparation method 2 (SPM2)

Sample preparation is identical to that followed in SPM1 to the end of centrifugation. Following centrifugation the sample is subjected to filtration (step 6).

Once filtered, this sample (from either SPM1 or SPM2) can be used for oligosaccharide analysis by HPAE-PAD. Oligosaccharide analysis by CGC-FID requires further treatment (see 7. below) as carbohydrates do not have an appreciable vapour pressure, and they require derivatization.

7. 100 μl of the solution is transferred to a gas chromatographic autosampler vial (12 × 32 mm) and lyophilized. The lyophilized sample is then treated with 0.5 ml of N-trimethylsilyl imidazole in pyridine (Tri-Sil Z, Pierce Chemical Company, Rockford, IL; Sylon TP, Supelco Inc., Bellefonte, PA). The vials are then capped, and the solutions heated at 70–80 °C for 0.5–1 h (heating block; Denville Scientific Inc., Metuchen, NJ).

5.5 Oligosaccharide fingerprinting employing high performance anion exchange liquid chromatography with pulsed amperometric detection (HPAE-PAD)

5.5.1 Sample analysis by high performance anion exchange liquid chromatography

A number of published HPAE-PAD protocols exist for oligosaccharide fingerprinting of fruit juice, honey and maple syrup for authenticity. Each of these methods will be presented in chronological order.

Fruit juice analysis by HPAE-PAD. The oligosaccharide fingerprint method for food authenticity determination was first introduced in 1991 by Swallow *et al*. In this pioneering work the authors employed extensive sample clean-up with charcoal celite in order to remove most of the monosaccharides from the sample. The reasoning behind this extensive sample clean-up and dual column usage was the need for high theoretical plate numbers for oligosaccharide separation.

Following clean-up, these samples were analysed on a Waters 625 (Waters Chromatography, Milford, MA) metal-free gradient HPLC system. The carbohydrates in the sample were separated using two Carbo Pac PA1 (Dionex Crop., Sunnyvale, CA) pellicular anion exchange columns (4 × 250 mm) connected in series. A 100 µl sample loop was utilized for the analysis and the flow rate was set at 0.7 ml/min. The carbohydrates were detected by a PAD (pulsed amperometric detector; Waters Model 464) with dual gold electrode and triple pulsed amperometry at a sensitivity of 50 µA. The electrode was maintained at the following potentials and durations: $E_1 = 50$ mV ($T_1 = 0.299$ sec); $E_2 = 600$ mV ($T_2 = 0.299$ sec); $E_3 = -800$ mV ($T_3 = 0.499$ sec). A post column delivery system delivering 300 mM NaOH at a flow rate of 0.8 ml/min was used to minimize baseline drift. The authors used an ion pac column (Dionex ATC-1) prior to the LC pumps in order to remove trace amounts of carbonate from the mobile phase which can otherwise affect solute retention times.

The linear gradient elution program employed is shown in Table 5.3. According to the authors, a single chromatographic run can reveal the addition of high fructose syrup (HFS), medium (MIS) and total invert sugar (IS) to fruit juices. The authors point out that liquid sodium hydroxide must be used in the preparation of the mobile phase due to high carbonate concentration in sodium hydroxide pellets. Further research by Low and Swallow (1991) showed that charcoal–celite sample treatment could be eliminated while maintaining oligosaccharide separation and sensitive detection.

Because of the lengthy analysis times involved (177 min, including column re-equilibration), White and Cancalon (1992) introduced a number of modifications to the previously described methodology. These modifications (Table 5.4) included the use of a column switching valve between the two analytical columns which resulted in monosaccharide and some sucrose elimination prior to the detector; more rapid linear gradient delivery of a higher molarity sodium acetate solution; and more rapid delivery of 300 mM sodium hydroxide. They showed that for detecting medium invert sugar adulteration of orange juice these modifications resulted in reduced analysis times (120 min, including column

Table 5.3 Gradient elution program for the detection of MIS in orange juice via fingerprint oligosaccharide analysis (Swallow et al., 1991)

Time (min)	Composition		
	% A	% B	% C
0.00	100	0	0
7.00	100	0	0
23.00	97	3	0
53.00	0	100	0
63.00	0	100	0
66.00	0	0	100
123.00	0	0	100
125.00	100	0	0
177.00	100	0	0

Where: %A: 100 mM NaOH; %B: 100 mN NaOH/100 mM NaOAc; %C: 300 mM NaOH.

Table 5.4 Gradient elution program for the detection of MIS in orange juice via fingerprint oligosaccharide analysis (White and Cancalon, 1992)

Time (min)	Composition		
	% A	% B	% C
0.00	75	25	0
4.00	75	25	0
20.00	0	100	0
30.00	0	100	0
30.10	0	0	100
90.10	0	0	100
90.20	75	25	0
120.00	75	25	0

Where: %A: 100 mM NaOH; %B: 100 mM NaOH/200 mM NaOAc; %C: 300 mM NaOH.

re-equilibration) while maintaining oligosaccharide resolution and detection limits.

Further modifications of the original method were made by Wisenberger *et al.* (1992). These modifications consisted of a different linear gradient elution program (Table 5.5), an alternate thermostabilized (25 °C) stationary phase (Carbo Pac PA 100; Dionex), higher molarity sodium hydroxide solution, and slightly modified pulsed amperometric detector settings. In addition the authors used a Dionex HPLC system. The authors report that these modifications resulted in a sample analysis time of 75 min with no loss in either the resolution or sensitivity of the fingerprint oligosaccharides when the method was used to detect the adulteration of orange juice with medium invert sugar.

Table 5.5 Gradient elution program for the detection of MIS in orange and grapefruit juice via fingerprint oligosaccharide analysis (Wiesenberger *et al.*, 1992)

Time (min)	Composition		
	% A	% B	% C
0.00	100	0	0
2.00	100	0	0
5.00	90	10	0
20.00	65	35	0
35.00	16	84	0
36.00	0	100	0
41.00	0	100	0
42.00	0	0	100
57.00	0	0	100
57.20	100	0	0
75.00	100	0	0

Where: %A: 100 mM NaOH; %B: 100 mM NaOH/100 mM NaOAc; %C: 300 mM NaOH.

The modification which resulted in fastest analysis times for detecting beet medium invert sugar addition to orange juice while maintaining oligosaccharide resolution and sensitive detection was that introduced by Dionex in 1992 (Dionex, 1992). A single Carbo Pac PA100 column is used in conjunction with a non-linear gradient elution program. The non-linear gradient was as follows: 0–1 min 99% A and 1% B; then 1–20 min to 100% B; this was followed by re-equilibration for 10 min under initial conditions. This modified system has been used to detect the addition of a variety of inexpensive sweeteners to citrus juice products (Hammond, 1993).

Our research group (Wudrich et al., 1993) also introduced a modified gradient elution program for the detection by HPAE-PAD of total invert sugar (beet or cane) produced either by acid or enzymatic hydrolysis; high fructose corn syrup; and medium invert sugar (beet or cane). The total analysis time between sample injections was 96 min and the linear gradient elution program employed is shown in Table 5.6. This program afforded oligosaccharide fingerprint separation for each of these inexpensive sweeteners in a single chromatographic run with detection limits for each sweetener of ~5%.

Maple syrup analysis by HPAE-PAD. In addition to fruit juice authenticity analysis, HPAE-PAD has been extended to honey and maple syrup authenticity determination (Swallow and Low, 1990, 1994; Stuckel and Low, 1995). For maple syrup analysis the authors used the sample clean-up method SPM2 and the gradient elution program outlined in Table 5.7 to detect medium invert sugar and high fructose syrup addition. For honey a number of modifications to the HPAE-PAD methodolgy were required before adulterant detection could be monitored.

Table 5.6 Gradient elution program for the detection of IS, MIS, and HFS in citrus juices via fingerprint oligosaccharide analysis (Wudrich et al., 1993)

Time (min)	Composition		
	% A	% B	% C
0.00	100	0	0
6.59	100	0	0
7.00	95	5	0
23.00	89	11	0
30.00	60	40	0
35.00	60	40	0
45.00	10	90	0
50.00	10	90	0
53.00	0	0	100
68.00	0	0	100
68.10	100	0	0
96.00	100	0	0

Where: %A: 100 mM NaOH; %B: 100 mM NaOH/100 mM NaOAc; %C: 300 mM NaOH.

Table 5.7 Gradient elution program for the detection of MIS and HFS in maple syrup via fingerprint oligosaccharide analysis (Stuckel and Low, 1995)

Time (min)	Composition		
	% A	% B	% C
0.00	100	0	0
8.00	100	0	0
40.00	0	100	0
41.00	0	0	100
70.00	0	0	100
71.00	100	0	0
116.00	100	0	0

Where: %A: 100 mM NaOH; %B: 100 mM NaOH/100 mM NaOAc; %C: 300 mM NaOH.

Honey analysis by HPAE-PAD. Because of the high concentration of monosaccharides present in honey (Table 5.2) Swallow and Low (1994) found that the majority of these carbohydrates had to be removed prior to oligosaccharide analysis. Monosaccharides were removed from the sample employing a modified charcoal/celite procedure. One gram (1.00 g) of honey was diluted in 19.00 g of deionized water and stirred with 4.0 g of activated charcoal, 50–200 mesh (Fischer Scientific Co., Edmonton, AB), for 17 h at 4 °C. After mixing, the sample was placed on a 3.0 cm × 40 cm column containing 4.0 g of activated charcoal-celite (50+50, w/w; Fischer Scientific Co.). More than 95% of the monosaccharides can be removed from the honey sample (estimated by HPLC) by washing with 1 l of 0.1% (v/v) ethanol at room temperature at a flow rate of 10.0 ml/min maintained by vacuum. The oligosaccharides were then eluted from the column with 500 ml of a 60 °C solution of 50% (v/v) ethanol at the same flow rate. The filtrate was dried at 35 °C in a rotary evaporator and dissolved in 10 ml of deionized water. Following filtration, oligosaccharide analysis was accomplished using the HPAE-PAD methodology previously presented (i.e. columns, detector settings etc.) using the gradient elution programs shown in Table 5.8 (for total invert sugar) and 5.9 (for high fructose corn syrup).

5.5.2 Interpretation of HPAE-PAD results

Fruit juices. In general, orange and grapefruit juice contain the least number and lowest concentration of 'natural' oligosaccharides with apple and pineapple juice containing the highest levels (both in number and concentration). In order to illustrate the efficacy of the HPAE-PAD fingerprint oligosaccharide method for fruit juice authenticity analysis grapefruit juice is used as a practical example. The following fingerprint oligosaccharide result interpretation for pure grapefruit juice and the same juice intentionally adulterated with IS, MIS and HFS is based on our published methodology (Low and Wudrich, 1993).

Table 5.8 Gradient elution program for the detection of IS in honey via oligosaccharide fingerprinting (Swallow and Low, 1994)

Time (min)	Composition		
	% A	% B	% C
0.00	75	25	0
4.00	75	25	0
20.00	0	100	0
30.00	0	100	0
30.10	0	0	100
90.10	0	0	100
90.20	75	25	0
120.00	75	25	0

Where: %A: 100 mM NaOH; %B: 100 mM NaOH/100 mM NaOAc; %C: 300 mM NaOH.

The carbohydrate profile (HPAE-PAD) of a pure grapefruit juice is shown in Figure 5.3. The large offscale peaks in this chromatogram occurring in the 10.5 to 28 min region correspond to glucose, fructose and sucrose.

Because the major carbohydrate profile/approximate ratio for grapefruit and other citrus juices is glucose : fructose : sucrose/1 : 1 : 2, the inexpensive sweetener which best matches this profile/ratio is MIS. However, our experience in the analysis of commercial samples has shown that HFS and IS have also been used to adulterate citrus juices.

Figures 5.4, 5.5, and 5.6 are HPAE-PAD profiles of three commercial inexpensive sweeteners/syrups, MIS, HFS and IS respectively. Each of these materials contain a large number of oligosaccharides (retention times >40 min) of varying concentrations. Our research has shown that each of these materials

Table 5.9 Gradient elution program for the detection of HFS in honey via oligosaccharide fingerprinting (Swallow and Low, 1994)

Time (min)	Composition		
	% A	% B	% C
0.00	100	0	0
5.00	100	0	0
37.00	0	100	0
38.00	0	0	100
67.00	0	0	100
68.00	100	0	0
113.00	100	0	0

Where: %A: 100 mM NaOH; %B: 100 mM NaOH/250 mM NaOAc; %C: 300 mM NaOH.
Note: only one Carbo Pac PA1 pellicular anion exchange column is used and the sample injection volume is 600 µL.

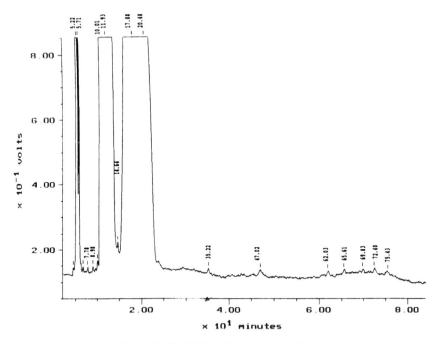

Figure 5.3 HPAE-PAD of pure grapefruit juice.

contain unique oligosaccharides which can be used as 'fingerprints' to detect their addition to foods rich in carbohydrate such as fruit juices. It is important to note that the production of the fingerprint oligosaccharides present in MIS is due to acid catalysed formation during processing (Wudrich et al., 1993; Cancalon, 1993). Therefore, a false positive (only for MIS) may be observed if a citrus concentrate has been excessively heated. Research results from our laboratory have shown that the heating of a citrus juice concentrate also increases the levels of 5-hydroxymethyl-2-furaldehyde (HMF), 3-deoxyglycosulose, and 2-furaldehyde. Hammond (1996) has also noted a linear correlation between HMF and the total combined peak area of the fingerprint oligosaccharides. Therefore, the presence of these heat generated compounds in addition to the four fingerprint oligosaccharides in a citrus juice concentrate may be useful in distinguishing heat abuse from intentional adulteration. With respect to fresh pressed and pasteurized citrus juice no false positive occurs as extensive heating of a single strength juice does not result in the formation of the characteristic four fingerprint oligosaccharides present in medium invert sugar. In our laboratory we have also found that capillary gas chromatography (section 5.6.3) can be useful for distinguishing a heat abused citrus concentrate from one intentionally adulterated with medium invert sugar.

Chemical conversion of beet/cane sucrose to MIS (Fig. 5.4) results in the formation of four fingerprint oligosaccharides which have retention times in

OLIGOSACCHARIDE ANALYSIS

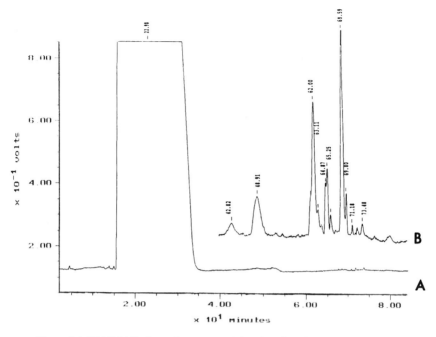

Figure 5.4 HPAE-PAD of pure beet sucrose (A) and medium invert sugar (MIS; B).

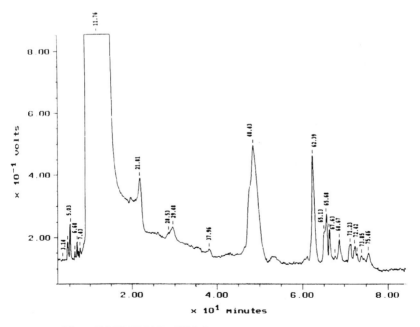

Figure 5.5 HPAE-PAD of high fructose corn syrup (HFS; HFS 55).

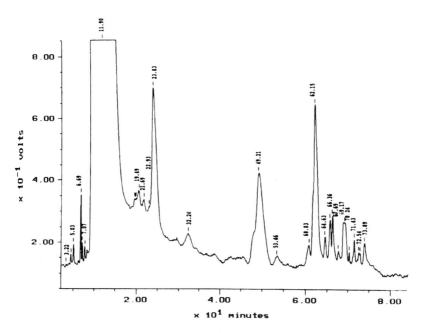

Figure 5.6 HPAE-PAD of total invert sugar (IS).

the 62–72 min region. The chemical structures of two of these fingerprint compounds have been elucidated (Swallow and Low, 1993) and have been shown to be 'kestose' oligosaccharides (fructose linked to a sucrose molecule). The latest eluting (retention time of ~70 min) of the four fingerprint oligosaccharides is O-α-D-fructofuranosyl-(2 → 6)-β-D-fructofuranosyl-α-D-glucopyranoside. This carbohydrate had not previously been isolated or identified in nature and we therefore named it kelose. The second of the four fingerprint oligosaccharides in medium invert sugar was also isolated and structurally identified as O-β-D-fructofuranosyl-(2 → 6)-β-D-fructofuranosyl-α-D-glucopyranoside, commonly known as 6-kestose.

The fingerprint oligosaccharides formed during the enzymatic production of high fructose syrups (HFS; Fig. 5.5) include maltose (retention time of ~48 min), isomaltose (~22 min), maltotriose (~62 min), maltotetraose (~65.5 min) and other glucose polymers. Total chemical inversion of sucrose to produce IS results in the formation (Fig. 5.6) of two major oligosaccharides with retention times of 49–51 and 62–64 min, respectively. The structures of these two compounds have not yet been determined but unpublished results from our laboratory indicate that they are disaccharides.

Oligosaccharide profiles of a grapefruit juice sample intentionally adulterated with each of the previously described inexpensive sweeteners/syrups are shown in Figs 5.7–5.9. Each of these chromatograms are vastly different from those obtained for pure juices (Fig. 5.3). The presence of pronounced fingerprint

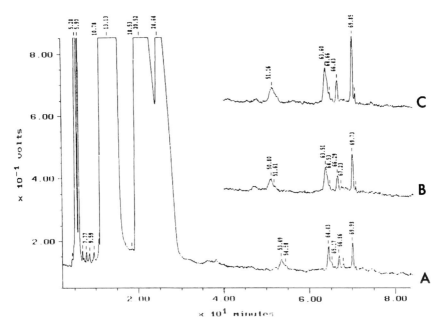

Figure 5.7 HPAE-PAD of pure grapefruit juice intentionally adulterated with 5% (A), 10% (B) and 15% (C) MIS.

oligosaccharide peaks which elute in the 48–71 min region of the chromatogram clearly indicate whether a juice has been adulterated. Oligosaccharide fingerprint results for fruit juices show that adulteration levels of 10% are readily detected and that 5% detection levels are possible.

To detect citrus juice adulteration with inexpensive sweeteners in our laboratory we use kelose as the fingerprint oligosaccharide for medium invert sugar; maltose and maltotriose for high fructose syrup from starch; and the two unidentified oligosaccharides with retention times of ~49–51 and 62–64 min, respectively for total invert sugar. Results from HPAE-PAD analysis of 76 pure and intentionally adulterated orange juice (similar results were obtained for grapefruit juice) samples representing the major world producing regions are shown in Table 5.10 (Wudrich, 1993).

Due to differences in column packing, solvent delivery, and HPLC equipment, the elution times of these fingerprint oligosaccharides may vary. Therefore, on each individual HPAE-PAD system standards must always be run, and a database of both pure and intentionally adulterated samples established.

Considerable misinformation has been circulating on both the commercial membrane filtration of inexpensive sweeteners to remove trace oligosaccharides, and the production (without membrane filtration) of 'oligosaccharide free' inexpensive sweeteners. Although commercial membrane filters with a molecular mass cutoff of 500 daltons (trisaccharides have an approximate molecular

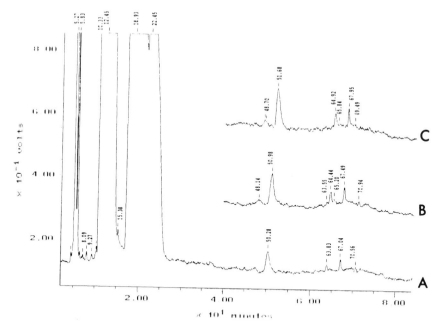

Figure 5.8 HPAE-PAD of pure grapefruit juice intentionally adulterated with 5% (A), 10% (B) and 15% (C) HFS.

weight of 594 g/mole) are available, filtration rates are very slow and it is impossible to remove trisaccharides and tetrasaccharides from disaccharides (such as sucrose). In addition the so-called 'oligosaccharide free' inexpensive sweeteners produced chemically without membrane filtration still contain detectable levels of fingerprint oligosaccharides (unpublished results).

Maple syrup. Pure, non-aged maple syrup has monosaccharide concentrations <11–12% (Morselli, 1975; Stuckel and Low, 1996), however, aged (>10 months) syrup may contain monosaccharide concentrations up to 25% (Low, 1993). This fact coupled with the high cost differential between pure maple syrup and inexpensive sweeteners has made this food product susceptible to adulteration with MIS and HFS.

In our research 80 pure maple syrup samples representative of world producing regions were analysed by HPAE-PAD for their oligosaccharide content. Chromatographic results showed few detectable oligosaccharide peaks after ~35 min in the pure samples while several peaks were observed for HFS. On the basis of all 80 pure maple syrups the fingerprint oligosaccharide chosen to detect HFS addition was unique and had a relative retention time of ~39 min (tentatively identified as maltopentaose). This oligosaccharide was determined to be unique because it was absent (peak area <30 000 µV × sec) in each of the pure maple syrup samples analysed. The mean peak area for the fingerprint

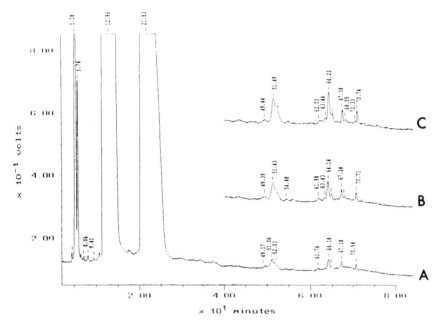

Figure 5.9 HPAE-PAD of pure grapefruit juice intentionally adulterated with 5% (A), 10% (B) and 15% (C) IS.

oligosaccharide rose to $1.14 \, \mu V \times sec \times 10^6$ for maple syrup samples intentionally adulterated with 10% HFS (Stuckel and Low, 1995).

A similar unique fingerprint oligosaccharide was also used to detect MIS adulteration of pure maple syrup. On the basis of all 80 pure maple syrups the fingerprint oligosaccharide chosen to detect MIS addition was unique and had a relative retention time of ~40 min. This oligosaccharide was determined to be unique because it was absent (peak area $<30\,000 \, \mu V \times sec$) in each of the pure maple syrup samples analysed. The mean peak area for this fingerprint oligosaccharide rose to $0.59 \, \mu V \times sec \times 10^6$ for maple syrup samples intentionally adulterated with 10% MIS (Stuckel and Low, 1995).

Table 5.10 Mean HPAE-PAD fingerprint oligosaccharide areas for 76 pure orange juice samples intentionally adulterated with MIS, HFS and IS at levels of 5, 10 and 20%

Sample	Adulterant area ($\mu V \times sec \times 10^6$)				
	MIS	HFS(I)	HFS(II)	IS(I)	IS(II)
Pure	0.37	0.33	0.23	0.22	0.25
5% Adulteration	0.83	0.87	0.90	0.91	0.90
10% Adulteration	1.66	1.60	1.63	1.76	1.87
20% Adulteration	3.02	2.96	3.02	3.36	3.54

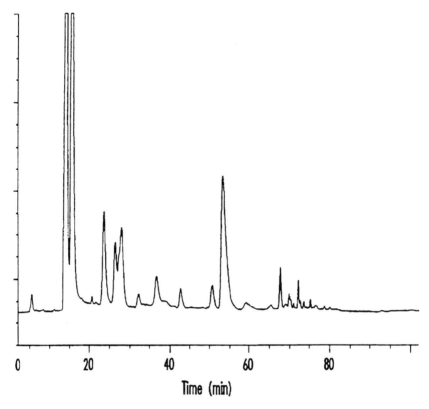

Figure 5.10 HPAE-PAD of pure honey.

Honey. The detection of honey adulteration is much more difficult than that in fruit juices due to the high level of monosaccharides and the presence of a large number of 'naturally' occurring oligosaccharides in the pure food. Upon removal of the majority (~95%) of these monosaccharides by charcoal/celite chromatography the natural oligosaccharide pattern of honey can be readily discerned by HPAE-PAD (Fig. 5.10).

Both IS and HFS can be used to adulterate honey because each of these materials approximates honey with respect to major carbohydrates profile/ratio (~1:1 ratio of fructose to glucose). HPAE-PAD analysis of an intentionally adulterated sample of honey with IS at a level of 10% is shown in Fig. 5.11. The fingerprint carbohydrate which can be used to detect this sweetener has a retention time of ~21 min. (labelled I in Fig. 5.11). This unidentified oligosaccharide was not present in 22 of 44 pure honey samples analysed by HPAE-PAD and was present in low concentrations (mean area of 1.13 $\mu V \times sec \times 10^6$ (Swallow and Low, 1994). These results show that HPAE-PAD can be used to detect the addition of commercially produced IS to honey at the 10% level.

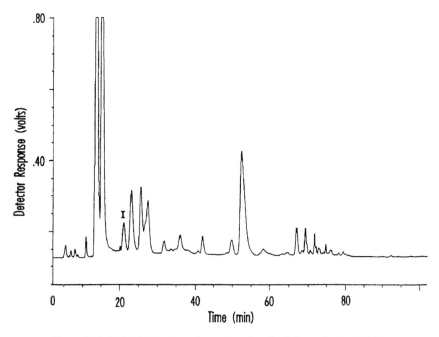

Figure 5.11 HPAE-PAD of pure honey intentionally adulterated with 10% IS.

A different gradient elution program (Table 5.8) is used for the detection of HFS addition to honey. Under these gradient elution conditions the fingerprint oligosaccharide (tentatively identified as a branched DP-6 oligosaccharide) for the detection of HFS adulteration of honey has a retention time of ~40.9 min (labelled III in Fig. 5.12). This oligosaccharide was not detected in 4 of the 44 pure honeys analysed by HPAE-PAD and was present in low concentrations (mean area of $2.07\,\mu V \times \sec \times 10^5$) in the remaining samples. Intentional adulteration of these pure honey samples with 10% HFS raised the mean area of peak III to $1.97\,\mu V \times \sec \times 10^6$ (Swallow and Low, 1994). These results show that HPAE-PAD can be used to detect the addition of commercially produced HFS to honey at the 10% level.

5.5.3 HPAE-PAD conclusions

Methodology has been developed for the analysis of a variety of foods rich in carbohydrates such as fruit juices, honey and maple syrup to detect adulteration with inexpensive sweeteners. The disadvantages of this technique include the specialized equipment required to perform the analysis such as a metal-free gradient high performance liquid chromatographic system equipped with a pulsed amperometric detector and lengthy analysis times. However, a single analysis can result in the detection of MIS, IS and HFS in a sample.

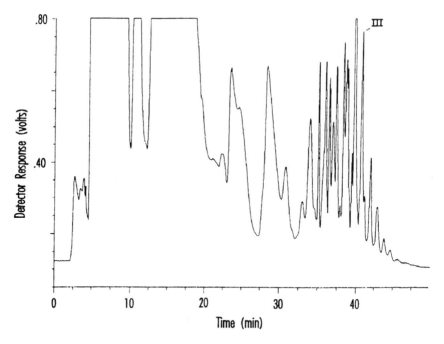

Figure 5.12 HPAE-PAD of pure honey intentionally adulterated with 10% HFS (HFS 55).

5.6 Oligosaccharide fingerprinting employing capillary gas chromatography with flame ionization detection (CGC-FID)

5.6.1 Sample analysis by capillary gas chromatography with flame ionization detection

Following the derivatization technique the trimethylsilylated carbohydrates are analysed using a gas chromatograph (i.e. Hewlett-Packard, Model 6890; Hewlett-Packard, Mississauga, ON) equipped with a flame ionization detector and if possible an autosampler (Hewlett-Packard, Model 7673A). Analysis is carried out in the splitless mode. Two CGC-FID programs can be used and the choice is dependent on the carrier gas employed.

1. If the carrier gas used is ultra high purity (UHP) helium or UHP nitrogen: oven temperature program of 210 °C for 15 min followed by a 1 °C/min gradient to a temperature of 290 °C, maintained for 20 min. Carrier gas flow is maintained at a constant velocity of 27 cm/sec. Injector temperature and detector temperatures are kept at 250 °C and 300 °C, respectively. Detector make-up gas is UHP nitrogen at a flow rate of 30 ml/min.
2. If the carrier gas used is UHP hydrogen: an oven temperature program of 210 °C for 10 min followed by a 1 °C/min gradient to a temperature of 248 °C; this temperature was maintained for 30 sec and was followed by a 30 °C/min gradient to a final temperature of 290 °C, maintained for 15 min.

Carrier gas is maintained at a constant flow of 1.4 ml/min. The total linear velocity should be ~38 cm/sec. Injector temperature and detector temperatures are kept at 250 °C and 300 °C, respectively. Detector make-up gas is UHP nitrogen at a flow rate of 30 ml/min.

Depending on the sensitivity of the CGC-FID system the sample volume injected can be either 1.0 µl or 2.0 µl.

Fingerprint oligosaccharide separation in literature is based on the use of either of the following capillary columns: J&W DB-5 (0.25 mm × 30 m; 0.25 µm film thickness) [J&W Scientific, Folsom, California, USA]; or an HP-5 (0.32 mm × 30 m; 0.25 µm film thickness) [Hewlett-Packard, Wilmington, Delaware, USA].

5.6.2 Interpretation of CGC-FID results

Fruit juices. Oligosaccharide fingerprinting by CGC-FID has been successfully used to detect IS and HFS in orange (Low, 1995, 1996), and pineapple (Low *et al.*, 1994) juices. Unpublished results from our laboratory and others have shown that this methodology may be used to detect IS, MIS, HFS and HIS in a variety of fruit juices such as cranberry, grape, strawberry, grapefruit etc. Due to the complexity of the oligosaccharide profile for apple juice, and the fact that HPAE-PAD methodology is not applicable to readily detect adulteration in this food, apple juice will be used a practical example for CGC-FID oligosaccharide fingerprint analysis.

The major carbohydrates present in apple juice are sorbitol, glucose, fructose, and sucrose with fructose predominating and resulting in a fructose/glucose ratio greater than 2.0 (Brause and Raterman, 1982; Low, 1996). Three commercially available commercial sweeteners which approximate the carbohydrate composition of apple juice are beet/cane invert sugar (IS; 1:1 ratio of fructose to glucose), high fructose corn syrup (HFCS; specifically HFCS 55; 55% fructose and 40% glucose), and high fructose inulin syrup (HIS; fructose to glucose ratios of 8:1 to 3:1).

To illustrate how difficult it is to detect an adulterated apple juice, two samples were intentionally spiked with 5, 10 and 15% high fructose corn syrup. These samples were then subjected to a wide range of chemical analyses and the results are shown in Table 5.11. Comparison of these intentionally adulterated samples with the authentic range for pure apple juice clearly indicates how difficult it is to identify an adulterated juice.

The CGC-FID carbohydrate profile for a pure apple juice with hydrogen as the carrier gas is shown in Fig. 5.13. Pure apple juice samples contain a number of oligosaccharides as indicated by the peaks eluting in the 15–40 min region of the chromatogram. The large offscale peak at ~29 min corresponds to sucrose. It is important to note the lack of appreciable peaks in the chromatographic regions of 19.4–20, 29.4–30, and 38.5–45 min, respectively.

Table 5.11 Chemical composition data for selected pure and adulterated apple juice samples[a]

Sample	So (%)	G (%)	F (%)	S (%)	F/G	CGA (ppm)	M (g/100 mL)	C (g/100 mL)	K (ppm)	Ca (ppm)	Mg (ppm)	$\delta^{13}C$
PAJ-1	0.42	1.75	6.66	1.70	3.81	97	420	22	1000	78	60	−25.6
PAJ-1 + 5% HFS	0.40	1.70	6.62	1.66	3.89	95	390	21	959	71	57	−24.9
PAJ-1 + 10% HFCS	0.36	1.66	6.90	1.49	4.16	88	351	20	858	70	52	−23.9
PAJ-1 + 15% HFCS	0.30	1.46	7.14	1.44	4.89	80	360	20	888	65	55	−23.1
PAJ-2	0.39	2.00	6.16	1.89	3.08	47	440	20	1006	83	49	−25.9
PAJ-2 + 5% HFCS	0.33	1.93	6.37	1.80	3.30	46	444	19	1011	82	44	−25.2
PAJ-2 + 10% HFCS	0.34	1.80	6.50	1.68	3.61	44	400	19	929	80	44	−24.5
PAJ-2 + 15% HFCS	0.28	1.67	6.63	1.60	3.97	44	379	20	897	75	39	−23.7
Literature values	0.16–1.20[b,c]	0.89–3.99[b,c]	3.00–10.5[b,c]	0.88–5.62[b,c]	>20[c]	<30[b]	200–900[b,c]	685–1510[b,c]	30–120[b,c]	40–70[b,c]	<−20.2[d]

[a] All samples were at 11.5 °Brix. Abbreviations: C, citric acid; Ca, calcium; CGA, chlorogenic acid; F, fructose; G, glucose; HFCS, high fructose syrup; K, potassium; M, malic acid; Mg, magnesium; PAJ, pure apple juice; S, sucrose; So, sorbitol.
[b] Brause and Raterman, 1982.
[c] Mattick, 1988.
[d] Doner and Phillips, 1981.

Detection of total invert sugar. The fingerprint oligosaccharides used to detect the addition of invert sugar to apple juice have approximate retention times of ~39.1 (FPI) and ~39.8 (FPII) min, respectively (Fig. 5.14). The CGC-FID analysis of 123 pure apple juice samples gave mean peak heights/areas for FPI and FPII of <2 pA/94 µV × sec and <2 pA/94 µV × sec, respectively. Spiking of the pure apple juice samples with 10% IS raised these means to 5.3 pA/~3000 µV × sec and 23.5 pA/~10 000 µV × sec, respectively (Low, 1996).

The approximate area/height ratio of peaks FPI and FPII has been shown to range from ~3 to ~5.5 depending on the CGC-FID equipment employed (Low, 1995, 1996; unpublished results). The chemical structures represented by these peaks have not yet been elucidated. However, oligosaccharide standards were compartmentalized via CGC-FID analysis with disaccharides having elution times ranging from 16–50 min, and tri- and tetrasaccharides had elution times >50 min. The two fingerprint peaks for invert sugar are therefore most likely to be disaccharides.

Detection of adulteration by total invert sugar vs heat effect on juices. The commercial process used to produce total invert sugar (IS) involves beet/cane sucrose hydrolysis with hydrochloric acid. This process results in the formation of the fingerprint oligosaccharides which are used to detect the addition of this

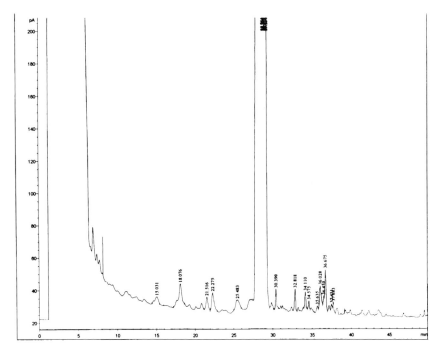

Figure 5.13 CGC-FID of pure apple juice (PAJ).

material to fruit juices. Therefore, the question arises as to the relationship between heat treatment of juice concentrates and the production of these fingerprint oligosaccharides. We recently participated in an interlaboratory study with 11 other laboratories which was designed to answer this important question for apple juice. These results have not been published, however, results from 11 of the 12 laboratories indicate that there is a clear distinction between a juice sample intentionally adulterated compared with one which was heat abused.

Twenty-four samples analysed in this study comprised: two pure apple juice concentrates (AJC); two AJC adulterated with 10% IS; two AJC heated at 105 °C/5 min; two AJC heated at 105 °C/10 min; two AJC heated at 105 °C/15 min; two AJC heated at 120 °C/5 min; two AJC + 10% IS heated at 105 °C/5 minutes; two AJC + 10% IS heated at 105 °C/10 min; two AJC + 10% IS heated at 105 °C/15 min; two AJC + 10% IS heated at 120 °C/5 min; two AJC + 20% IS; and two AJC + 30% IS. The results from the analysis of these samples by CGC-FID by our laboratory were very clear. The pure apple juice samples and those heated at 105 °C/5 min contained no detectable levels (peak height <2 pA) of the fingerprint oligosaccharides used to detect total invert sugar. The intentionally adulterated apple juice concentrate samples showed each of the fingerprint oligosaccharides (FP) with FP1: ~38.6 min/peak height

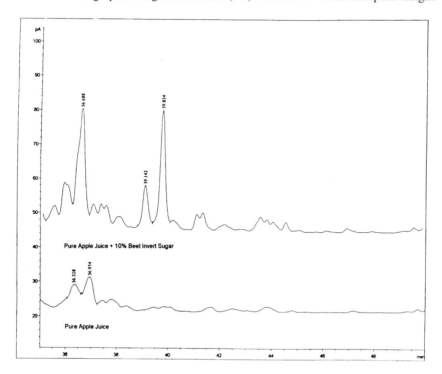

Figure 5.14 Expansion of the 35–50 min region of the CGC-FID chromatograph of PAJ and PAJ + 10% IS.

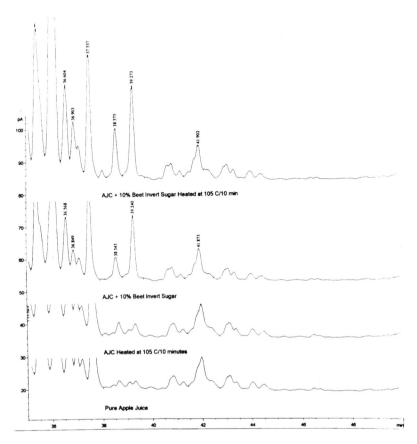

Figure 5.15 Expansion of the 35–50 min region of the CGC-FID chromatogram of PAJ, PAJ heated at 105°C/10 min, PAJ + 10% IS, and PAJ + 10% IS heated at 105°C/10 min.

of 6.2 pA for 10% IS (~12 pA for 20% IS; and ~17 for 30% IS); and FP2: ~39.3 min/peak height of 19 pA for 10% IS (~38 for 20% IS; and ~62 for 30% IS). The remainder of the heated apple juice concentrates all had detectable levels of the two fingerprint oligosaccharides (peak heights of 5–14 for FP1 and ~4–12 pA for FP2, with the peak area/height for FP1 always being greater than that of FP2). The intentionally adulterated and then heated samples had peak height ratios (FP1 to FP2) ranging from 1.74 to 2.07. Therefore, the heat vs total invert sugar question for apple juice can be answered by both the presence and respective ratio of the two fingerprint oligosaccharide peaks in the sample. (Fig. 5.15).

Detection of high fructose corn syrup. The fingerprint oligosaccharide used as a marker for high fructose corn syrup had retention times of ~41.2 (FPIII) and ~44.6 min (FPIV), respectively (Fig. 5.16). The identity of this fingerprint

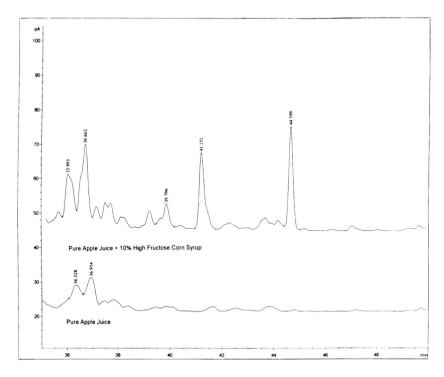

Figure 5.16 Expansion of the 35–50 min region of the CGC-FID chromatogram of PAJ and PAJ + 10% HFS.

oligosaccharide was determined by standards to be isomaltose, and was confirmed by HPAE-PAD (Swallow and Low, 1990).

The presence of two peaks for isomaltose by capillary gas chromatography with flame ionization detection (CGC-FID) can be explained by the fact that reducing carbohydrates can exist in a number of tautomeric forms. For D-glucose the two major tautomeric forms are β-D-glucopyranose ($\sim$62%) and α-D-glucopyranose ($\sim$38%) (i.e. a 1.6:1 ratio). The addition of a glucose moiety at the 6-position of glucose (i.e. isomaltose, O-α-D-glucopyranosyl-(1 $\rightarrow$ 6)-D-glucopyranose) makes the β-tautomer less stable and results in an approximately 1.3:1 area ratio of the β-(FPIV) to α-tautomer (FPIII).

The CGC-FID analysis of 123 pure apple juice samples gave mean peak heights/areas for FPIII and FPIV of <2 pA/not detected (area <1000 μV × sec) and <2 pA/139 μV × sec, respectively. Spiking of the pure apple juice samples with 10% HFS raised these means to 22.5 pA/$\sim$13 000 μV × sec and 26.7 pA/$\sim$17 000 μV × sec, respectively (Low, 1996).

A recent AOAC peer validation (McLaughlin et al., 1997) study done on the CGC-FID method in which my laboratory participated indicated that detection limits for HFS of 2% were possible.

Detection of high fructose syrup from inulin. The recent (Low and Hammond, 1996) finding of commercial apple juice intentionally adulterated with high fructose syrup from inulin resulted in significant interest from industry, government and legal bodies (Butler, 1996). High fructose syrup from inulin contains two unique fingerprint oligosaccharides which have been used as markers to detect its addition to apple juice. These fingerprint oligosaccharides have relative retention times of ~20.0 (FPV) and ~30.1 (FPVI) min, respectively (Fig. 5.17). In addition to these two oligosaccharides two others have also been shown to be present in much lower concentrations in commercial HIS samples, and have approximate retention times of 25.7 and 39.1 min, respectively (Low and Hammond, 1996).

One hundred and sixty-seven pure apple juice concentrates representing numerous global apple production/processing regions have been analysed for

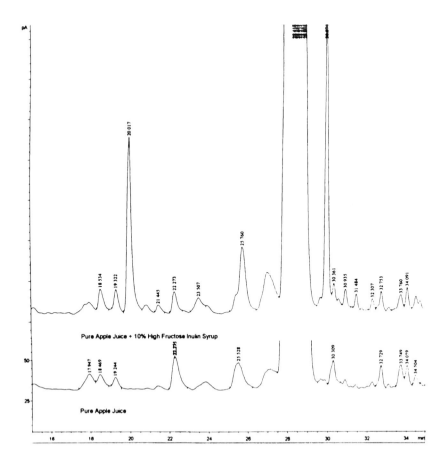

Figure 5.17 Expansion of the 15–35 min region of the CGC-FID chromatogram of PAJ and PAJ + 10% HIS.

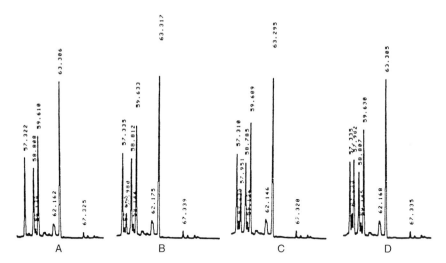

Figure 5.18 Expansion of the 56–70 min region of the CGC-FID chromatogram of pure honey (A), and the same sample intentionally adulterated with 5% (B), 10% (C) and 15% (D) IS.

the presence of oligosaccharides FPV and FPVI. None of the 167 pure apple juice samples analysed contained detectable (peak height >2 pA) levels of the two fingerprint oligosaccharides (FPV and FPVI) which we use as markers for HIS adulteration. The intentional spiking of pure apple juice samples with 10% HIS raised the peak height/area of FPV and FPVI to 71.1 pA/0.66 × 10^6 pA × sec, and 145.6 pA/0.64 × 10^6 pA × sec, respectively (Low, 1996; Low and Hammond, 1996).

As was the case with HFS, the recent AOAC peer validation study on the CGC-FID method indicated that detection limits for HIS of 2% were possible. Our research group is currently working on the isolation and structural identification of these two fingerprint oligosaccharides.

Honey. Originally (Low and Sporns, 1988) CGC-FID methodology was developed in our laboratory to relate the di- and trisaccharide composition of honey to its hydrolase enzyme content and activity. Based on this fundamental research we developed a CGC-FID method to detect IS adulteration of honey (Low and South, 1995).

In this work 68 pure honey samples from various world producing regions and 5 commercial total invert sugar samples were analysed by CGC-FID for their oligosaccharide content and concentration. It was noted that IS contained two fingerprint oligosaccharide peaks (previously identified as FPI and FPII) oligosaccharide markers with retention times of $\sim$57.6 (FPI) and $\sim$57.9 (FPII), respectively (Fig. 5.18; helium as carrier gas). Although these fingerprint oligosaccharides were not detectable in 46 of the pure honey samples they were present in low concentrations in the remaining 24 samples (overall mean of $\sim$300 μV × sec for FPI and $\sim$600 μV × sec for FPII). When spiked with 10%

IS the mean for the 68 honey samples rose to ~2500 µV × sec for FPI and ~10 000 µV × sec for FPII (Low and South, 1995).

The authors point out that honey is pasteurized, stored and may be heat abused which could result in the synthesis of the two fingerpoint oligosaccharides used to detect IS addition. Subsequent unpublished results from our laboratory and those of Dr Alan Brause (Analytical Chemical Services of Columbia, Columbia, Maryland, USA) indicate that honey HMF levels correlate closely to the production of these two fingerprint oligosaccharides. Therefore if Codex (1969) guidelines of honey HMF levels <40 ppm are followed, false positive IS adulteration results are avoided.

5.6.3 Other applications of oligosaccharide fingerprinting by CGC-FID

Unpublished results from our laboratory indicate that CGC-FID may also be useful in distinguishing a heat abused citrus juice concentrate from one which has been intentionally adulterated with MIS. Under isothermal conditions of 245 °C a commercial orange juice concentrate (C) and the same concentrate heated at 90 °C/3 min (B) shows three oligosaccharide peaks in the 39–44 min region of the chromatogram. The same concentrate intentionally adulterated with 10% MIS (A) shows six oligosaccharide peaks in the 39–44 min region of the chromatogram (Fig. 5.19; helium as carrier gas).

Oligosaccharide profiling by CGC-FID can also be used to monitor the enzymatic treatments/processes used by manufacturers to produce a fruit juice. For example, the treatment of macerated apple/apple pomace with cellulase to increase both juice and yield and the soluble solids content of the final juice, results in the formation of cellobiose (α- and β-anomers with relative retention times of 30.4 and 36.7 min, respectively) which can be readily detected by CGC-FID (Fig. 5.20). This enzymatic treatment is not allowed in Europe and is currently under discussion in both Canada (Agriculture and Agri-Food Canada, personal communication) and the United States (Food and Drug Administration, personal communication). We are currently applying our developed CGC-FID methodology to maple syrup adulteration with IS, MIS, HFS and beet/cane sucrose.

5.6.4 CGC-FID conclusions

This method is extremely useful for detecting inexpensive sweetener adulteration in honey, maple syrup, and a wide variety of fruit juices. Sample preparation is facile and analysis times (employing hydrogen as carrier) are reasonably rapid (~66 min). Currently, some controversy exists in the application of CGC-FID to detect apple juice adulteration with IS. This controversy is based on the formation of the IS fingerprint oligosaccharides in an apple juice concentrate which has been severely heat treated. As was shown in section 5.6.2 CGC-FID

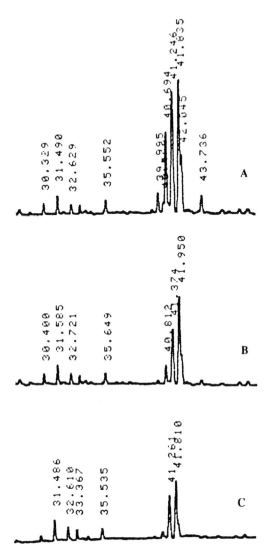

Figure 5.19 Expansion of the 28–47 min region of the CGC-FID chromatogram of pure orange juice (POJ; C), POJ heated at 90°C/3 min (B) and POJ + 10% MIS (A).

results can clearly discern an adulterated sample from one which has been heat abused or one which has been adulterated and then heat abused.

This method like all others in this area should be used in conjunction with other analytical authenticity tests. Where possible a CGC-FID result indicating adulteration should be analytically confirmed. However, due to the sensitivity of this test and its ability to detect adulterants which are basically 'invisible' by other authenticity tests it may need to stand alone.

5.7 Oligosaccharide fingerprinting employing capillary zone electrophoresis

5.7.1 Introduction

In general terms, capillary electrophoresis (CE) is the electrophoretic separation of a substance from a complex mixture within a narrow bore tube. In this system a small volume of sample (100 pl to 50 nl) is introduced into one end of a fused silica capillary tube containing an appropriate electrolyte (buffer). The component molecules migrate along the length of the capillary tube at different rates under the action of an applied electric field. Analysis times can be as short as 15 sec and are rarely longer than 20–30 min. Individual molecules are detected as they approach the other end of the capillary tube by employing one of many available on-line detection schemes.

Electro-osmosis occurs (in uncoated) silica capillary tubes because the inner wall of the capillary bears acidic silanol groups that become negatively charged when in contact with many of the buffers commonly employed in CZE. These

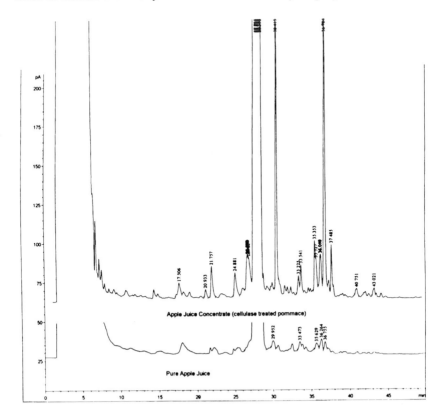

Figure 5.20 CGC-FID of pure apple juice and pure apple juice containing cellulase treated pomace.

negative charges are immobile, however, they attract hydrated counterions from the electrolyte solution to form an electric double layer between the surface of the capillary and the electrolyte solution, which is characterized by the zeta potential. When an electrical potential is applied to the capillary the hydrated counterions are mobilized and this mobility results in a bulk of the solvent (electro-osmotic flow) moving towards the cathode.

Capillary zone electrophoresis (CZE) is the most commonly employed technique in CE. Solute separation is based on their interaction in the electrolyte solution with the applied electric field/electro-osmotic flow and ionic species separate on the basis of their charge-to-mass ratio. Cations are eluted first, followed by all neutral molecules (typically these are unresolved), followed by anions.

Capillary zone electrophoresis affords high resolution (up to 1 million theoretical plates in a 60 cm capillary column), fast analysis times, minimal electrolyte/solvent usage, minimal sample requirements, and is non-destructive. For these reasons we have been actively involved in the development of CZE oligosaccharide fingerprinting methodology for food adulteration detection.

5.7.2 Sample preparation for CZE

Carbohydrate analysis by CZE has been challenging due to the general neutrality of these compounds. The first reported carbohydrate separation by CZE was by Honda *et al.* (1989) which required pre-column derivation to yield an anionic borate complex. This work was followed by a simple and rapid carbohydrate derivative/derivatization procedure using aminonaphthalene sulphonic acids by Jackson (1990) and Chiesa and O'Neill (1994). These authors were successful in separating glucose polymer (up to DP 15) standards on fused silica capillary columns. Unfortunately, the derivatizing agents used by these authors were extremely expensive.

Our research group has been active in the development of an oligosaccharide fingerprinting CZE method employing commercial inexpensive derivatizing reagents such as 2-amino-4-fluorobenzoic acid (AF) and 3-amino-4-hydroxybenzenesulphonic acid (AHBA).

The derivatization procedure involves the treatment of 100 µl of 5.5 Brix fruit juice, honey or maple syrup sample with 15 µl of 1 M sodium cyanoborohydride in dimethyl sulphoxide and 15 µl of a 0.80 M AF or AHBA in 15% acetic acid. The resulting solution/suspension was heated at 75 °C for 15 min. The solution was then freeze dried and diluted with 200 µl of 15 mM borate buffer (pH 10; CZE electrolyte). Derivatized carbohydrate analysis was performed on a Waters Quanta 4000 capillary electrophoresis system (Waters Chromatography, Milford, MA) employing a 60 cm×50 µm fused silica capillary column (Polymicro Technologies Inc., Phoenix, AZ). The system was equipped with a positive power supply and the applied voltage was 20 kV. Sample introduction into the capillary was by a 30 sec hydrostatic injection from a height of 10 cm.

The electrolyte solution for derivatized carbohydrate separation was 15 mM boric acid (Aldrich Chemical Co., Milwaukee, WI) which was adjusted to pH 10 with sodium hydroxide. Direct UV detection was employed for the derivatized carbohydrates at 254 nm. The electro-osmotic flow was calculated to be $5.73 \times 10^{10-4}$ cm/Vs.

The ionic strength (i.e citric acid concentration in citrus juice) effect of the sample on solute electrophoretic mobility can be a significant problem in fruit juice analysis (Swallow and Low, 1994). In our laboratory we minimize this effect by the addition of caffeine as our internal standard.

5.7.3 Interpretation of CZE results

The order of elution of derivatized glucose standards was based on molecular weight with the largest having the shortest relative electrophoretic migration time (Fig. 5.21). Baseline resolution of glucose standards to DP 16 was observed with run times <15 min. Detection limits (based on a sample signal 3× baseline noise) for maltose and maltotetraose were 10^{-9} M. Application of this methodology for the qualitative and quantitative analysis of oligosaccharides in high fructose corn syrup correlated well with HPAE-PAD results (Table 5.12; Ferley and Low, 1997).

We are currently in the process of applying this methodology to HFS, MIS and IS adulteration of citrus juices and HFS, MIS and IS adulteration of apple juice, honey and maple syrup (Ferley and Low, 1997).

5.7.4 CZE conclusions

The development of capillary electrophoretic methodology for authenticity determination is still in its infancy. However, CE methods afford excellent solute resolution and rapid analysis times and appear to be conducive to oligosaccharide fingerprinting.

5.8 Conclusions

Currently, a major problem facing the fruit juice, honey and maple syrup industries is the undeclared addition of inexpensive sweeteners/syrups to these foods. The sophisticated adulteration of foods high in carbohydrate with materials which contain the same major carbohydrate profile as the pure food have made adulteration detection extremely difficult.

Fruit juices, honey, maple syrup, and inexpensive sweeteners/syrups (i.e. HFS, MIS and IS) each contain a complex mixture of oligosaccharides. These carbohydrates are either present in the raw material or arise during production/processing. HPAE-PAD, CGC-FID and CZE analysis reveal the presence of unique oligosaccharides in each inexpensive sweetener/syrup which can be used as 'fingerprints' for their addition to pure foods.

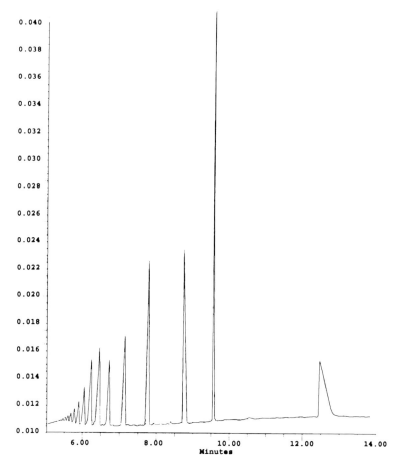

Figure 5.21 CZE (AHBA derivative) of glucose polymers (DP 15 to DP 1).

Table 5.12 Oligosaccharide concentration (g/L) in high fructose corn syrup 55 as determined by HAPE-PAD and CZE (AHBA derivative) (Ferley and Low, 1997)

Oligosaccharide	HPAE-PAD	CZE
Isomaltose	16.6	16.3
Isomaltotriose	0.25	0.24
Maltose	10.7	10.6
Maltotriose	0.83	0.85
Maltotetraose	0.22	0.21
Maltopentaose	0.36	0.34
Maltohexaose	0.29	0.27
Maltoheptaose	0.14	0.14

Oligosaccharide methodology has been applied to the adulteration of orange, grapefruit, pineapple, and apple juice. In addition this method has been modified to detect the adulteration of honey and maple syrup. Detection limits for fingerprint oligosaccharide methods are ~5%.

References

Asami, T., Ohyama, T., Minamisawa, K. and Tsukihashi, T. (1989) New tuber Yacon containing large amounts of frutooligosaccharides. *Nogyo Oyobi Engei*, **64**, 1033–6.
Brause, A.R. and Raterman, J.M. (1982) Verification of authenticity of apple juice. *J. Assoc. Off. Anal. Chem.*, **65**, 846–9.
Butler, D. (1996) Apple juice alert goes out. *Food News*, **23**, 1.
Cancalon, P.F. (1993) Oligosaccharide generation in acidic sugar media. *J. Assoc. Off. Anal. Chem. Int.*, **76**, 584–90.
Chiesa, C. and O'Neill, R.A. (1994) Capillary zone electrophoresis of oligosaccharides derivatized with various aminonapthalene sulfonic acids. *Electrophoresis*, **15**, 1132–40.
Codex Alimentarius Commission (1969) Recommended European Regional Standards for Honey. Rome: Joint F.A.O./W.H.O. Food Standards Programme, CAC/RS 12-1969.
Dionex (1992) Analysis of fruit juice adulterated with medium invert sugar from beets. *Dionex Application Note* 82.
Doner, L.W. and Phillips, J.C. (1981) Detection of high fructose syrup in apple juice by mass spectrometric $^{13}C/^{12}C$ analysis: Collaborative study. *J. Assoc. Off. Anal. Chem.*, **64**, 85–90.
Ferley, S. and Low, N.H. (1997) Detection of inexpensive sweetener addition to orange juice by capillary zone electrophoresis. *J. Agric. Food Chem.* (under review).
Florida Department of Agriculture (1996) *Florida Agricultural Statistics: Citrus Summary 1995–96*. Florida Agricultural Statistics Services, Orlando.
Gott, B. (1983) Microseris scapigera: A study of a staple food of Victorian Aborigines. *Australian Aboriginal Studies*, No. 2.
Hammond, D.A. (1993) Going sweet on ion chromatography. *Lab. Products Technol.*, April, 26.
Hammond, D.A. (1996) Authenticity of fruit juices, jams and preserves, in *Food Authentication*, (eds P.R. Ashurst and M.J. Dennis), Chapman & Hall, London, pp. 15–59.
Hassid, W.Z., and Ballou, C.E. (1957) Oligosaccharides, in *The Carbohydrates*, (ed. W. Pigman), Academic Press., New York, pp. 478–535.
Helferich, B., Bohn, E. and Winkler, S. (1930) Ungesatatigte derivative von gentiobiose und cellobiose. *Chem. Ber.*, **63**, 989–92.
Honda, S., Iwase, S. and Makino, A. (1989) Simultaneous determination of reducing monosaccharides by capillary zone electrophoresis as the borate complexes of N-2-pyridylglycamines. *Anal. Biochem.*, **176**, 72–7.
Jackson, P.E. (1990) The use of polyacrylamide-gel electrophoresis for the high-resolution separation of reducing saccharides labelled with the fluorophore 8-amino-napthalene-1,3,6-trisulfonic acid. Detection of picomolar quantities by an imaging system based on a cooled charged-coupled device. *J. Biochem.*, **270**, 705–13.
Krol, B. (1978) Side reactions of acid hydrolysis of sucrose. *Acta Aliment. Pol.*, 4, 373–80.
Low, N.H. (1993) Maple syrup adulteration detection methods Part I. *Maple Syrup Digest*, **5A**, 13–16.
Low, N.H. (1995) Apple and orange juice authenticity analysis by capillary gas chromatography with flame ionization detection. *Fruit Processing*, **11**, 362–7.
Low, N.H. (1996) Determination of fruit juice authenticity by capillary gas chromatography with flame ionization detection. *J. Assoc. Off. Anal. Chem. Int.*, **79**, 724–37.
Low, N.H., Brause, A. and Wilhelmsen, E. (1994) Normative data for commercial pineapple juice from concentrate. *J. Assoc. Off. Annal. Chem. Int.*, **77**, 965–75.
Low, N.H. and Hammond, D.A. (1996) Detection of high fructose syrup from inulin in apple juice by capillary gas chromatography with flame ionization detection. *Fruit Processing*, **4**, 135–9.
Low, N.H. and South, W. (1995) Determination of honey authenticity by capillary gas chromatography. *J. Assoc. Off. Anal. Chem. Int.*, **78**, 1106–13.

Low, N.H. and Sporns, P. (1988) Analysis and quantitation of minor di- and trisaccharides in honey, using capillary gas chromatography. *J. Food Sci.*, **53**, 558–61.
Low, N.H. and Swallow, K.W. (1991) Nachweis des zusatzes teilinvertierter saccharose zu orangensaft mit hilfe der HPLC. *Flussiges Obst*, 1, 13–18.
Low, N.H. and Wudrich, G.G. (1993) Detection of inexpensive sweetener addition to grapefruit juice by HPLC-PAD. *J. Agric. Food Chem.*, **41**, 902–909.
Mattick, L.R. (1988) An evaluation of the methodology for determining the authenticity of apple juice and concentrate, in *Adulteration of Fruit Juice Beverages* (eds S. Nagy, J.A. Attaway and M.E. Rhodes), Marcel Dekker, New York, pp. 175–93.
McCance, R.A. and Lawrence, R.D. (1929) The carbohydrate content of foods – inulin and the fructosans. Medical Research Council, *Special Report Series, No. 135*, p. 58.
McLaughlin, M., Brause, A.R. and Low, N.H. (1997) Peer validation study on the Low II method for the detection of high fructose corn syrup and hydrolyzed inulin syrup in apple juice. *J. Assoc. Off. Anal. Chem. Int.* (under review).
Morselli, M.F. (1975) Chemical composition of maple syrup. Maple Research Data No. 1. University of Vermont, Burlington, Vermont.
Reeve, A. (1992) Starch hydrolysis: Process and equipment, in *Starch Hydrolysis Products: Worldwide Technology, Production, and Applications* (eds F.W. Schenck and R.E. Hebeda). VCH Publishers, New York, pp. 79–120.
Stuckel, J.G. (1995) Chemical composition of, and adulteration detection in, maple syrup. MSc. thesis, University of Saskatchewan, Saskatchewan, Canada.
Stuckel, J.G. and Low, N.H. (1995) Maple syrup authenticity analysis by anion-exchange chromatography with pulsed amperometric detection. *J. Agric. Food Chem.*, **43**, 3046–51.
Stuckel, J.G. and Low, N.H. (1996) The chemical compostion of 80 pure maple syrup samples produced in North America. *Food Research Int.*, **29**, 373–9.
Swallow, K.W. and Low, N.H. (1990) Analysis and quantitation of the carbohydrates in honey using high-performance liquid chromatography. *J. Agric. Food Chem.*, **38**, 1828–32.
Swallow, K.W. and Low, N.H. (1993) Isolation and identification of oligosaccharides in a commercial beet medium invert sugar. *J. Agric. Food Chem.*, **41**, 1587–92.
Swallow, K.W. and Low, N.H. (1994) Determination of honey authenticity by anion-exchange liquid chromatography. *J. Assoc. Off. Anal. Chem. Int.*, **77**, 695–702.
Swallow, K.W., Low, N.H. and Petrus, D.R. (1991) Detection of orange juice adulteration with beet medium invert sugar using anion-exchange liquid chromatography with pulsed amperometric detection. *J. Assoc. Off. Anal. Chem.*, **74**, 341–5.
Swinkles, J.J. (1985) Sources of starch: Its chemistry and physics, in *Starch Conversion Technology* (eds G.M.A. Van Beynum and J.A. Roels), Marcel Dekker, New York, pp. 15–46.
Szejtli, J., Henriques, R.D. and Castineira, M. (1970) Mechanism of the acid hydrolysis of saccharose and raffinose. *Acta Chim. Acad. Sci. Hung.*, **66**, 213–27.
United States Department of Agriculture (1996) *Horticultural Products Review*.
White, D.R. and Cancalon, P.F. (1992) Detection of beet sugar adulteration of orange juice by liquid chromatography/pulsed amperometric detection with column switching. *J. Assoc. Off. Anal. Chem. Int.*, **75**, 584–7.
Wiesenberger, A., Kolb, E., Haug, M. and Gurster, D. (1992) Detection of the addition of beet medium invert sugar to OJC by HPLC. Application of the method for GJC and experiences with quantitative evaluation. *Fruit Processing*, **2**, 137–9.
Wudrich, G.G. (1993) Detection of inexpensive sweetener addition to citrus juice products by high performance liquid chromatography. MSc. thesis, University of Saskatchewan, Saskatchewan, Canada.
Wudrich, G.G., McSheffrey, S. and Low, N.H. (1993) Liquid chromatographic detection of a variety of inexpensive sweeteners added to pure orange juice. *J. Assoc. Off. Anal. Chem. Int.*, **76**, 342–54.

6 Enzymic methods of food analysis
G. HENNIGER

6.1 Introduction

Enzymatic analysis is an independent and separate branch of analytical chemistry. It has become one of the most important methodologies used in food analysis and allows the quick and reliable determination of many food ingredients.

Methods of enzymatic analysis are being increasingly used today in the investigation of foodstuffs; these methods were originally used in biochemical research laboratories. However, today the necessary reagents are commercially available and enzymatic measurement techniques are used in industry and in official/governmental laboratories (food inspection) for routine analysis. The sample material for enzymatic analysis is not restricted to foodstuffs.

Enzymatic analysis as an analytical technique has been known for a long time. Enzymatic analysis means:

1. analysis using enzymes as reagents to measure all substances capable of being transformed by enzymes (these substances are metabolites in living organisms and they are the substrates of the enzyme):

or

2. the determination of enzyme activities. The activity of an enzyme is measured by the rate of consumption of the reactants (substrates), or by the formation of the end products.

Enzyme activity measurements are used in food chemistry to check, for example, for sufficient heating during pasteurization or sterilization procedures. This relies on the fact that enzymes are thermally labile ingredients (or additives) of certain foodstuffs. An example of this is the measurement of diastase in honey where a lack of diastase activity can show that the honey was overheated during processing. Measurement of the enzymes phosphatase, catalase, xanthine oxidase and peroxidase in milk enables a conclusion to be drawn regarding the duration and intensity of heating of a milk sample as there are differing temperature sensitivities of the various enzymes. Enzyme measurements can also be used to differentiate fresh from frozen meat: enzymes bound to cell structures, e.g. in mitochondria, are liberated when frozen meat is thawed. Enzyme activities can indicate e.g. proteolytic activity during the ripening of cheese and furthermore, enzyme activities in foodstuffs can provide an indication of undesirable microbial contamination.

The measurement of enzyme activities will not be specifically dealt with in this chapter. Such measurements are carried out under established conditions set by convention:

- buffer: type, concentration = ionic strength, addition of salts, pH;
- temperature: 25 °C, 30 °C, 37 °C;
- type of substrates: natural (which is often not standardized and difficult to handle), synthetic (e.g. in the analysis of carbohydrate/starch and protein converting enzymes: p-nitrophenyl derivatives of oligoglucosides, resp. peptides or proteins, amides and esters);
- concentrations of substrate and coenzyme in the assay system in order to have a saturation of the enzyme and to measure a zero-order kinetics;
- amount/activity of auxiliary and indicator enzymes, if necessary;
- direction of measurement (forward reaction in the direction of the equilibrium or backward reaction 'against' the equilibrium, because enzymes often catalyse equilibrium reactions);
- addition of activators, e.g. metal ions/salts, or reagents for enzyme activity protection: EDTA, thioglycol, glutathione, but also serum albumin or sucrose;
- type of measurement: initial reaction rate, measurement at defined/fixed times, 1-point (end-point) kinetics;
- measuring principle: e.g. photometry, fluorometry, potentiometry.

Problems of such systems include:

- the quantitative isolation of enzymes from the sample without loss of activity and the selection of a solution which is suited for the measuring principle (deproteinization with acid or CARREZ reagents is not allowed because enzymes are proteins);
- the origin/raw material (species) dependent optimization of the reaction conditions (after the isolation and characterization of the respective enzyme);
- the low stability of enzymes in diluted solutions, especially in the diluted sample solution and the assay system, dependent e.g. on pH and temperature;
- limited range of linearity with respect to amount of enzyme and time of reaction;
- inhibitors in the sample solution which reduce the catalytic activity (e.g. heavy metal ions, phenolic compounds);
- insufficient repeatability (one sample, one operator, same equipment, analyses one after the other) and reproducibility (one sample, different operators, analyses in different laboratories with different equipment). Results with an error of analysis of up to ±20% may be observed.

Enzymatic activities from micro-organisms in general, and from starter cultures in particular, as well as technical enzymes from plants, animal organs and micro-organisms are often used for food production. After heating of food, i.e.

after denaturation of enzymes, they cannot be measured by their catalytic activity. They can only be measured by their protein composition e.g. by enzyme immunoassays.

Enzymes can bind to compounds and substances which have very different structures and still maintain their activities. Very sensitive detection systems have been developed which incorporate the highly specific antigen-antibody reaction (enzyme immunoassay). The measurements are based on the activity of the enzymes.

In choosing a suitable analysis method a number of criteria should be met regarding expenditure and quality of results. This list can also serve as a description of enzymatic methods or as a check list in comparing various types of analytical methodologies.

1. The assay should be specific/selective for the substance to be determined. Specificity of the determination means that the result obtained is accurate. Enzymatic methods may replace traditional chemical methods, and may be used to check specificity of other analytical systems, e.g. chromatographic techniques.
2. The sample preparation should not be complicated so as to minimize errors (e.g. loss of analyte). Practically clear and almost colourless solutions or extracts are needed for the measuring principle (photometry). The risk of losing the analyte and thereby obtaining wrong, i.e. too low results is minimal.
3. The performance of the assay, the handling of the equipment and the actual measurement should be simple enough for semi-skilled workers to perform. The performance of enzymatic determinations is very simple and involves pipetting of solutions into a cuvette, obtaining photometric readings and the calculation of results.
4. The method should be reliable and when comparing analysis results from day to day or from laboratory to laboratory, the method should show a good reproducibility. The scattering of the results around the mean should be low. Photometric readings, also used in other analytical methodologies, contribute mostly to the scattering of results, followed by the random error of pipetting the sample solution. The precision of results from enzymatic analysis is good. Coefficients of variation of between 0.5 and 2% can be achieved by 'normal' analysts. Results from recovery experiments often give data of 100 ±0.5 to 2% recovery of added standard material.
5. For the determination of low concentrations of substances in the sample, the assay should possess a high sensitivity. The sample solution should preferably be usable for the determination without prior concentration. The pipetting scheme mostly used for enzymatic determinations allows a large sample volume (approx. $\frac{1}{3}$ of the total volume), so that measurements in the range of mg/l sample solution can be done with an error of analysis of approx. ±10%. The sensitivity of enzymatic methods is sufficient.

6. The method should be reliable: foodstuff samples contain heterogeneous mixtures of substances, the determination should show no interference from other ingredients or additives in the sample. In enzymatic analysis interference by the sample matrices are seldom observed when analysing foodstuffs. This can be easily explained because the nature of the enzymes in the assay is the same as of the enzymes e.g. in human metabolism. If a sample can be eaten or drunk without interference in the human body, it can often be analysed with enzymes without any interference. The sample matrix often determines the type of sample preparation only.
7. The analysis should be rapid because determinations are often used for production monitoring. The time requirement must be low, since the production process must be actively regulated as a result of the measurement. Enzymatic methods are quick. The time needed depends on the enzyme characteristics, temperature and reaction conditions, as well as of the availability of suitable equipment and good experience in handling. Sample preparation for enzymatic analysis is also fast.
8. The cost for analysis including reagents, equipment and personnel should be economically viable. Analysis costs are usually dominated by the costs for personnel, followed by the costs for equipment (purchasing, repairing and servicing). The costs for disposables and reagents are relatively low in enzymatic analysis; the equipment is often available in the laboratories and therefore, there are often no additional costs. Semi-skilled staff are able and authorized (see 10 below: danger from reagents) to carry out enzymatic determinations.
9. For large numbers of samples the determination must be suitable for mechanization or automation. Instruments for processing large numbers of samples should be commercially available. Enzymatic determinations can be run in series in the case of the analysis of one analyte in different samples, and also in the case of different analytes in one sample because the equipment and the procedures are always the same or very similar, there is only a difference in the reagents. A large number of determinations can be processed by rationalization, mechanization or automation of the methods. Labour saving devices like dispensers, piston type pipettes and disposable cuvettes are available at low costs. Different types of automated equipment either continuously working systems or descrete analysers are commercially available.
10. In consideration of discussions on the problems of work and environmental safety, the reagents used in the analytical laboratory should be safe. The disposal of waste should not be problematic. The reagents used for many enzymatic determinations are, with the exception of methanol and isopropanol in the determination of cholesterol, not hazardous materials in the usual sense of hazardous substances regulations and chemical properties. Only the general safety measures that apply to all chemical substances should be adhered to. After use, the reagents can be disposed with

laboratory waste or household waste water although the local regulations have to be observed.

These requirements are to some extent mutually exclusive, so that compromises often have to be made. Very rapid test methods are frequently only qualitative, meaning that they result only in approximate values of yes/no decisions. Exact results often are expensive in terms of time and labour, e.g. when performing multiple determinations. Furthermore, highly sensitive test methods are often very non-specific and very specific methods are often less sensitive. With the application of enzymatic analysis methods, such compromises are frequently no longer necessary. The reason for this is that enzymes as ingredients of living cells are the result of millions of years of evolution, and enzymatic analysis is the transfer of an enzyme catalysed reaction from the living cell into a cuvette.

6.2 Specificity of enzymatic methods

The greatest advantage of using enzymatic methods is their high specificity. This characteristic is dictated by the enzymes used in analysis. Enzymes often react in a highly specific way, but not all enzymes can be expected to be absolutely specific. (Nevertheless, a high specificity can only be found within natural occurring macromolecules like enzymes, antibodies, DNA and RNA. Chemical reactions and adsorption/desorption (=chromatographic) materials are often not specific.) The use of an enzyme with a known non-specificity need not necessarily be a disadvantage. Often a method needs several coupled reactions. In this situation only one of the enzymes in the reaction chain has to be specific to render the overall reaction specific. Nevertheless, the specificity of enzymes should always be known when they are used for analytical purposes.

6.2.1 *Highly ('absolutely') specific enzymes*

Ideally, the enzymes used for analysis are absolutely specific and convert only the analyte. Highly specific enzymes are shown in Table 6.1. An example of a highly specific enzyme is glutamate dehydrogenase. This enzyme not only reacts specifically with L-glutamate in the forward reaction but also with ammonia in the reverse reaction. Another example is glucose-6-phosphate dehydrogenase (G6P-DH), an 'absolutely' specific enzyme. This enzyme exclusively converts glucose-6-phosphate (G-6-P) with a measurable velocity. On the other hand, hexokinase (HK) is a 'less' specific enzyme, it converts, with approximately the same speed of reaction, the hexoses D-glucose, D-fructose and D-mannose in the presence of ATP to their corresponding sugar-6-phosphates.

The determination of D-glucose (Table 6.2) uses the enzyme mixture HK/G6P-DH. Since the hexokinase reaction is not photometrically measurable (reactants and products do not differ in their light absorption) it is necessary to

Table 6.1 Highly specific enzymes in food analysis

Enzyme	Used for the determination of
Ascorbate oxidase (AAO)	L-ascorbic acid
Citrate lyase (CL)	Citric acid
Citrate synthase (CS)	Acetic acid
Creatine kinase (CK)	Creatine/creatinine
Formate dehydrogenase (FDH)	Formic acid
Gluconate kinase	D-Gluconic acid
Glucose-6-phosphate dehydrogenase (G6P-DH)	D-Glucose, D-fructose, D-sorbitol (HK procedure), maltose, sucrose, starch
Glutamate dehydrogenase (GlDH)	L-Glutamic acid, ammonia, urea
3-Hydroxybutyrate dehydrogenase (HBDH)	D-3-Hydroxybutyric acid
Nitrate reductase (NR)	Nitrate
Oxalate decarboxylase (OxDC)	Oxalate
Phosphoglucose isomerase (PGI)	D fructose
Urease	Urea

couple an indicator reaction to it. It is logical to choose the G6P-DH reaction, since of the sugar-6-phosphates formed in the less specific hexokinase reaction, only glucose-6-phosphate is converted in a highly specific reaction. Thus it is possible to determine D-glucose with absolute specificity .

6.2.2 *Less specific enzymes*

Specificity of a method and accuracy of the results are also guaranteed when the less specific reaction of an enzyme is combined with a highly specific enzyme/reaction as shown in the determination of D-glucose. Other examples of such combinations are shown in Table 6.3.

6.2.3 *Relative specificity of enzymatic procedures*

Specificity is also guaranteed if the substrate of a non-specific reaction is not contained in the sample (Table 6.4). An example of this is the enzymatic

Table 6.2 Reaction scheme for the determination of D-glucose

D-Glucose		D-glucose-6-phosphate
D-Fructose + ATP	$\xrightarrow{\text{hexokinase}}$	D-fructose-6-phosphate + ADP
D-Mannose		D-mannose-6-phosphate
D-Glucose-6-phosphate + NADP$^+$	$\xrightarrow{\text{G6P-DH}}$	D-gluconate-6-phosphate + NADP + H$^+$

Table 6.3 Less specific enzymes combined with highly specific enzymes in food analysis

Enzymes	Used for the determination of
Acetyl-CoA synthetase (ACS) with L-MDH and CS	Acetic acid
Hexokinase (HK) with G6P-DH	D-Glucose
Hexokinase (HK) with PGI and G6P-DH	D-Fructose
Sorbitol dehydrogenase (SDH) with HK, PGI and G6P-DH	D-Sorbitol

Table 6.4 Relative specificity with less specific enzymes in food analysis

Enzyme	Used for the determination of	Relatively specific in the absence of
Cholesterol oxidase (Chol OD)	Cholesterol	Phytosterols
Alcohol dehydrogenase (ADH)	Ethanol	n-Propanol, n-butanol
Galactose dehydrogenase (Gal DH)	D-galactose	L-Arabinose
Succinyl-CoA synthase (SCS)	Succinic acid	Itaconate

determination of ethanol. The enzymes, alcohol dehydrogenase (ADH) and aldehyde dehydrogenase (AlDH) also convert n-propanol and n-butanol, but these alcohols are usually not present in foodstuffs or if present, are only in negligible (low) concentrations.

6.2.4 Enzymes with slow side activity

Often the non-specific reactions are considerably slower than the specific reaction, thus the main and side reactions can be monitored separately either by using different amounts of enzyme in the assays or in other cases, the small side reaction ('creep' reaction) is compensated for by graphical or mathematical extrapolation. Examples are shown in Table 6.5.

6.2.5 Group-specific enzymes

An additional type of specificity of enzymes used for food analysis is group-specificity, found primarily in oligo- and polysaccharide converting enzymes (Table 6.6). The enzymes recognize a common group in various different

Table 6.5 Enzymes with a side activity in food analysis

Enzyme	Used for the determination of	Side activity to
Glycerokinase (GK)	Glycerol	Dihydroxyacetone
L-Malate dehydrogenase (L-MDH)	L-Malate	L-Tartrate
Aldehyde dehydrogenase (Al-DH)	Acetaldehyde	Other aldehydes
Sulphite oxidase	Sulphite	Other sulphur containing compounds

Table 6.6 Group-specific enzymes in food analysis

Enzyme	Used for the determination of
Amyloglucosidase (AGS)	Starch
Esterase	Triglycerides
β-Fructosidase	Sucrose
α-Galactosidase	Raffinose
β-Galactosidase	Lactose, lactulose
α-Glucosidase	Maltose
Lipase	Triglycerides
Phospholipase	L-α-Lecithin

Table 6.7 Stereo-specific enzymes in food analysis

Enzyme	Used for the determination of
D-Isocitrate dehydrogenase (ICDH)	D-Isocitric acid
D-Lactate dehydrogenase (D-LDH)	D-Lactic acid
L-Lactate dehydrogenase (L-LDH)	L-Lactic acid
D-Malate dehydrogenase (D-MDH)	D-Malic acid
L-Malate dehydrogenase (L-MDH)	L-Malic acid

molecules and convert this group. For example, β-fructosidase cleaves β-fructosidic bonds (e.g. in sucrose) and α-glucosidase cleaves α-glucosidic bonds (e.g. in sucrose and maltose) with the uptake of water and the liberation of the monomeric sugar(s).

6.2.6 Stereo-specific enzymes

Another important type of specificity is stereo-specificity (Table 6.7). The enzymes convert stereo-specifically only one of the two stereo-isomeric forms of the substrate. L-Lactate dehydrogenase (L-LDH), which occurs in animals and humans, exclusively converts L-lactate. Micro-organisms can also possess D-lactate dehydrogenase which reacts with the D-form of lactic acid.

A determination can therefore be termed specific:

- if the assay system contains an enzyme with high ('absolute') specificity,
- if a less specific enzyme is combined with a highly specific enzyme,
- if the specific reaction is considerably faster than the non-specific reaction,
- if the substrate of the non-specific reaction is not present in the sample.

Enzymatic analysis is very specific because enzymes are the most specific reagents known. This is still true after the development of 'modern' analytical equipment. The reaction principles used in enzymatic analysis have been developed in nature in a million years of evolution process.

Specificity of the method guarantees accuracy of results. Enzymatic analysis not only replaces less specific chemical methods, but often is also used as a reference method by which the results from instrumental analyses are verified. Another technical term for specificity (used, for example, in biochemistry) is selectivity (used, for example, in organic chemistry). The meaning is always the same: the reaction with one and only one analyte.

The specificity of enzymes and the mechanisms of enzymatic catalysis can easily be explained.

Enzymes are, as a rule, much larger than the substrates they convert. With high molecular weight substrates like carbohydrates and especially proteins, the enzymes act on only a 'small' localized group of the macromolecule and not on the whole molecule. (An example is the hydrolysis of peptide bonds in proteins by proteases.) These properties lead to the concept of an active centre as the part of the enzyme surface responsible for the binding and catalytic activity of the substrate.

The active centre of the enzyme may be a crevice, cavity, pocket, hole, or depression on the surface of the enzyme-protein. Each particular enzyme has its own typical three-dimensionally structured active centre whose surface is constructed from 10 to 20 amino acids, with a corresponding diameter of about 10 to 40 nm. The shape of the active centre in its transitional state corresponds to the shape of the substrate as a negative impression. The transitional state is characterized by its particular spatial arrangement and its particular charge distribution.

The active centre and the substrate must fit together. This theory, proposed by Emil Fischer in 1894, is known as the Lock and Key Theory. This theory could not only explain the enzyme's specificity, but also the mode of action of inhibiting substances.

The Fischer hypothesis has been supported by X-ray crystallographic studies, e.g. by Phillips at Oxford University in the 1960s (Dickerson, E. and Geis, I., *The Structure and Action of Proteins*, 1971). The weakness in Fischer's theory is that it describes the enzymes as rigid, inflexible molecules; that problem had been solved by Koshland's Induced Fit Theory: the structure of the enzyme is altered by the substrate.

Today the mode of action of enzymes is well understood. The charged groups on the enzyme cause an electrostatic attraction in its immediate vicinity. This is also true for hydrophobic bonds. The substrate attaches to the enzyme. Then the interactions of the electromagnetic fields cause a deformation of the electron clouds, and this produces a distortion of the bonds. As the substrate molecules deform, the resonance structures are often altered. Planar bonds (formed through resonance) within the molecule are distorted from the plane and this causes the electronic clouds to further distort, thus weakening the bonds. Binding forces between enzyme and substrate induce a distortion of the binding angles in the substrate.

Many reactions proceed slowly in water. Because of the dipolar nature of water and its high dielectric constant, charged molecules are kept apart from each other. Water also acts as a good insulator. The interior of the enzyme as well as the active centre contain many hydrocarbon residues of the non-polar amino acids. The low dielectric constant and poor insulating properties affect very strong electric forces on the reactants. This leads to an acceleration of the reaction. The active centre of the enzyme can be regarded as the 'organic solvent' of the living cell.

M.J.S. Dewar (*Enzyme*, **36**, 8–20; 1986) described the mode of enzyme action as 'the correct' substrate fitting exactly into the active centre of the enzyme. The substrate must be of the correct size so that all water molecules are 'squeezed out' when the substrate is bound to the active centre. A substrate which is 'too large' does not fit into the active centre and therefore no reaction takes place. A substrate which is 'too small' is not large enough to push out all the water molecules from the active centre. These water molecules hinder or impair the enzymatic reaction.

Enzymatic reactions may be explained without the supposition of 'special strength'. Enzymatic catalysis is, for example, acid-base-catalysis as known from organic chemistry (saponification of esters), or covalent catalysis: a highly reactive intermediate is formed from the enzyme and 'its' substrate, e.g. a Schiff base from carbonyl- and amine-groups. The 'life-force' (*vis vitalis*) was previously believed to be responsible for enzyme actions. The laws of electrostatic electricity are sufficient for the total understanding of enzyme catalysis. The substrate is bound to the enzyme, thereby extracting it from the aqueous milieu and taken into the interior of the enzyme where a strong electronic environment affects the reactants. The 'activated' substrate can either react with the enzyme surface, or react with other molecules which have also been attached to the enzyme.

6.3 The analytes

Enzymatic analysis is independent of the sample material. All types of food, also products from the non-food area as well as biological samples can be analysed by means of enzymes. The matrix of the sample material determines sample preparation only. Enzymatic analysis depends on the availability and suitability of purified enzymes and protocols for sample preparation and performance of assay which have been proven in routine analysis.

Enzymes are isolated from living animal, plant and microbial materials. These contain enzymes for synthesis (anabolism) as well as for catabolism of metabolites. If it is possible to isolate such an enzyme in a sufficient purity, and a suitable measuring equipment is available, these compounds can be determined enzymatically.

The limitations of enzymatic analysis with respect to the analytes is very often the reason for giving information about the individual analytes determined in food analysis.

6.3.1 *Acetaldehyde*

A test, based on ultra-violet (UV) light absorption, exists for the analysis of alcoholic beverages (beer, brandy, champagne and wine), of fruit juices, dairy products (yoghurt, cheese and butter). In the analysis of wine, the procedure measures 'total acetaldehyde' which is the sum of free acetaldehyde and on sulphite bound acetaldehyde.

Acetaldehyde occurs in nature in all organisms, even if in very small quantities because it is a product of many metabolic processes. Acetaldehyde is also an intermediate product in alcoholic fermentation. (Its content indicates the fermentative production of ethanol.) When fermentation processes are used for production of food, the content of acetaldehyde increases under certain conditions. Furthermore, acetaldehyde is formed by oxidation of ethanol in the presence of oxygen from air.

Acetaldehyde is an important flavour compound. High concentrations as in beer indicate the presence of off-flavour compounds and are not favourable because of a bad odour. In wine production, acetaldehyde is bound to sulphite in order to improve the taste. On the other hand, acetaldehyde is the most common aldehyde found in dairy products (yoghurt, cheese, etc.). Low contents are responsible for a desirable specific flavour and high contents are responsible for flavour defects.

Principle of determination

$$\text{Acetaldehyde} + \text{NAD}^+ + \text{H}_2\text{O} \xrightarrow{\text{AlDH}} \text{acetic acid} + \text{NADH} + \text{H}^+$$

6.3.2 *Acetic acid*

Acetic acid is one of the most important organic acids which determine flavour/taste and smell of foodstuffs. It occurs, for example, in beer, dough, bread, fruits and wine. Acetic acid may be contained in must and is formed during fermentation processes in the production of wine. The limit of concentration may be regulated by food inspection. Acetic acid is an important metabolic analyte, it is not only an end product of fermentation processes but also the oxidation product of acetaldehyde and ethanol.

It is important to be able to measure acetic acid in the analysis of beer and wine, of vinegar, of fruit and vegetable products (juice, canned food, mixed pickles, ketchup), meat products (sausages), pickled fish and dairy products (cheese), of sour dressings, sauces and mayonnaise, of bakery products (baking agents, dough, bread and cake). In the analysis of wine, the main part of the 'volatile acids' is acetic acid.

Acetic acid is the main component of vinegar. Acetate is an ingredient of spices (additives) for the preparation of sausages.

Principle of determination

$$\text{Acetate} + \text{ATP} + \text{CoA} \xrightarrow{\text{ACS}} \text{acetyl-CoA} + \text{AMP} + \text{pyrophosphate}$$

$$\text{Acetyl-CoA} + \text{oxaloacetate} + \text{H}_2\text{O} \xrightarrow{\text{CS}} \text{citrate} + \text{CoA}$$

$$\text{Malate} + \text{NAD}^+ \xrightleftharpoons{\text{MDA}} \text{oxaloacetate} + \text{NADH} + \text{H}^+$$

6.3.3 *Ammonia*

Ammonia, an important component in the nitrogen cycle, occurs as a result of many chemical and biochemical processes such as in the decomposition or digestion of organic material. Ammonia may be a major component of off-flavours. High concentrations of ammonia can indicate the (microbial) decomposition of substances like milk, meat and seafood and also the presence of faeces, urine and micro-organisms in water.

Ammonium salts may be used for the adulteration of fruit juices (increase of the formol value). Ammonium carbonate is used for the production of baked goods (baking powder), caramels, desserts and puddings. Ammonium salts are used as dough conditioners and flavour enhancer. Ammonium chlorides are contained in licorice in quite large amounts.

A test based on UV absorption may be carried out for its analysis in baking agents and bakery goods, fruit juice, meat and meat products, milk, cheese, seafood and licorice.

Principle of determination

$$\text{2-Oxoglutarate} + \text{NADH} + \text{NH}_4^+ \xrightarrow{\text{GlDH}} \text{L-glutamate} + \text{NAD}^+ + \text{H}_2\text{O}$$

6.3.4 L-Ascorbic acid ('vitamin C')

The enzymatic method for the analysis of L-ascorbic acid in fruit, berry and vegetable products (juices, nectars and drinks, canned and frozen food, potato products, sauerkraut, tea), beer, wine, milk and dairy products, meat products (sausages), baby food, dietetic food and flour relies on a colorimetric determination.

L-Ascorbic acid is present in all animal and plant cells but in varying amounts. Humans cannot themselves synthesize L-ascorbic acid and are therefore dependent on external sources (vitamin C), the principal dietary sources of which are potato, fruit and vegetables. The human body is not able to store L-ascorbic acid so it must be consistently supplied with the food.

For technological reasons, L-ascorbic acid may be used as an antioxidant and as an acidifier in the manufacture of foodstuffs. L-ascorbic acid delays or inhibits browning reactions caused by enzymes in fruit and fruit products. The use of L-ascorbic acid in the production of beer and wine may be regulated on national or international (European) basis. L-ascorbic acid is added to meat products to reduce the quantity of nitrite necessary for pickling, or to flour to improve baking qualities. It may also be added to fruit juices to compensate its loss during production. On the other hand, L-ascorbic acid is a relatively sensitive substance; its determination is thus suitable not only for use in the assessment of the quality of processing food from fruit and vegetable raw materials, but also for the control of storage.

Principle of determination

$$\text{L-Ascorbate (x-H}_2) + \text{MTT}^+ \xrightarrow{\text{PMS}} \text{dehydroascorbate (x)} + \text{MTT-formazan}^- + \text{H}^+$$

$$\text{L-Ascorbate} + \tfrac{1}{2}\text{O}_2 \xrightarrow{\text{AAO}} \text{dehydroascorbate} + \text{H}_2\text{O}$$

6.3.5 L-Aspartic acid

A UV test is used for the analysis of fruit and vegetable products (juices, soft drinks), sugar beets and sugar refinery products, fruit juice, soft drinks, beer and wine.

L-Aspartic acid is a non-essential amino acid which occurs in all proteins. L-Aspartic acid may be added to food in order to improve the biological quality, or as a nutrient and dietary supplement.

Principle of determination

$$\text{L-Aspartate} + \text{2-oxo-glutarate} \xrightleftharpoons{\text{GOT}} \text{oxaloacetate} + \text{L-glutamate}$$

$$\text{Oxaloacetate} + \text{NADH} + \text{H}^+ \xrightarrow{\text{MDH}} \text{L-malate} + \text{NAD}^+$$

6.3.6 Cholesterol

Cholesterol is the most important of the animal sterols. It is an important component of the cell membranes of the higher species of organism and a precursor to a whole series of steroid hormones. High levels of cholesterol in serum are regarded as a risk factor for cardiovascular disease. There is, or may be, a positive correlation between the consumption of food rich in cholesterol and mortality from this disease.

Cholesterol is found in all animal fats. It is an important component of egg yolk. Due to the relative constancy of its content, it is frequently used in the determination of the egg content of foodstuffs like bakery goods, noodles and egg liqueur with egg yolks (16 g egg yolk contains 200 mg cholesterol).

Cholesterol is important nutritionally as it is partly ingested in the form of food.

A colorimetric method is used for analysis of egg, egg products and in products containing egg (bakery goods, egg liqueur, ice-cream, mayonnaise, noodles), of dietetic food, meat products, fats (butter, plant fats) and oils.

Principle of determination

$$\text{Cholesterol} + \text{O}_2 \xrightarrow{\text{Chol-OD}} \Delta^4\text{-cholestenone} + \text{H}_2\text{O}_2$$

$$\text{Methanol} + \text{H}_2\text{O}_2 \xrightarrow{\text{catalase}} \text{formaldehyde} + 2\,\text{H}_2\text{O}$$

$$\text{Formaldehyde} + \text{NH}_4^+ + 2\,\text{acetylacetone} \rightarrow \text{lutidine dye} + 3\,\text{H}_2\text{O}$$

6.3.7 Citric acid

Citric acid is a very important metabolite in animals, plants and micro-organisms, it is one of the most important fruit acids. It is manufactured biotechnologically on a large scale and is used also as an additive/acidulant in foods, such as in the production of processed cheese and sausages, soft drinks, candies, canned fruit and vegetables. Citric acid may be added to wine and juice. The addition may be regulated on a national and international (European) basis. (The unallowed addition to orange juice can be recognized by determining the ratio citrate to D-isocitrate.) Citric acid is an acidulant for pH adjustment and control, a buffer, a complexing agent for metal ions (responsible, for example, for turbidities) and an emulsifier. It enhances and modifies flavour and taste of food products. The addition of citric acid may improve the stability of goods, it improves technological processes in the production of fruit and vegetable products, and it prevents plant material from enzymatic browning reactions at low pH values.

A UV test is employed for the analysis of citric acid in beer and wine, beverages (soft drinks, lemonades, cola light drinks, tea), bakery goods (baking agents, dough and bread), candies and chewing gum, dietetic food, honey, soft drinks, of fruit, berry and vegetable products (juices, nectars, drinks, jam, canned and frozen food, sugar beets and sugar refinery products), meat products (additives, sausages) and dairy products (aqueous butter phase, cheese, processed cheese).

Principle of determination

$$\text{Citrate} \xrightarrow{CL} \text{oxaloacetate} + \text{acetate}$$

$$\text{Oxaloacetate} + \text{NADH} + \text{H}^+ \xrightarrow{MDH} \text{L-malate} + \text{NAD}^+$$

$$\text{Pyruvate} + \text{NADH} + \text{H}^+ \xrightarrow{LDH} \text{L-lactate} + \text{NAD}^+$$

(LDH is added to the assay system for the case that oxaloacetate is chemically or enzymatically decarboxylated with the formation of pyruvate before it is converted to malate by means of MDH. The amount of NADH formed in the MDH and LDH reactions is stoichiometric to the amount of citrate in the sample solution.)

6.3.8 *Creatine/creatinine*

Creatinine in soups is measured using a UV detection method to indicate the meat content. The concentration of creatinine may be regulated by food regulation.

Principle of determination

$$\text{Creatinine} + \text{H}_2\text{O} \xrightarrow{\text{creatininase}} \text{creatine}$$

$$\text{Creatine} + \text{ATP} \xrightarrow{GK} \text{creatine-phosphate} + \text{ADP}$$

$$\text{ADP} + \text{PEP} \xrightarrow{PK} \text{ATP} + \text{pyruvate}$$

$$\text{Pyruvate} + \text{NADH} + \text{H}^+ \xrightarrow{LDH} \text{L-lactate} + \text{NAD}^+$$

6.3.9 *Ethanol*

Ethanol occurs in nature in practically all organisms, even if in very small quantities. It is the end product of alcoholic fermentation and a 'desired' component of alcoholic beverages (the monetary value-determining component) but also the 'undesirable' component in non-alcoholic and low-alcoholic beverages. The content of ethanol in alcoholic beverages and also in chocolates containing brandy and spirits is of interest because of duties and taxes, and in the case of non- and low-alcoholic beverages because of national regulations.

Ethanol in fruits indicates ripeness, the presence of a high amount of ethanol in fruit products like fruit juices indicates that the components used for production may have decomposed or deteriorated. Ethanol in sugar solutions, curd and kefir indicates the presence of yeasts. Ethanol in fish indicates fish spoilage. Ethanol is contained in fermented food products (vegetables like sauerkraut and mixed pickles, soy sauce). Ethanolic products are added to desserts in order to improve the taste.

The UV test is used for the analysis of alcoholic beverages (beer, champagne, liqueur and wine), of low-alcohol and non-alcoholic beverages (e.g. beer, soft drinks, lemonades, malt beverages), of fruit and vegetable products (juices, nectars, jam, fermented products), chocolate and confectionery products, dairy products (kefir), bakery products (baking agents, dough, bread), desserts, fish, honey, ice-cream and vinegar.

Principle of determination

$$\text{Ethanol} + \text{NAD}^+ \xrightarrow{\text{ADH}} \text{acetaldehyde} + \text{NADH} + \text{H}^+$$
$$\text{Acetaldehyde} + \text{NAD}^+ + \text{H}_2\text{O} \xrightarrow{\text{AlDH}} \text{acetic acid} + \text{NADH} + \text{H}^+$$

6.3.10 *Formic acid*

Formic acid, the oxidation product of methanol and formaldehyde, occurs as a metabolite in many biochemical reactions, but its concentration is invariably very low. Formic acid is a small part of the 'volatile acids' in wine, it is contained in beer and may be formed in baking processes. Formic acid is measured in sugar factories because it is the oxidation product of formaldehyde used for disinfection. Formic acid is formed from sugar under alkaline conditions of sugar juice purification.

As formic acid in low concentrations has both a bactericidal and fungicidal effect, it can be used as a food preservative in the production of juices and mixed pickles (when legal aspects have to be taken into account).

Moulds tend to produce formic acid as a metabolite; hence, determination of formic acid can give an indirect indication as to the properties such as the degree of decomposition of samples (e.g. tomato products).

The method of detection involves a UV test for the analysis of beer, fruit and vegetable products (juices, canned and frozen food, jam), bakery goods (ginger bread), vinegar, honey, wine, fish, meat and sugar refinery products.

Principle of determination

$$\text{Formate} + \text{NAD}^+ + \text{H}_2\text{O} \xrightarrow{\text{FDH}} \text{bicarbonate} + \text{NADH} + \text{H}^+$$

6.3.11 D-Fructose ('fruit sugar')

The analysis of fructose in beverages (soft drinks, lemonades), chocolate, sweets and candy, baby food, diabetic food, dietetic food, fruit, berry and vegetable products (juices, nectars, drinks, jam, canned and frozen food, potato products), honey, ice-cream, sugar refinery products, invert sugar, and wine is by an enzymic method using UV detection. (D-Fructose is mostly determined together with D-glucose and/or with sucrose.)

Free D-fructose is mostly found in plants, often together with D-glucose and also sucrose, where it is a very important sugar component particularly of fruit. D-Fructose is a component of di-, tri- and oligo-saccharides (sucrose, lactulose; raffinose; oligo-β-fructosans) and is also a component of the polysaccharide inulin. D-Fructose can be produced by hydrolysis (inversion) of sucrose and separation from D-glucose, or, better, by hydrolysis of inulin from chicory. D-Fructose replaces sucrose in the case of diabetic food.

D-Fructose, D-glucose and sucrose are the most important sweeteners in foodstuffs. D-Fructose is sweeter than sucrose, and in the case of wine the taste of D-fructose is more pleasant than the taste of D-glucose. When reduced-calorie dietetic foods are required sucrose should be exchanged for D-fructose.

D-Fructose is mostly measured together with D-glucose, because the sum of both are often compared with the results from the analysis of 'reducing sugars'. The ratio D-glucose to D-fructose may be used to find out adulteration of orange juice, for example, by the addition of glucose syrup. Normally, the ratio D-glucose to D-fructose is approx. 0.90 to 0.92 in the case of orange juice.

Principle of determination

$$\text{D-Fructose} + \text{ATP} \xrightarrow{\text{HK}} \text{F-6-P} + \text{ADP}$$

$$\text{F-6-P} \xrightarrow{\text{PGI}} \text{G-6-P}$$

$$\text{G-6-P} + \text{NADP}^+ \xrightarrow{\text{G6P-DH}} \text{D-gluconate-6-P} + \text{NADPH} + \text{H}^+$$

6.3.12 D-Gluconic acid/D-glucono-δ-lactone

A UV test for the analysis of D-gluconic acid in wine, fruit juice, meat products (sausages), and dairy products (cheese) is used.

D-Gluconic acid is the oxidation product of D-glucose and a product of D-glucose metabolism of a number of micro-organisms. D-Gluconic acid in wine indicates a high quality of the product and it may be added to beverages in order to eliminate calcium turbidity.

D-Glucono-δ-lactone is used in the manufacture of foodstuffs, such as sausages and cheese, because of the slow release of the acid from the lactone and a slow pH reduction. In this specific use, legislative restrictions may apply.

Principle of determination

$$\text{D-Glucono-}\delta\text{-lactone} + H_2O \xrightarrow{pH\ 10} \text{D-gluconate}$$

$$\text{D-Gluconate} + ATP \xrightarrow{\text{gluconate kinase}} \text{D-gluconate-6-P} + ADP$$

$$\text{D-Gluconate-6-P} + NADP^+ \xrightarrow{\text{6-PGDH}} \text{ribulose-5-P} + NADPH + H^+ + CO_2$$

6.3.13 D-Glucose ('grape sugar')

UV test for the analysis of bakery goods, beer, diabetic and dietetic food, meat products.

D-Glucose only is very seldom measured outside clinical chemistry. The determination of D-glucose in foodstuffs is normally carried out together with the determination of D-fructose (see above), maltose and sucrose (see below): beverages (soft drinks, lemonades), of chocolate, sweets and candy, fruit, berry and vegetable products (juices, nectars, drinks, jam, canned and frozen food, potato products), honey, ice-cream, sugar refinery products and wine. D-Glucose and D-fructose are the main components of the 'reducing sugars'.

D-Glucose occurs widely in the animal and plant kingdoms. It is an essential component of carbohydrate metabolism and occurs frequently in free form along with D-fructose and sucrose. However, the more important forms are those of di-, tri-, oligo- and polysaccharides (lactose, maltose, sucrose; raffinose; dextrins, starch; cellulose).

D-Glucose is produced by hydrolysis of starch with the enzyme amyloglucosidase. D-Glucose is an ingredient of glucose syrup which is used to sweeten beverages and other products. The addition to wine and fruit juice is often prohibited by regulations. Such addition can be recognized by measuring D-glucose and D-fructose and the calculation of the glucose/fructose ratio.

D-Glucose, D-fructose and sucrose are the most important sweeteners in foodstuffs. D-Glucose and D-fructose are the ingredients of inverted sugar. The method of analysis, requiring UV measurement only, is carried out in bakery goods, beer, diabetic and dietetic and meat products.

Principle of determination

$$\text{D-Glucose} + ATP \xrightarrow{HK} \text{G-6-P} + ADP$$

$$\text{G-6-P} + NADP^+ \xrightarrow{\text{G6P-DH}} \text{D-gluconate-6-P} + NADPH + H^+$$

6.3.14 L-Glutamic acid

L-Glutamic acid is a non-essential amino acid and occurs in large quantities in most proteins. Its sodium salt (mono-sodium glutamate, MSG) is often used as a taste improver and taste intensifier for a series of foodstuffs. L-Glutamate is a nutrient, dietary supplement and salt substitute.

L-Glutamic acid is measured by a colorimetric method for the analysis of beer, cheese, meat products (sausages, soups), fish, fruit and vegetable products (e.g. tomato products), sugar refinery products, soups, spices and sauces.

Principle of determination

$$\text{L-Glutamate} + \text{NAD}^+ + \text{H}_2\text{O} \xrightarrow{\text{GlDH}} \text{2-oxoglutarate} + \text{NADH} + \text{NH}_4^+$$
$$\text{NADH} + \text{INT} + \text{H}^+ \xrightarrow{\text{diaphorase}} \text{NAD}^+ + \text{formazan}$$

6.3.15 Glycerol

Glycerol is a by-product of alcoholic fermentation and its fatty acid esters (glycerides) occur frequently in nature.

Glycerol is a component of high quality wines. It often determines the 'body' of wine and has a strong impact on its taste. Glycerol is considered to be one of the most important compounds in wine. The addition of glycerol to wine is an adulteration (such as the production of a quality wine from a table wine) and therefore not allowed. Addition of glycerol can be detected by the measurement of ethanol and glycerol and the calculation of the ethanol/glycerol ratio.

Glycerol is used as a moisturiser in, for example, marzipan, tobacco and cosmetics, in order to maintain a certain moisture content and to prevent drying-out.

Glycerol is measured by a UV based test in the analysis of beer, honey, juice, marzipan, sweets, vinegar, distilled liquors/spirits and wine.

Principle of determination

$$\text{Glycerol} + \text{ATP} \xrightarrow{\text{GK}} \text{L-glycerol-3-P} + \text{ADP}$$
$$\text{ADP} + \text{PEP} \xrightarrow{\text{PK}} \text{ATP} + \text{pyruvate}$$
$$\text{Pyruvate} + \text{NADH} + \text{H}^+ \xrightarrow{\text{LDH}} \text{L-lactate} + \text{NAD}^+$$

6.3.16 Guanosine-5'-mono-phosphate

Guanosine-5'-monophosphate is commercially used as a flavour enhancer and is measured in soups and spice blends by a UV based method.

Principle of determination

$$\text{G-5-MP} + \text{ATP} \xrightarrow{\text{G-5-MP kinase}} \text{guanosine-5'-di-phospate (GDP)} + \text{ADP}$$
$$\text{GDP} + \text{ADP} + 2\,\text{PEP} \xrightarrow{\text{PK}} \text{GTP} + \text{ATP} + 2\,\text{pyruvate}$$
$$2\,\text{Pyruvate} + 2\,\text{NADH} + 2\,\text{H}^+ \xrightarrow{\text{LDH}} 2\,\text{L-lactate} + 2\,\text{NAD}^+$$

6.3.17 D-3-Hydroxybutyric acid

D-3-Hydroxybutyric acid is found in blood and most animal organs. In untreated diabetes, it is excreted in the urine.

An increase in the content of D-3-hydroxybutyric acid in eggs can be detected six days after chickens have been fertilized. This increase continues even after the embryo has died. D-3-Hydroxybutyric acid is thus a typical indicator for fertilized and incubated eggs. The use of these eggs for the production of human food is either generally forbidden or in some countries permitted for up to six days incubation. For the identification of incubator rejected eggs the upper limit is 10 mg/kg in the dry matter of the unmodified egg product (approx. 2.5 mg/kg liquid whole egg) as regulated by the European Council Directive from 1989.

A colorimetric method is used for the analysis of D-3-hydroxybutyric acid in liquid egg, and egg products.

Principle of determination

$$\text{D-3-Hydroxybutyrate} + \text{NAD}^+ \xrightarrow{\text{3-HBDH}} \text{acetoacetate} + \text{NADH} + \text{H}^+$$

$$\text{NADH} + \text{INT} + \text{H}^+ \xrightarrow{\text{diaphorase}} \text{NAD}^+ + \text{formazan}$$

6.3.18 *D-Isocitric acid*

D-Isocitric acid is part of the citric acid cycle and hence occurs in all animals and plants. The contents are usually very low. Juices of different berries or fruits contain D-isocitrate in varying amounts. The ratio citrate to D-isocitrate, however, may be quite constant in a certain range. The determination of D-isocitric acid has become of importance in the analysis of fruit juices, especially of orange juice, for the detection of adulteration by citric acid. (The chemical D-isocitrate is quite expensive, citrate is inexpensive and also free from D-isocitrate.) The determination of D-isocitrate may be performed to assist the calculation of the fruit juice content of fruit-juice-containing beverages.

D-Isocitric acid occurs in free and ester form, but also in lactone form; alkaline saponification is often thus necessary as sample preparation.

Isocitric acid is measured by a UV test in the analysis of fruit juice, fruit juice beverages and fruit products.

Principle of determination

$$\text{D-Isocitrate} + \text{NADP}^+ \xrightarrow{\text{ICDH}} \text{2-oxoglutarate} + \text{CO}_2 + \text{NADPH} + \text{H}^+$$

6.3.19 *D-Lactic acid*

D-Lactic acid is formed only by some micro-organisms, e.g. by *Lactobacillus lactis*, *Lb. bulgaricus* and *Leuconostoc cremoris*. The presence of D-lactic acid may indicate a microbial contamination/bacterial spoilage and its presence may be a parameter for the evaluation of the freshness of pre-packed heated meat products. (L-Lactic acid is often the final product in the metabolism of living organisms.)

The stereo-specific measurement of the lactate forms is of considerable interest, e.g. in the manufacturing of sour milk products in order to assess the activity of micro-organisms forming either L-lactate, D-lactate or the racemic D-/L-lactate.

D-Lactate should not be contained in baby food because of no or only slow metabolization and excretion.

D-Lactic acid is measured by a UV test in the analysis of sour milk products, fruit and vegetable juices, meat products, baby food, beer and wine. (D-Lactate is very often measured together with L-lactate.)

Principle of determination

$$\text{D-Lactate} + \text{NAD}^+ \xrightarrow{\text{D-LDH}} \text{pyruvate} + \text{NADH} + \text{H}^+$$
$$\text{Pyruvate} + \text{L-glutamate} \xrightarrow{\text{GPT}} \text{L-alanine} + \text{2-oxoglutarate}$$

6.3.20 *L-Lactic acid*

L-Lactic acid is often a final product in the metabolism of all living organisms.

The stereo-specific measurement of the lactate forms is of great interest, for example, in the manufacturing of sour milk products in order to assess the activity of micro-organisms. L-Lactate only is formed by certain organisms such as *Streptococcus cremoris, S. lactis, S. thermophilus, Lactobacillus casei, Leuconostoc citrovorum* and all kinds of Bifido bacteria. Other micro-organisms, e.g. Lactobacillus acidophilus and Lactobacillus helveticus, produce the racemic D-/L-lactic acid.

Lactate in sugar refinery products indicates a loss of sugar. L-Lactate indicates the microbial quality of eggs: the upper limit is 1000 mg/kg in the dry matter of the untreated egg product (approx. 250 mg/kg liquid whole egg) as regulated by the European Council Directive 1989. L-Lactate is formed in beer after an infection or in a second fermentation (Belgian beers) and in wine by malo-lactic fermentation.

A UV based test is employed for the analysis of L-lactic acid in sour milk products, cheese, milk and milk powder, fruit and vegetable products, sugar refinery products, bakery goods (baking agents, dough and bread), honey, ice-cream, meat products, seafood, salads and dressings, eggs and egg products, beer and wine. (L-Lactate is often measured together with D-lactate.)

Principle of determination

$$\text{L-Lactate} + \text{NAD}^+ \xrightarrow{\text{L-LDH}} \text{pyruvate} + \text{NADH} + \text{H}^+$$
$$\text{Pyruvate} + \text{L-glutamate} \xrightarrow{\text{GPT}} \text{L-alanine} + \text{2-oxoglutarate}$$

6.3.21 *Lactose*

Lactose is an important carbohydrate component of mammalian milk. As it is relatively insoluble in cold water, it can easily be isolated from whey.

Lactose, milk powder and whey powder may be added to food in order to improve 'body' and taste (lactose may accentuate flavours) as well as to improve browning reactions. (In contrast to sucrose, lactose is not fermented by yeasts as used for the production of bakery goods.) Lactose is also added to improve texture of foodstuffs by increasing total solids without becoming too sweet. (The sweetness of lactose is lower than that of sucrose.) Lactose is an indirect indicator for the addition of milk products to food. (An addition that may be regulated by food regulations.) The measurement of lactose is also of nutritional interest because on the one hand lactose improves the utilization of calcium, on the other hand people may suffer from lactose intolerance because of the lack of the enzyme β-galactosidase.

Lactose measurement is important in the analysis of milk and dairy products (cheese, processed cheese), in foodstuffs containing milk and milk products (baby food, bread and bakery goods, dietetic food, milk chocolate, ice-cream, meat products).

Principle of determination

$$\text{Lactose} + H_2O \xrightarrow{\beta\text{-galactosidase}} \text{D-glucose} + \text{D-galactose}$$

either via D-galactose:

$$\text{D-Galactose} + NAD^+ \xrightarrow{\text{Gal-DH}} \text{D-galactonic acid} + NADH + H^+$$

or via D-glucose:

$$\text{D-Glucose} + ATP \xrightarrow{HK} \text{G-6-P} + ADP$$

$$\text{G-6-P} + NADP^+ \xrightarrow{\text{G6P-DH}} \text{D-gluconate-6-P} + NADPH + H^+$$

6.3.22 Lactulose

Lactulose is formed from lactose when milk is heated (sterilized, pasteurized, UHT heated). It is measured in commercial milk to differentiate between pasteurized milk (e.g. less than 50 mg lactulose/l milk) and UHT or sterilized milk (e.g. higher than 600 mg lactulose/l milk). It is expected that the European Union will use this measurement for the classification of heat-treated milks.

UV test for the analysis of milk.

Principle of determination

$$\text{Lactulose} + H_2O \xrightarrow{\beta\text{-galactosidase}} \text{D-glucose} + \text{D-fructose}$$

$$\text{D-Fructose} + ATP \xrightarrow{HK} \text{F-6-P} + ADP$$

$$\text{F-6-P} \xrightarrow{PGI} \text{G-6-P}$$

$$\text{G-6-P} + NADP^+ \xrightarrow{\text{G6P-DH}} \text{D-gluconate-6-P} + NADPH + H^+$$

6.3.23 L-α-Lecithin

Lecithin (phosphatidyl choline) is a fraction of lecithins which is a mixture of phospholipids. (Note: soybean lecithin contains approx. 20% of phosphatidyl choline which is the biochemically defined 'lecithin'. Egg lecithin contains approx. 70% phosphatidyl choline. This fact has to be considered when planning experiments and interpreting data from analysis.) It is used in food production because of its emulsifying characteristics. Lecithin is added to bakery goods, chocolate, margarine, instant products.

The measurement of lecithin may be carried out for the calculation of the egg content of foodstuffs. The measurement may also be suited to differentiate buttermilk and skimmed milk. A UV measurement is used for lecithin analysis in these products.

Principle of determination

$$\text{L-}\alpha\text{-Lecithin} + H_2O \xrightarrow{\text{phospholipase C}} \text{1,2-diglyceride} + \text{phosphoryl choline}$$
$$\text{Phosphoryl choline} + H_2O \xrightarrow{\text{alkaline phosphatase}} \text{choline} + \text{phosphate}$$
$$\text{Choline} + \text{ATP} \xrightarrow{\text{choline kinase}} \text{phosphoryl choline} + \text{ADP}$$
$$\text{ADP} + \text{PEP} \xrightarrow{\text{PK}} \text{ATP} + \text{pyruvate}$$
$$\text{Pyruvate} + \text{NADH} + H^+ \xrightarrow{\text{LDH}} \text{L-lactate} + NAD^+$$

6.3.24 D-Malic acid

D-Malic acid is almost unknown in nature; it is a metabolite produced by some micro-organisms alone. Analysis of freshly pressed fruit juices shows the presence of D-malic acid at the detection limit. Hence, a legal limit for fruit and fruit juices could be recommended as 10 mg/l.

D-Malic acid is a component of (racemic) D-/L-malic acid as prepared chemically. As natural products are practically free from D-malic acid, the detection of D-malic acid indicates that racemic-malic acid has been added to a product.

The addition of D-/L-malic acid to wine may be allowed, although regulated by law. The addition to fruit juice is forbidden. Malic acid is measured by a UV based method in the analysis of fruit and vegetable products, fruit juice and wine.

Principle of determination

$$\text{D-Malate} + NAD^+ \xrightarrow{\text{D-MDH, decarb.}} \text{pyruvate} + CO_2 + \text{NADH} + H^+$$

6.3.25 L-Malic acid

L-Malic acid, a component of the citric acid cycle, is an important substance in animals, plants and micro-organisms; it is one of the most important fruit acids and is the dominant acid in apples.

The addition of racemic malic acid may be allowed, e.g. in the case of wine, but its addition may be forbidden, such as in the case of fruit juices. Stereospecific L-malic acid can be produced biotechnologically/enzymatically using fumarase to attach water to fumaric acid (when synthesized chemically, racemic D-/L-malic acid is produced).

Microbial decomposition of L-malic acid leads to the formation of L-lactate; this can be a desirable reaction in, for example, the production of wine (malolactic fermentation).

L-Malic acid is an acidulant stronger than citric acid, and with a comparable longer taste retention. It is used in the production of beverages, fruit fillings, jellies, puddings, and also delicatessen products. Measurement of L-malic acid in these products relies on a UV based enzyme test.

Principle of determination

$$\text{L-Malate} + \text{NAD}^+ \xrightarrow{\text{L-MDH}} \text{oxaloacetate} + \text{NADH} + \text{H}^+$$

$$\text{Oxaloacetate} + \text{L-glutamate} \xrightarrow{\text{GOT}} \text{L-aspartate} + \text{2-oxoglutarate}$$

6.3.26 *Maltose*

Maltose is formed from starch in seed and malt by enzymatic hydrolysis (amylases). Glucose-syrup ('starch-sugar') produced from starch by means of amyloglucosidase also contains maltose. (Maltose is often measured together with sucrose, D-glucose and D-fructose.)

Maltose is a principal component of beer wort. It is also contained in bread and instant food and the analysis method is used to measure maltose in baby food, chocolate, liquid sugar as well as other foods.

Principle of determination (*based on UV measurements*)

$$\text{Maltose} + \text{H}_2\text{O} \xrightarrow{\alpha\text{-glucosidase}} 2 \text{ D-glucose}$$

$$\text{D-Glucose} + \text{ATP} \xrightarrow{\text{HK}} \text{G-6-P} + \text{ADP}$$

$$\text{G-6-P} + \text{NADP}^+ \xrightarrow{\text{G6P-DH}} \text{D-gluconate-6-P} + \text{NADPH} + \text{H}^+$$

6.3.27 *Nitrate*

Nitrates are important nitrogen-containing salts in nature. The nitrogen required, for example, by plants for the formation of proteins is taken up almost entirely in the form of nitrate (liquid and solid manure, fertilizer). Nitrate and nitrite occur in human food as intentional additives or as undesirable contaminants, especially from water used for the production of food. The nitrate content of water depends largely on the amount of fertilizer used in plant production. Nitrate may be stored in plants and its content is quite high in the case of salad, beetroots (red beets), radish and celery, spices, and spinach. The average daily human intake of nitrate is e.g. 500 mg (250 g fresh vegetables), FAO/WHO accept a daily intake of 5 mg sodium nitrate per kg body weight.

Nitrate in foodstuffs is also of nutritional importance due to its reduction to nitrite and the subsequent formation of compounds that can attach to haemoglobin; it can also form nitrosamines that are known to be carcinogenic. The intake of nitrate has to be low in the case of baby food.

The nitrate concentration in beer is determined by that in the water and hops used. Nitrate in wine may indicate the (unauthorized) addition of tap water.

A test for the analysis of nitrate in baby food, fruit and vegetable products (juices), meat and meat products (sausages), of dairy products (cheese, processed cheese, whey powder), beer and wine, as well as water is based on UV measurements.

Principle of determination

$$\text{Nitrate} + \text{NADPH} + \text{H}^+ \xrightarrow{\text{NR}} \text{nitrite} + \text{NADP}^+ + \text{H}_2\text{O}$$

6.3.28 *Oxalic acid*

Oxalic acid occurs in high concentrations in some plant leaves. Its content is quite high in spinach, rhubarb, beetroot and other vegetables (which may also be rich in minerals and vitamins). It is also a product of human metabolism. In pathological conditions, it is excreted with the urine when its presence can lead to the formation of calcium oxalate stones in the ureter and kidneys.

Oxalate interferes with the resorption of calcium and is thus of some considerable nutritional importance.

Oxalate in beer causing a Gushing effect, is also responsible for turbidity and sediments.

A UV test is used for the analysis of oxalate in fruit and vegetable products, beer, chocolate.

Principle of determination

$$\text{Oxalate} \xrightarrow{\text{oxalate decarboxylase}} \text{formate} + \text{CO}_2$$

$$\text{Formate} + \text{NAD}^+ + \text{H}_2\text{O} \xrightarrow{\text{FDH}} \text{bicarbonate} + \text{NADH} + \text{H}^+$$

6.3.29 *Pyruvate*

A UV test is used for the analysis of beer, cheese, milk, sugar refinery products and wine.

Pyruvate is an important metabolite in metabolism. It is formed from proteins, and carbohydrates. Pyruvate has been measured in milk for the assessment hygienic/bacteriological quality and cleanliness of milk production. The measurement, based on a UV test is simple, quick, inexpensive and can be run automatically (continuous flow analyser). It can be used to measure pyruvate levels in beer, cheese, milk, sugar products and wine.

Principle of determination

$$\text{Pyruvate} + \text{NADH} + \text{H}^+ \xrightarrow{\text{L-LDH}} \text{L-lactate} + \text{NAD}^+$$

6.3.30 *Raffinose*

Raffinose is a trisaccharide consisting of D-galactose, D-glucose and D-fructose. It occurs in sugar beets in a relatively high content, and interferes with the crystallization of sucrose and accumulates in molasses. Raffinose is also contained in other plants, e.g. in soybean. Its presence in food is an indirect proof of adulteration, such as by the unallowed addition of soybean products.

A UV test is used for the analysis of sugar refinery products, soybean and cereal flour.

Principle of determination

$$\text{Raffinose} + \text{H}_2\text{O} \xrightarrow{\alpha\text{-galactosidase}} \text{D-galactose} + \text{sucrose}$$

$$\text{D-Galactose} + \text{NAD}^+ \xrightarrow{\text{Gal-DH}} \text{D-galactonic acid} + \text{NADH} + \text{H}^+$$

6.3.31 *D-Sorbitol/Xylitol*

A colorimetric method for the analysis of sorbitol in fruit products (juices, jam), bakery goods, chewing gum, chocolate and confectionery products, diabetic food (honey, jam, ice-cream), dietetic food, low-calorie beverages, chewing gum, sweets and candy.

D-Sorbitol, the reduction product of D-fructose, occurs extensively in fruits, e.g. in apples, pears and prunes. It is used in the food technology industry as a moisturizing agent and as a sugar substitute for diabetic products as, in contrast to D-glucose, insulin is not necessary for its metabolism. Addition and content of D-sorbitol in diabetic food may be regulated by food inspection.

D-Sorbitol is contained in apple juice, but only in traces or not at all in grape juice, wine and wine vinegar. Its presence can be used to indicate adulteration of wine by the addition of apple products.

Xylitol is a sugar alcohol that occurs frequently in fruits, vegetables and mushrooms. Xylitol is a sweetener and a sugar substitute; xylitol is stable, shows no browning reactions and is not fermented by cariogenic bacteria in the mouth. Its use for food production is regulated.

Principle of determination

$$\text{D-Sorbitol} + \text{NAD}^+ \xrightarrow{\text{SDH}} \text{D-fructose} + \text{NADH} + \text{H}^+$$

$$\text{Xylitol} + \text{NAD}^+ \xrightarrow{\text{SDH}} \text{D-xylulose} + \text{NADH} + \text{H}^+$$

$$\text{NADH} + \text{INT} + \text{H}^+ \xrightarrow{\text{diaphorase}} \text{NAD}^+ + \text{formazan}$$

6.3.32 Starch

Starch is an important ingredient in food and animal feed; it is also an important energy source in human diet. When hydrolysed enzymatically or chemically (by acid), oligoglucosides and D-glucose are formed.

Starch is an important energy reservoir in plants, and seeds like barley, corn, maize, rice and wheat contain up to 75% starch. These plant materials are of considerable importance in human and animal nutrition. Starch content is measured because it contributes mainly to the caloric value of food like bakery goods, cereals or potato products. The starch content in foodstuffs may be controlled by legislation (e.g. meat products like sausages and hamburgers) because starchy products are of lower value than meat, it may also be of interest because of agricultural or customs duties and taxes.

A UV test is used for the analysis of starch in baby food, cereal products, flour, bakery goods, dairy products (yoghurt, instant desserts, drinks), dietetic food, fruit and vegetables (apples, potato), meat products (sausages, hamburger), as well as for the determination of partially hydrolysed starch (glucose-syrup, starch-sugar) in beverages, energy drinks and jam and of dextrins in beer.

Principle of determination

$$\text{Starch} + (n-1)\,\text{H}_2\text{O} \xrightarrow{\text{AGS}} n\,\text{D-glucose}$$

$$\text{D-Glucose} + \text{ATP} \xrightarrow{\text{HK}} \text{G-6-P} + \text{ADP}$$

$$\text{G-6-P} + \text{NADP}^+ \xrightarrow{\text{G6P-DH}} \text{D-gluconate-6-P} + \text{NADPH} + \text{H}^+$$

6.3.33 Succinic acid

As a metabolite of the citric acid and glyoxylate cycles, succinic acid occurs widely in animals, plants and micro-organisms.

Succinic acid is a specific indicator of microbial decomposition in eggs and egg products. The upper limit is regulated in Europe: 25 mg/kg of egg product dry matter.

Analysis of succinic acid indicates the degree of maturity of fruit and cheese and a test based on UV measurement is used for its analysis in relevant products.

Principle of determination

$$\text{Succinate} + \text{ITP} + \text{CoA} \xrightarrow{\text{SCS}} \text{IDP} + \text{succinyl-CoA} + P_i$$

$$\text{IDP} + \text{PEP} \xrightarrow{\text{PK}} \text{ITP} + \text{pyruvate}$$

$$\text{Pyruvate} + \text{NADH} + \text{H}^+ \xrightarrow{\text{LDH}} \text{L-lactate} + \text{NAD}^+$$

6.3.34 Sucrose ('sugar')

Sucrose has a central position in plant metabolism. Its isolation from cane and beet is of great economic interest. When sucrose is hydrolysed/inverted, D-glucose and D-fructose are formed.

Sucrose is an important ingredient of foodstuffs. It is not only of monetary value, but also the most important carbohydrate sweetener.

Sucrose is often measured together with D-glucose and D-fructose, and a test using UV measurements is used for its analysis in sugar refinery products (beet and cane sugar), baby food, bakery goods, beverages (soft drinks, lemonades, cola drinks), chocolate, sweets and candies, dairy products (desserts, cheese, sweetened condensed milk), dietetic food, eggs and egg products, food dressings, fruit, berries and vegetable products (juices, jam, tomato products, canned and frozen food), potato for the production of chips, honey, ice-cream, meat products (sausages), mustard, spices (meat additives, pepper), beer and wine.

Principle of determination

$$\text{Sucrose} + \text{H}_2\text{O} \xrightarrow{\beta\text{-fructosidase}} \text{D-glucose} + \text{D-fructose}$$

$$\text{D-Glucose} + \text{ATP} \xrightarrow{\text{HK}} \text{G-6-P} + \text{ADP}$$

$$\text{G-6-P} + \text{NADP}^+ \xrightarrow{\text{G6P-DH}} \text{D-gluconate-6-P} + \text{NADPH} + \text{H}^+$$

6.3.35 Sulphite

Sulphur dioxide, sulphurous acid and its salts (sulphites) occur in very low concentrations in nature. However, they have been used for a very long time in the industrial production of foodstuffs ('sulphuring'). Sulphite is used as an antimicrobial agent (yeasts, mould, bacteria), a browning inhibitor (fruits and vegetables) and an antioxidant (beer, seafood). Sulphite is regarded as being poisonous for cells; in metabolism, it is rapidly oxidized to sulphate and excreted.

The sulphite content in foodstuffs is legally prescribed in a number of countries and its presence has usually to be declared on the label.

A UV test for the analysis of sulphite in fruit and vegetable products (juices, jam, potato products, raisin, dried fruits, canned and frozen vegetables), beverages, fish and fish products, salad dressings, spices, vinegar, beer and wine, is available.

Principle of determination

$$\text{Sulphite} + O_2 + H_2O \xrightarrow{\text{sulphite OD}} \text{sulfate} + H_2O_2$$
$$H_2O_2 + NADH + H^+ \xrightarrow{\text{NADH POD}} 2\,H_2O + NAD^+$$

6.3.36 Urea

Urea is the most important decomposition product of human protein metabolism. Its measurement in body fluids gives an indication of the state of protein balance in muscle cells and of the protein supply, such as of cows when urea is measured in milk.

Urea is sometimes added (illegally) to meat products in order to indicate a higher content of muscle protein (which is determined via nitrogen) than is actually present (adulteration).

An enzymic test using UV measurements is used for the analysis of bakery goods, meat products, milk and seafood.

Principle of determination

$$\text{Urea} + H_2O \xrightarrow{\text{urease}} 2\,NH_3 + CO_2$$
$$\text{2-Oxoglutarate} + NADH + NH_4^+ \xrightarrow{\text{GlDH}} \text{L-glutamate} + NAD^+ + H_2O$$

6.4 Sample preparation

6.4.1 Sample preparation methods

Sample preparation refines the sample, so that the analysis can be carried out with no interferences.

In classical chemical analysis this means the removal of interfering components in the sample, or the quantitative isolation of the substance of interest. These techniques are necessary because chemical reactions are usually group reactions and not substance specific. There are also practically no chemical reactions without side reactions. Side reactions are dependent on the reaction conditions and the composition of the sample and in order to optimize these chemical methods, sample preparation has continually been improved with the objective of removing further interfering substances from the sample material. Thus in classical analysis there is frequently no 'universal' method suitable for a wide variety of sample materials, rather there are 'individual' methods dependent on the nature of the sample.

Such isolation and removal operations are not necessary in enzymatic analysis because enzymes convert only the substance to which they are specific. Enzymatic analysis uses photometric measurements to follow the reaction and photometric measurements only require solutions which are 'practically' clear and 'almost' colourless. Practically clear and almost colourless means that a slight opalescence or coloration of the sample solution does not cause interference. Sample preparation for enzymatic analysis is very easy (Table 6.8). The ability of 'losing' the analyte and thereby obtaining low result, is minimal.

6.4.2 Special sample preparations

Dependent on the characteristics of the particular analyte.

L-Ascorbic acid. L-Ascorbic acid (vitamin C) is not stable especially when in aqueous solution. For example an aqueous L-ascorbic acid solution shows a drop in concentration of L-ascorbic acid to about half within 2 h in the presence of oxygen. In fruit juice L-ascorbic acid is oxidized until the gaseous and dissolved oxygen is exhausted or until there is no remaining L-ascorbic acid.

To solubilize pure L-ascorbic acid powder or vitamin C tablets, to dilute the sample and to extract vitamin C from the sample, aqueous meta-phosphoric acid solution is recommended as a stabilizing solvent.

Starch. Starch is a macromolecule consisting of D-glucose subunits which are joined by α-1,4- and α-1,6-bonds. Amylose and amylopectin are starches found in plants, and glycogen is an animal starch. These substances called 'insoluble starches' are not soluble in water. Thus starch must be solubilized or transformed into a soluble form before its determination.

Starch may be solubilized by autoclaving. A better alternative is to use a mixture of hydrochloric acid and dimethylsulphoxide (DMSO). DMSO solubilizes starch without destruction of the molecules and without generating free D-glucose. These conditions hydrolyse sucrose totally and split about 10% of lactose and maltose. Therefore a sample blank (without amyloglucosidase) is necessary (in the absence of maltose), or the sample must be extracted with an (40%: w/v) ethanol/water mixture (in the presence of maltose) to remove 'low molecular weight' sugars.

The use of 'technical enzymes' in the sample preparation (to solubilize starch) is strongly advised against. The numerous enzymatic activities present in a technical enzyme may lead to other uncontrolled enzymatic reactions occurring. Also, (depending on the sample matrix) unwanted reactions may proceed which may interfere in the determination. Thus wrong results (either too low or too high) are obtained.

Due to analytical principles.

L-Glutamic acid, D-sorbitol/xylitol. In the determination of L-glutamate, D-sorbitol and xylitol the indicator reaction is colorimetric (it uses INT – Iodonitro-tetrazolium-chloride).

High concentrations of reducing substances (e.g. L-ascorbic acid, or sulphurous acid) interfere in the assay. The reducing substances react with INT and are the cause of a creep reaction. Treating the sample with hydrogen peroxide at an alkaline pH eliminates the interference.

Table 6.8 Sample preparation for enzymatic food analysis

The sample is	Necessary sample preparation techniques	Reason for sample preparation
Liquid, clear, colourless or slightly coloured, neutral	Use sample directly for the assay, if necessary dilute	There must be a coenzyme excess in the assay system to convert the analyte quantitatively; measurements must be done in the linear part of the calibration curve; the Lambert–Beer law must be valid; the measured signal (= absorbance difference) must be high enough to minimize the error of analysis
Liquid, coloured	Decolorize, if necessary dilute (decolorization is often not necessary after dilution)	The dyestuffs in coloured samples may absorb light at the measurement wavelength thus exceeding the optimal range of measurement of absorbances; changes of the pH in the coloured sample solution may be responsible for creep reactions before the start enzyme is added
Liquid, turbid	Filter or centrifuge, treat with CARREZ reagents, if necessary dilute	Turbidities interfere in the photometric measurement because of refraction, scattering and absorption of light
Liquid, acidic or alkaline	Neutralize or adjust to the pH of the assay system, if necessary dilute (often neutralization is not necessary after dilution)	Enzymatic reactions occur in buffered solutions with pH values optimal for the enzymes and/or the equilibrium of the reaction; the buffer capacity of the assay system must be sufficient to maintain the optimum pH necessary for the enzymatic reaction; a creep reaction may be caused when a coloured acid sample is introduced into an alkaline assay system
Liquid, emulsion	Treat with acid or CARREZ reagents, filter or centrifuge, dilute if necessary	The emulsion of the sample may be responsible for turbidity in the assay system (see 'liquid, turbid' samples)
Pasty or solid	Homogenize or blend, extract, filter or centrifuge, if necessary dilute	Solid samples can be analysed with only a few analytical techniques, therefore the analyte is extracted from the sample material; samples must be ground to obtain representative results; in the analysis of plant materials, plant cells are broken down when the particle size is smaller than 0.2 mm and can be extracted quantitatively
Pasty or solid, containing fat	Homogenize or blend, extract, remove fat in the cold, filter or centrifuge; if applicable use CARREZ-clarification, if necessary dilute	Fat in the sample may be responsible for turbidity in the assay system (see 'liquid, turbid' samples)
Pasty or solid, containing protein (enzymes)	Homogenize or blend, extract, treat with acid or CARREZ reagents, remove fat in the cold, filter or centrifuge, if necessary dilute	Protein in the sample may be responsible for turbidity in the assay system (see 'liquid, turbid' samples); enzymes in 'living' sample materials have to be denaturated otherwise there may be interferences in the enzymatic assay
Containing CO_2	Degas (stir, shake or filter, use an ultrasonifier bath), or alkalinize, if necessary dilute	Solutions containing carbon dioxide cannot be accurate pipetted

D-Fructose, sucrose. An excess of free D-glucose interferes in the estimation of D-fructose and sucrose. If the ratio of D-glucose to D-fructose (or to sucrose) is 10 : 1 or greater, then the free D-glucose must be removed, e.g. by oxidation with glucose oxidase (GOD) and catalase.

In the presence of the enzymes glucose oxidase (GOD) and catalase, D-glucose is oxidized to D-gluconate. After the oxidation of the bulk of the D-glucose, the enzymes GOD and catalase are inactivated in a boiling water bath. The reaction need not proceed quantitatively because the remainder of the D-glucose is compensated for in a reagent blank assay.

6.5 Principles and measuring technique

Running an enzymatic determination is simple. In most cases it is merely pipetting buffer and coenzyme solutions, enzyme suspension (or solution), redistilled water and the sample solution into a cuvette (Table 6.9), following the reaction and calculating the result from the measured signals (see 6.5.6).

6.5.1 *Reaction mechanism*

Enzymatic reactions proceed according to very exact and well-known reaction mechanisms, and they proceed without any unknown side reactions. This is not always true for conventional chemical conversions. There are no 'main reactions' in chemical methods which do not have any side reactions; the mechanism of the main reaction is known but the mechanisms of the simultaneous side reactions are still unclear. The type of side reaction and its extent not only

Table 6.9 Example of protocol for an enzymatic determination

Pipette into cuvettes	Blank (ml)	Sample 1 (ml)	Sample 2 (ml)	Sample 3 (ml)
Buffer	1.000	1.000	1.000	1.000
Coenzyme(s)	0.100	0.100	0.100	0.100
Redist. water	2.000	1.900	1.800	–
Sample	–	0.100	0.200	2.000
Auxiliary enzyme	0.020	0.020	0.020	0.020
Mix and measure the absorbances A_1. Start the reaction by the addition of				
Starting enzyme	0.020	0.020	0.020	0.020
Mix. After the reaction is complete measure absorbances A_2				

Notes:
1. The (reagent) blank includes the light absorbance of the starting enzyme, as well as (traces) of the analyte in the reagents.
2. Assay 'sample 1' together with the blank is 'single' determination.
3. Assays 'sample 1' and 'sample 2' are a double determination under 'different conditions': the measured absorbance differences have to be proportional to the sample volumes. This is the simplest way to detect interferences of the assay by ingredients of the sample, as well as of gross errors in analysis.
4. Assay 'sample 3' shows the situation for trace level compound analysis: the sample volume is greatly increased.

depends on the particular reaction conditions, but also on the individual ingredients of the different samples. Furthermore, there may also be 'reverse reactions', e.g. in titration procedures, and in this case, the 'quality' of the results depends on the speed of work. An example is the determination of free and total sulphite in wine. Another example for a chemical determination is the reductometric determination of sugars (e.g. Fehling, Luff–Schoorl, Lane–Eynon methods). The difficulties encountered in these conventional methods are compensated for by the use of 'scientific' correction factors or by standardization of the method (e.g. by regulating the reaction conditions).

Enzymatic reactions proceed stoichiometrically. A general reaction equation is,

$$A + B \xrightarrow{enzyme} C + D$$
$$\text{substrate} \quad \text{coenzyme} \quad \quad \text{product} \quad \text{coenzyme'}$$

One mole of the substrate (i.e. the substance which is converted by the enzyme, the analyte in analysis) reacts with one mole of a coenzyme to produce (under the catalytic influence of the enzyme) one mole of the product and one mole of the 'altered' coenzyme (i.e. either oxidized or reduced NAD and NADP, for example).

Coenzymes are essential for many enzymatic reactions. (Coenzymes are often precursors of vitamins.) A number of different definitions have been used in the past for the term 'coenzyme'. Today, a coenzyme is defined as a transport metabolite. For example, reduced nicotinamide adenine dinucleotide (NADH) 'transports' hydrogen and adenosine-5'-triphosphate (ATP) 'transports' a reactive phosphate group. In the living cell the coenzyme, which has been modified by one reaction, is reconverted into its original form by the following reaction. This is in contrast to the substrate: the product formed in a reaction is the substrate for the following reaction.

A number of enzymatic reactions utilize the coenzymes NAD or NADP and a large number of other reactions can be coupled to a NAD(P)-dependent reaction. (Figure 6.1 shows the structural formula of NAD(P) in the oxidized form.) Generally, the term 'indicator enzyme' is chosen to represent the coupled NAD(P) dependent enzymes.

Examples of coupled reactions are the determination of D-glucose and glycerol. In the D-glucose assay, hexokinase is used as the first enzyme and glucose-6-phosphate dehydrogenase is the indicator enzyme. In the determination of glycerol, pyruvate kinase is an auxiliary enzyme and lactate dehydrogenase is the indicator enzyme.

The mechanisms of these indicator reactions are very well known: in enzymatic reactions utilizing the coenzymes NAD and NADP, a hydride ion is transferred from the substrate and added to the C_4-position of the pyridinium ring system. This is known as a nucleophilic addition reaction and a quinine system is formed from the aromatic system. Figure 6.2 shows the reaction mechanism in a simplified form.

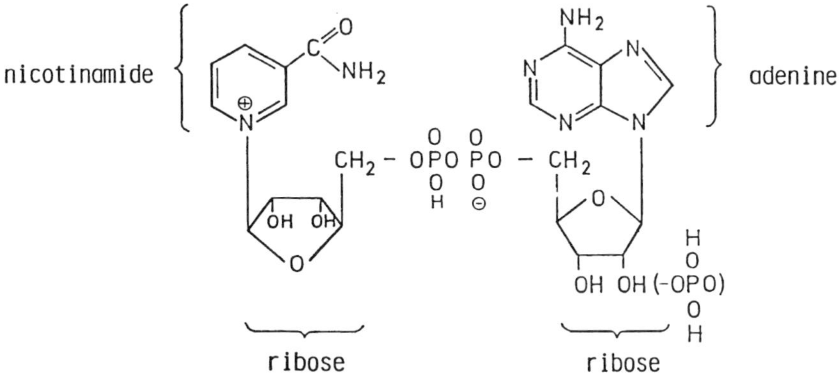

Figure 6.1 Structure of nicotinamide-adenine di-nucleotide (-phosphate) NAD(P) in the oxidized form.

Enzymatic reactions proceed quantitatively without unknown side reactions and with known reaction mechanisms. The substrate is totally converted and produces no by-products. The amount of NAD(P) or other light absorbing substances formed or decomposed is proportional to the amount of analyte. Thus the reaction proceeds stoichiometrically.

A well-known exception is the determination of acetic acid. Acetic acid is totally converted in the ACS catalysed reaction forming acetyl-CoA. Acetyl-CoA then reacts with oxaloacetate which has been formed in a previous indicator reaction utilizing L-malate dehydrogenase (L-MDH), and which is a 'true' equilibrium reaction and this must be taken into account when calculating the results.

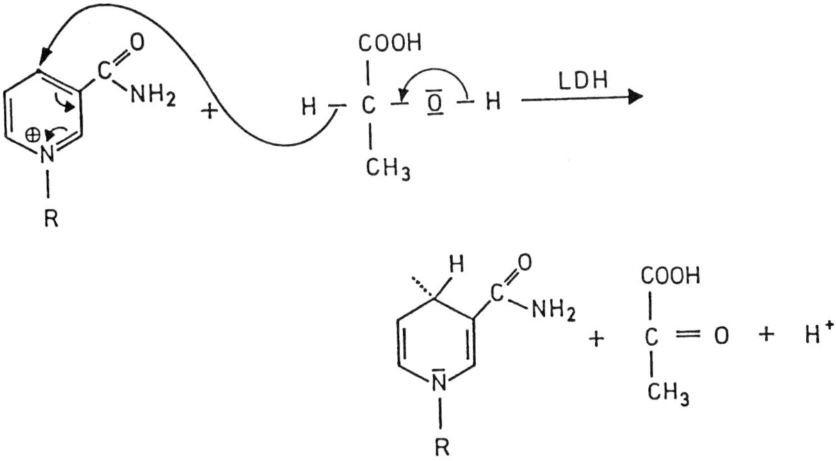

Figure 6.2 Mechanism of the lactate dehydrogenase (LDH) reaction.

6.5.2 Reaction equilibrium

Like all chemical reactions, enzyme catalysed reactions all have a position of equilibrium. Catalysts, and therefore also enzymes (bio-catalysts), have no influence on the position of this equilibrium, they only affect the speed at which the equilibrium is reached. Enzymes obey the rules of thermodynamics and kinetics. Enzymatic reactions are also subject to the laws of mass action.

For the general equation,

$$A + B \xrightarrow{enzyme} C + D$$

the equilibrium constant is:

$$K = \frac{[C] \times [D]}{[A] \times [B]}$$

The equilibrium constant K characterizes each chemical reaction. Reactions whose equilibrium lies to the left of the reaction have a very small constant K. K is large, however, when the equilibrium lies to the right of the reaction.

If the equilibrium lies almost totally to the right, that is on the side of the reaction products, the reaction proceeds quickly and largely quantitatively. In contrast, reactions whose equilibriums lie 'somewhere' in the middle between the reactants and the products (i.e. with K ~ 1) or lie on the left of the reaction, proceed only partially in the desired direction. If the concentration of the reactant (analyte) is to be analysed, then the conversion must be quantitative. To 'change' the reaction to the desired direction, the ratio of the concentrations of the reactants to products must be altered. Raising the concentration of B, or lowering the concentration of the products C and/or D reduces the concentration of substance A for a given equilibrium constant. It follows that the reaction then proceeds fully to the right.

Practically all biochemical reactions are reversible, at least 'microscopically'. Therefore, the assay conditions must be chosen so that in the assay the equilibrium is shifted quantitatively in favour of the products of the reaction.

An example is the determination of lactic acid using the enzyme LDH:

$$\text{Lactate} + NAD^+ \xrightarrow{LDH} \text{pyruvate} + NADH + H^+$$

Lactic acid (lactate) is oxidized by means of the coenzyme NAD (nicotinamide adenine dinucleotide), pyruvic acid (pyruvate) and NADH are formed. The equilibrium of this reaction ($K = 2.76 \times 10^{-12}$) lies largely on the side of lactate and NAD. Therefore, the determination of pyruvate by its reaction with NADH forming lactate presents no problems. To obtain a quantitative reaction of lactate with NAD to produce pyruvate, NADH and H^+, the concentration of the reactant NAD must be very high and the concentration of the end-products must be very low. In practice there is usually a large excess of NAD in the assay system.

There are several alternatives:

1. If the reaction proceeds in an alkaline medium, about pH = 10, the concentration of the H^+ ions is reduced to 10^{-10} mol/l.
2. Pyruvate may be removed from the equilibrium by reacting it with hydrazine to form hydrazone. The equilibrium for this reaction lies almost totally on the side of the hydrazone derivative.
3. Another way to remove pyruvate from the equilibrium is to react it further with L-glutamic acid to produce L-alanine and 2-oxoglutarate. This reaction is catalysed by glutamate pyruvate transaminase (GPT) and has the advantage of having fewer interfering addition products than the hydrazine reaction. Also the endpoint of the primary reaction is reached sooner.

$$\text{Lactate} + NAD^+ \xrightarrow{LDH} \text{pyruvate} + NADH + H^+$$

$$\text{Pyruvate} + \text{L-glutamate} \xrightarrow{GPT} \text{L-alanine} + \text{2-oxoglutarate}$$

4. The removal of NADH from the equilibrium (by reaction with tetrazolium salts and Diaphorase) also displaces the equilibrium.

Enzymatic reactions are equilibrium reactions. If enzymatic reactions are used in analysis, then test conditions must be chosen to allow the analyte to be quantitatively determined despite an unfavourable equilibrium, and to be measured within an acceptable time.

6.5.3 *Measurement of enzymatic reactions*

The light absorption characteristics of the coenzymes NAD and NADP (in the near UV range) are different for the reduced and oxidized forms. The reduced forms have an additional absorption peak with a maximum at 340 nm (Fig. 6.3). Because enzymatic reactions proceed stoichiometrically, and the absorption at 340 nm is characteristic of these coenzymes, the quantitative determination is possible, especially because the extinction coefficient used in the calculations has been determined with great accuracy.

Newer measurements show that the maximum absorbance is not at exactly 340 nm but nearer to 339 nm (339.2 nm). However, this does not affect the established value of the extinction coefficient used in the calculation.

Most enzymatic analytic procedures are based on the measurement of the increase or decrease in the absorbance (in the near UV range) of the coenzymes NADH and NADPH. There are also other methods, referred to as colorimetric tests. In this type of test, the concentration of the analyte is proportional to the formation of a product (dye or chromogen) which absorbs light in the visible spectrum. Examples of this are found in the determination of L-glutamic acid, D-3-hydroxybutyric acid, D-sorbitol, xylitol, L-ascorbic acid and cholesterol.

Enzymes are often very specific to a particular coenzyme. The living cell generally uses the NAD system for catabolic reactions (e.g. formation of NADH

172   ANALYTICAL METHODS OF FOOD AUTHENTICATION

in glycolysis) and uses the NADP system for synthetic (anabolic) reactions (e.g. consumption of NADPH in fatty acid and cholesterol synthesis). This means that when using enzymes as reagents for analysis the correct pyridine coenzyme must be used.

6.5.4 *Measurement techniques*

Many enzymatic measurements utilize the light absorption properties of NAD(P) in their oxidized and reduced forms. Monochromatic light is required for the photometric measurement and for the calculation of the results according to the Beer Lambert law. Monochromatic light can be obtained by using hydrogen, deuterium, halogen or tungsten lamps as light sources, and a filter or monochromator in order to get monochromatic light at a determined wave length.

The absorbance differences are measured at the maximum of the absorbance peak when a spectrophotometer or filter photometer is used, i.e. 340 nm. The molar extinction coefficient of NAD(P) is greatest at the absorption maximum, and this also gives the methods their great sensitivities. It is well accepted in photometry that the measurements are made at the light absorption maximum because small deviations, for example, in wavelength adjustment, can cause little or no error.

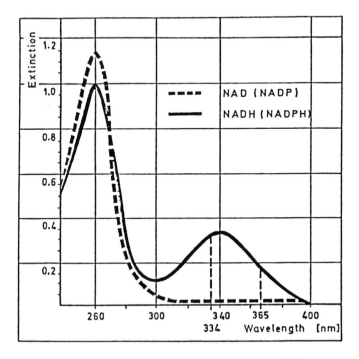

Figure 6.3 Absorption spectra of NAD(P) and NAD(P)H.

ENZYMIC METHODS OF FOOD ANALYSIS   173

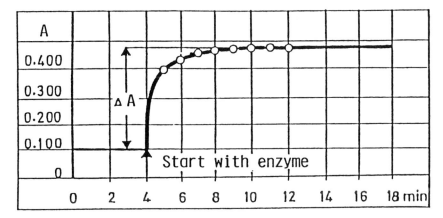

Figure 6.4 Reactions with NAD or NADP in the presence of dehydrogenases are redox reactions. The substrate to be determined is oxidized while the NADH or NADPH formed leads to an increase in light absorption. Examples are the determination of D-glucose, D-fructose, L- and D-lactic acid and ethanol.

6.5.5 *The reaction*

Figures 6.4 and 6.5 show typical reaction curves. At the beginning of the reaction the absorption curve is very steep, and the reaction is a zero-order reaction. As the reaction progresses it becomes slower and becomes a first-order reaction. The reaction becomes increasingly slower and the end of the reaction is

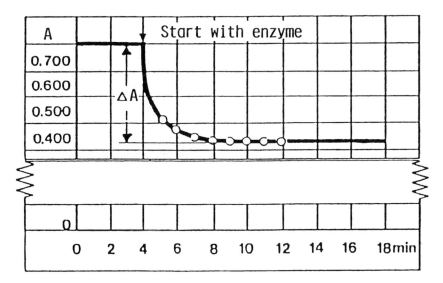

Figure 6.5 Reactions with NADH or NADPH also proceed in the presence of dehydrogenases. In these reactions the substrate to be determined is reduced and the coenzymes are oxidized. A decrease in light absorbance is observed. Examples are the determination of glycerol and citric acid.

reached asymptotically. After the measurement of initial absorbance A_1, the reaction is begun by addition of the starting enzyme. The final absorbance A_2 is read at completion of the reaction. The result is calculated using the Beer Lambert law from the differences in the two absorbance readings.

Occasionally the reaction does not come to a standstill and the absorbance continues to change constantly over time. This is because there is a side (or 'creep') reaction superimposed on the main reaction. The effects are caused, for example, by side activities or contamination activities of the enzymes employed if the substrates for these undesirable activities are contained in the sample material or in the assay system.

Contaminating activities in the enzyme preparation can be removed by the manufacturer, but side activities are characteristics of a particular enzyme and cannot be removed. Enzymatic activities present in the sample solution can be destroyed by heat inactivation or by deproteinization.

Since main reactions and side reactions proceed at different velocities, the 'creep' reactions can be compensated for in the calculation of the results. Fast main reactions and slow side reactions can be separated because the side reaction is still a zero-order reaction when the main reaction has finished.

'Creep' reactions proceed slowly and with a constant velocity. By plotting the absorbance values as a function of time, the creep reaction can be graphically eliminated by extrapolation of the absorbances back to the time the starting enzyme was added. A mathematic extrapolation is also possible. Calculate the absorbance change per minute (which has to be constant) and multiply by the total reaction time in minutes. Add (or subtract, depending on the reaction) the ΔA of the creep reaction to the ΔA of the total (combined) reaction.

Often the creep reaction is independent of the sample material but is caused by the reagents. In these cases, the creep reaction is identical in both the blank

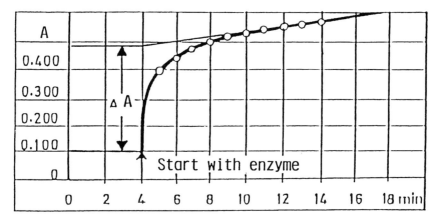

Figure 6.6 Reaction with creep reaction: sometimes enzymatic reactions do not stop and continue, e.g. because of slow side reaction. In this case, the absorbance difference for the main reaction is found by extrapolation of the absorbance back to the time of the beginning of the reaction.

assay and the sample assay and an extrapolation is not necessary. To compensate for the creep reaction, the absorbance of the sample assay is read against the reagent blank, or the absorbances of the sample assay and blank are read immediately one after each other.

Enzymatic reactions are chemical reactions and obey all the laws of chemistry and physics. When enzymatic systems are used in analysis the reaction conditions are always arranged so that the reaction is quantitative. Measurements are usually made photometrically.

6.5.6 Calculation

The photometric measurement is based on the difference between the light absorbance of the reactants of a chemical reaction and its products. The final calculation is based on the Beer Lambert law.

$$\Delta A = \varepsilon \times c \times d \tag{6.1}$$

The value of the absorbance (A) is proportional to the concentration (c) and the length of the light path (d). The proportionality factor, the molar extinction coefficient (ε) of the light absorbing substance, is in practice a constant for a particular wavelength.

The concentration of the absorbing substance is given by:

$$c = \frac{A}{\varepsilon \times d} \text{(mol/l)} \tag{6.2}$$

Concentration changes are measured:

$$c_1 - c_2 = \frac{A_1}{\varepsilon \times d} - \frac{A_2}{\varepsilon \times d} \text{(mol/1)} \tag{6.3}$$

If the reaction goes to completion, then $c_2 = 0$.

It must be remembered that the sample volume v is contained within the total volume V. A general formula to calculate the concentration of the analyte has been established which allows for variations to the sample volume and for the introduction of the molecular weight of the analyte (g/mol):

$$c = \frac{\Delta A \times V \times \text{Mwt}}{\varepsilon \times d \times v \times 1000} \text{[g/l]} \tag{6.4}$$

where, V = test volume (in ml), v = sample volume (in ml), Mwt = molecular weight of the analyte in g/mol, d = path length (in cm), ε = extinction coefficient of NADH or NADPH at 340 nm = 6.3 ($1 \times \text{mol}^{-1} \times \text{cm}^{-1}$).

Results from enzymatic determinations are calculated using photometric measurements and the Beer Lambert law. The extinction coefficients used in the formula have been very accurately determined. Empirical calculation factors are not recommended and the reference to a standard substance is not necessary. It is a question of an absolute method versus a relative measurement.

6.6 Interferences with the determination

6.6.1 *The nature of interference*

Food chemists/analysts are accustomed to interferences in traditional chemical analysis, and they are often referred to as 'matrix effects'. This term is sometimes used, to excuse or circumvent the 'poor' specificity of a method.

The sample materials which are investigated in food analysis are very diverse. The samples range from pure substances (e.g. sugar = sucrose) to the most complex mixtures (e.g. thickened sauces). The ingredients of food consist of (to varying degrees) carbohydrates, proteins, fat, water, minerals, additives, etc. Heating processes, like cooking, pasteurization and sterilization can introduce changes; for example, some compounds may be destroyed while others may be formed. Interferences can be expected in the analysis of these samples; for enzymatic analysis this is true for only a few examples. Generally, it can be said that samples which are metabolized by enzymes in man without interference are also determined without interference in the enzymatic assay system. This is because the nature of the enzyme in the assay is the same as in the human digestive tract.

Interferences to the enzymatic determination can be introduced by components in food, such as tannins in wine and phenolic compounds in apples, potatoes and hops. These interferences in the assay are easily recognized and usually removed either by suitable sample preparation techniques, or by the correct interpretation of the measurement signal (e.g. by extrapolation when a creep reaction is present).

Interferences are seldom seen in enzymatic analysis procedures. Compounds which are structurally similar to the analyte do not interfere because the enzymes, due to their specificity, convert only those compounds for which they are specific. Details of the specificity should always be stated in the method instruction sheet.

Additives such as preservatives or antioxidants could cause interferences in the assay. These substances hinder or prevent deterioration of foodstuffs caused partly by micro-organisms by inhibiting microbial enzymatic reactions. Addition of these substances in an amount which would interfere with an enzymatic determination is not feasible for reasons of cost, taste and toxicity. Many countries have regulations which limit the amount of preservative permitted in foodstuffs. In the development of enzymatic methods, the effect of additives (at least the most widely used ones) on the assay must be tested.

In food analysis, interferences may also be caused by endogenous colourings, by enzymes which are transferred with the sample into the reaction mixture, or by phenolic compounds in the sample, e.g. in the determination of sulphite. Suitable sample preparation removes dyes, enzymes or enzyme inhibitors.

6.6.2 *Recognizing interferences*

Interferences are easily detected. If the reaction is complete in the given time, this is an indication that the reaction is interference-free. It is, however, recommended to always repeat the measurement at the end of the enzymatic reaction to make certain that the reaction is complete. Usually an interference means that the reaction is slowed down and this is detected by the repeated measurement.

As a security measure, after the reaction has stopped the reaction can be re-started, either qualitatively or quantitatively, by the addition of standard material or a standard solution. A further change in the absorbance after the addition of standard is evidence that the absorbance A_2 can be confidently used to calculate the result.

If variable sample volumes are used in a series of determinations, and proportionality exists between sample volume and absorbance, this is further evidence that the reaction has no interferences (Table 6.9, assays 1 and 2). The explanation is simple: a large amount of sample in the assay also contains a large amount of 'enzyme-inhibitors' which (when all other factors in the assay are constant) will produce 'more interference' and increase the reaction time. If the measured signal is not proportional to the amount of sample (i.e. the absorbance difference for the larger sample is too low) then it is recommended that the measurement of the assay be repeated later (e.g. until proportionality with the signal is achieved).

Another way to detect interferences is the use of an internal standard. As well as a blank, a standard and a sample assay, an assay containing both sample and standard is done and the absorbances measured (Table 6.10).

A 'real' recovery experiment is also a means of detecting interferences and is even recommended when applying 'new' sample preparation techniques on unknown sample materials/matrices. The sample is assayed with and without the addition of standard and the standard must be quantitatively recovered. (The purity of the standard material must, of course, be established in a parallel experiment.)

Experience has shown that enzymatic reactions are usually free of interferences. Disturbances to the assay system can be recognized easily and can be eliminated with suitable sample preparation procedures.

6.7 Sensitivity

Enzymatic analysis covers an extraordinarily wide concentration range, which is dependent on the measurement technique employed, e.g. photometry, fluorimetry or luminescence techniques, and on the assay system.

Table 6.10 Example of protocol for an enzymatic determination with internal standard

Pipette into cuvettes	Blank (ml)	Sample (ml)	Standard (ml)	Sample and standard (ml)
Buffer	1.000	1.000	1.000	1.000
Coenzyme(s)	0.100	0.100	0.100	0.100
Water	2.000	1.900	1.900	1.800
Sample solution	–	0.100	–	0.100
Standard solution	–	–	0.100	0.100
Auxiliary enzyme	0.020	0.020	0.020	0.020
Mix and measure the absorbances A_1. Start the reaction by the addition of				
Starting enzyme	0.020	0.020	0.020	0.020
Mix. After the reaction is complete measure absorbances A_2				

Note: The recovery is calculated from the absorbance differences only:

$$\text{Recovery} = \frac{\Delta A_{\text{standard}+\text{sample}} - \Delta A_{\text{sample}}}{\Delta A_{\text{standard}}} \times 100 \ (\%)$$

The sensitivity of a method is the ability of a method to differentiate significantly between small differences of concentrations or contents. The detection limit is the smallest single result which is clearly detectable and different from the background noise. The detection limit may be defined as three standard deviations of the appropriate blank value. For practical reasons it may also be of interest to know the lowest content or concentration of an analyte in a sample solution which can be determined with acceptable precision. In practice, an 'acceptable precision' may be defined as a single measurement whose deviation from the correct value is, for example, $\pm 10\%$. (The definition could be set by other analysts to $\pm 50\%$ dependent on the analytical problem.)

In food analysis photometric measurements are normally used. Fluorometric measurements may cause problems because of the type of sample material/ sample matrix available, and usually the contents of analytes in foodstuffs is so high that luminescence techniques are more sensitive than required. Also luminescence techniques are not easy to handle. The sensitivity of photometric enzymatic substrate estimations is dependent on the assay system (sample and total volumes), the molecular weight of the analyte and on the extinction coefficients employed.

In general, the pipetting schemes are designed so that 1.000 ml of buffer, approximately 2.000 ml of double distilled water and 0.100 ml of test solution are pipetted into the cuvette. If the concentration of the analyte is below a certain limit the sample volume can usually be increased up to 2.000 ml, with a corresponding reduction in the volume of water (Table 6.9, sample assay 3).

The sensitivity of enzymatic systems can be calculated from 5 to 10 milli absorbance units using the Beer Lambert law. The detection limits are calculated from 10 to 20 milli absorbance units, and results with an error of analysis of approx. $\pm 10\%$ in a single determination from 50 milli absorbance units.

The sensitivity can be increased by a factor of approx. 3 if NADH is reacted with iodo-nitro-tetrazolium chloride (INT) in the presence of diaphorase. The reaction produces a formazan which can be measured in the visible spectrum at 492 nm ($\varepsilon = 19.9 \ [l \times mmol^{-1} \times cm^{-1}]$).

Examples for measurements where a high sensitivity in enzymatic analysis is needed.
Succinic acid in whole egg indicates the activity of micro-organisms. The upper limit of 25 mg/kg dry weight, (i.e. approx. 6 mg/kg liquid whole egg) is regulated in Europe. The sample solution is prepared by CARREZ clarification of a relatively large sample (5 g liquid whole egg/25 ml = 200 g/l) and a sample volume of 1.000 ml is used in the assay. The absorbance difference obtained for a concentration of 25 ppm succinate is about 0.090 absorbance units.

D-3-Hydroxybutyric acid is found **in a fertilised chicken egg** after 6 days incubation. The upper limit of 10 mg/kg dry weight, or 2.5 mg/kg liquid whole egg, is also regulated in Europe. The sample preparation is similar to succinic acid and involves CARREZ clarification. The absorbance difference obtained

Table 6.11 Sensitivity detection limit and results with an error of ±10% for enzymatic methods

Analyte	Sensitivity in mg/l	Detection limit in mg/l	Error of analysis ±10% in mg/l
Acetaldehyde	0.25	1	2.5
Acetic acid	0.1	0.15	1
Ammonia	0.02	0.08	0.2
L-Ascorbic acid	0.1	0.3	1
Cholesterol	7	20	35
Citric acid	0.25	0.5	2.5
Ethanol	0.1	0.5	1
Formic acid	0.05	0.2	0.5
D-Fructose	0.2	0.4	2
D-Gluconic acid	0.25	0.5	2.5
D-Glucose	0.2	0.4	2
L-Glutamic acid	0.06	0.2	0.6
Glycerol	0.1	0.4	1
D-3-Hydroxybutyric acid	0.04	0.2	0.4
D-Isocitric acid	0.25	1	2.5
D-/L-Lactic acid	0.15	0.3	1.5
Lactose	2	7	20
D-Malic acid	0.2	0.35	2
L-Malic acid	0.25	0.5	2.5
Maltose	1	2	5
Nitrate	0.1	0.15	1
Oxalic acid	0.4	1.6	4
Raffinose	3	5	30
Sorbitol	0.1	0.2	1
Starch	0.6	1.2	6
Succinic acid	0.15	0.6	1.5
Sucrose	1	2	10
Sulphite	0.1	0.3	1
Urea	0.04	0.15	0.4

for the upper concentration limit (10 ppm), when the sample volume is 1.000 ml, is about 0.035 absorbance units. In this case the formazan is measured (see above).

D-Isocitric acid is found **in fruit juices** (e.g. citrus juice) in only low concentrations. The determination is performed to ascertain if fruit juice or fruit drinks have been adulterated. D-Isocitrate is precipitated as the barium salt. After resolubilization with sodium sulphate, 1.000 ml of the sample solution is used in the assay.

The measurement of **nitrate** is of interest because of its questionable effect on human health. The enzymatic determination of nitrate gives impeccable results even in trace level amounts, if some (basic) points are noted:

- use paper and membrane filters which are free of nitrate,
- use adsorbants which are nitrate free,
- use double distilled water which is free of nitrate,
- make a reagent mixture (buffer, coenzyme) and do not use one tablet for each individual cuvette. The tablets may vary slightly and this will influence the A_1 value of each cuvette,
- measure A_2 and A_3 at exactly the given time for each cuvette.

Sulphite, used as a food additive, has recently been widely discussed. Sulphite can be determined enzymatically in trace level amounts if:

- a reagent mixture of buffer and NADH-tablet is made,
- the production of standard and test solution is made using only distilled water,
- the creep reaction at A_2 is treated correctly. (Ingredients in spices can produce a sample-dependent creep reaction.) The absorbance A_2 should be measured 10 min after the addition of sulphite oxidase and measured again 10 min later. The measurements should be made every 10 min until a constant absorbance change (over 10 min) is noted. The absorbance A_2 is then determined by extrapolation back to the time of the addition of the sulphite oxidase.

For analysis of ingredients which determine the monetary value of a substance an error in analysis of (say) $\pm 1\%$ is expected or even demanded. For trace level analysis an error in analysis of $\pm 10\%$ or even higher is acceptable and is in most cases realistic. Enzymatic methods show, as a rule, an adequate sensitivity, this is due to the test system and the extinction coefficients employed.

6.8 Conclusion

Enzymatic analysis, in contrast to traditional chemical or 'modern' instrumental analysis, fulfils most of the requirements of an optimal analytical method. It offers a unique set of advantages such as specificity and accuracy of results, ease of sample preparation and simple performance of assays, high precision, sufficient sensitivity, low costs and the use of safe reagents. The enzymatic determination of substrates may be the only methodology that contains methods which are reference, common and fast methods at the same time.

Enzymatic methods are mostly reference methods: enzymatic methods give accurate and precise results; they are often published by legal bodies (food inspection), standardization institutes and by analytical committees. Enzymatic methods are quick, economic and need no special laboratories and equipment. This is in contrast to the assumption that the performance of reference methods is time consuming and needs 'expensive' equipment. Furthermore, determinations should be carried out only in 'special' laboratories and specially educated staff is needed for performance.

Enzymatic methods are common methods: a great number of samples can be analysed without any bigger problems. No special education of technicians is necessary, the equipment needed for analysis is generally available in the laboratories. Enzymatic methods are performed not only in the laboratories of food inspection bodies but also in many laboratories of small, medium-sized and large companies. (In the case of chemical analysis, common methods are often 'modified' or 'simplified' reference methods adapted to individual sample matrices.)

Enzymatic methods are fast methods: enzymatic determinations, sample preparation included is not time consuming with relation to working time. (During the incubation times other work can be done.) Results from enzymatic analysis are 'much better' (because they are accurate) than those from rapid methods/screening methods, for example, which may give either yes/no decisions or semi-quantitative results.

The principle of enzymatic analysis should be free from disadvantages, because the system of measurement has been used successfully by the living cell for millions of years. Enzymatic analysis is the transfer of a reaction from the living cell into a cuvette. The possible weaknesses of enzymatic analysis lie solely in its use, i.e. equipment, reagents and analyst.

Further reading

Bergmeyer, H.U. (1983–1986) *Methods of Enzymatic Analysis*, Verlag Chemie Weinheim, Deerield Beach/Florida, Basel.
Desphande, S.S. (1996) *Enzyme Immunoassays from Concept to Product Development*, Chapman & Hall, New York.
Fennema, O.R. (1985) *Food Chemistry*, 2nd edn, Marcel Dekker, New York and Basel.
Passonneau, J.V. and Lowry, O.H. (1993) *Enzymatic Analysis: a Practical Guide*, Humana Press, Totowa, New Jersey.
Stauffer, C.E. (1989) *Enzyme Assays for Food Scientists*, an AVI book, Van Nostrand Reinhold, New York.

7 DNA/PCR techniques
W.S. DAVIDSON

7.1 The problem: species identification of raw and processed meat

The keys to identifying different meat species normally rely heavily on morphological characteristics. However, when an animal has been killed for food or sport, these markers are usually destroyed or intentionally removed from the animal. This presents a problem for many government agencies who require a means of identifying the species origin of meat or other biological specimens. For example, food inspectors check meat for accurate labelling to prevent the substitution of cheaper meats (e.g. horse for beef, or turkey for chicken). Some religious organizations demand proof that their meat does not contain any pork. Wildlife officers must be able to identify the species type of meat to determine if an animal has been hunted illegally (i.e. without a permit, out of season, or incorrect sex taken). Customs officers and conservation groups require methods for identifying restricted and endangered species or products derived from them. Even law enforcement agencies must be able to tell if a bloodstain or body part comes from a human or an animal. In addition, with the introduction of new strains or breeds of animals through genetic engineering (i.e. transgenic animals) producers want to have a technique that can verify their own particular variety. There is therefore a need for rapid, accurate and reliable methods for determining the species origin of a biological sample. In this chapter I shall briefly review some of the traditional procedures that rely on the properties of proteins to authenticate the species origin of raw and processed meat products and then describe in depth recently developed procedures that take advantage of the resiliency of DNA, the genetic material, in the sample. Other methods that use DNA and the polymerase chain reaction (PCR) to determine the sex of a sample, or to identify the strain or variety of a specimen will also be included.

7.2 Protein-based methods

7.2.1 *Gel electrophoresis*

Electrophoretic techniques are probably the most commonly used method for analysing the protein components of meat and meat products (AOAC, 1984). They are particularly useful for fresh or freshly frozen meat. Water soluble proteins are extracted and separated according to their charge, size or isoelectric point on gels made from polyacrylamide, starch or agarose or on cellulose acetate strips. The proteins are visualized using a nonspecific protein stain such as Coomassie blue or a more sensitive silver staining procedure. In some instances specific proteins, such as creatine kinase, are identified on the basis of their enzymatic activity. This method has been used successfully to differentiate between meat from game and domestic species (McCormick et al., 1992) and to identify raw beef, pork, chicken and turkey meats (Kim and Shelef, 1986). Isoelectric focusing (IEF) gels of sarcoplasmic proteins from fish muscle revealed that this is a reliable method for species identification although some intra-specific variation was noted (Lundstrom, 1981). The Institut Francais de la Recherche pour l'Exploitation de la Mer (IFREMER) has produced a catalogue of 34 fish families that contains the scientific names of the fish and drawings and photographs of water soluble proteins separated by IEF (Durand et al., 1985). Despite the widespread use of IEF of muscle proteins for species identification, it should be noted that this procedure was unable to distinguish between the closely related bluefin tuna (a regulated species with tight quota restrictions) and yellowfin tuna (an unregulated species with no quota restrictions) (Bartlett and Davidson, 1991).

Proteins are denatured by heating during the cooking or processing of meat. For most practical purposes this limits protein electrophoresis for species identification to fresh or freshly frozen meat samples although heat-stable proteins (e.g. parvalbumins) can be extracted and analysed by IEF in the presence of a denaturant such as urea. Heat-denatured proteins or proteins that have been extensively modified during processing can be analysed by an examination of the peptide profiles produced by specific proteases or by chemical reagents that cleave the protein at a particular residue (Sotelo et al., 1993). The major drawback to protein electrophoresis is the need for standards with which to compare the unknown. A comparison with diagrams of standard electropherograms can provide an initial indication of what species might be candidates as the species of origin but confirmation of the identity of the actual species by electrophoretic analysis requires reference samples to be run beside the unknown on the same gel. Another complication is that meat from different muscle types in the same organism may have different protein components and this can lead to additional bands or inconsistent banding patterns. A comprehensive review of electrophoretic methods is provided by Andrews in Chapter 8 of this book.

7.2.2 Immunology

Immunological methods for the species identification of meat and meat products rely on the ability to raise antibodies against soluble proteins. Simple Ouchterlony analyses using polyclonal antibodies usually suffer from cross-reaction of the homologous proteins from closely related organisms. More elaborate methods like enzyme-linked immunosorbent assays (ELISA) could tell pork from beef or mutton but could not distinguish beef from mutton, or chicken from turkey (Berger et al., 1988). Species identification of autoclaved meat samples using antisera to thermostable muscle antigens in an enzyme immunoassay has been tried but without much success (Kang'ethe and Gathuma, 1987). Similarly, it was found that heating raw or cured meat greatly reduced the ability of competitive and indirect ELISA procedures to determine the species origin of meats (Dincer et al., 1987). Hitchcock and Crimes (1985) reviewed some of the immunological techniques available for meat species identification and in Chapter 9 of this book this has been updated by Märtlbauer.

The major problem with immunological procedures is one of sensitivity. High titre antibodies are required and these must be produced for each species of interest. Epitopes that are available on proteins isolated from fresh or freshly frozen meat are often lost upon heating and this means that a different set of antibodies must be prepared for heat-stable antigenic determinants. Although immunological procedures often give quick, reliable results, they do not appear to be practical for identifying the species origin of many forms of meat based food.

7.2.3 *Liquid chromatography*

Liquid chromatographic separation of water soluble muscle proteins was investigated as a means of speeding up the process of protein analysis with respect to electrophoretic techniques. Ashoor et al. (1988) developed a rapid procedure for identifying raw beef, pork, lamb, chicken, turkey and duck. Cuts of meat from the same species had similar profiles that differed quantitatively from one another whereas meat from different animal sources had qualitative differences. These latter differences allowed the ratios of chicken to turkey in unheated poultry mixtures to be determined (Ashoor and Osman, 1988). Liquid chromatography overcomes the need to run standards beside the sample and adds the capability of estimating the amount of a contaminant in a sample but this procedure still cannot be used routinely with heat-treated samples.

7.2.4 *Need for alternative approach*

The disadvantages of each of the methods outlined above are a consequence of their dependence on the ability to characterize proteins. Many proteins are heat-labile, lose their biological activity soon after death and are subject to modification in different cell types. Their presence in a sample is a function of the cell type being examined. Moreover, electrophoretic, immunological and chromatographic methods do not detect all the differences that may occur between two proteins. For example, electrophoresis will only detect changes in the net charge of two proteins, and immunological techniques require different antigenic determinants to be present on the proteins under investigation. The difference in the structure of an homologous protein from one species to another resides in the organism's genetic material, namely its DNA. It is therefore preferable to examine the DNA in a biological sample if one wishes to know its species origin.

DNA is the same in every cell type of an individual. Thus, it does not matter if one isolates DNA from a bloodstain, a muscle extract or a liver sample. The information content is also greater in isolated DNA than in proteins. Information is lost during the progression of transcription of DNA to messenger RNA and its subsequent translation into protein. DNA is a remarkably stable molecule (Paabo *et al.*, 1989). It has been isolated from animal skins from the last century (Higuchi *et al.*, 1984) and from a 7000-year-old human brain (Paabo *et al.*, 1988). As will be illustrated below, DNA can be isolated from meat products that have undergone the most intense processing (e.g. extensive cooking and then autoclaving after canning in the case of tuna). Heat is the scourge of the protein-based procedures because it denatures the molecule under investigation. In contrast heat denaturation of DNA into its two constituent strands is a prerequisite step in all processes that involve DNA analysis.

Having been convinced of the advantages of DNA over proteins for the species identification of meats, the question then becomes what segment of the genome will best serve this purpose. Genomes have regions that reflect their evolutionary history in that they are common to a broad range of related species. Other regions are more variable and taken together combine to give an individual-specific 'genetic fingerprint' (Jeffreys *et al.*, 1985). The trick is to select a segment that is sufficiently variable that it differs from species to species but not so variable that every individual is different at this genetic locus. The segment of choice should also be present in multiple copies in the cell so that small quantities of extractable DNA are more likely to contain it. To date, the majority of work that has used DNA to identify the species origin of food or other biological specimens has employed repetitive genomic DNA (e.g. maxi-satellites or ribosomal DNA) or else a region of the mitochondrial genome that is present in several thousand copies per cell. Although these are still the regions of choice for most techniques, the advent of PCR permits the investigation of sequences of DNA that occur only once per genome (i.e. single copy DNA).

7.3 DNA-based methods

7.3.1 *Hybridization*

The first method that used DNA to identify the species origin of cooked meat involved DNA hybridization in a procedure known as a slot blot. In this procedure, DNA is extracted from the sample and the strands are separated by heating or using sodium hydroxide to break the hydrogen bonds that hold together the bases on the complementary strands. The single-stranded, denatured DNA is then bound to a membrane made from nitrocellulose or a modified nylon material. Under appropriate conditions of salt concentration and temperature, the single-stranded DNA covalently bound to the membrane can interact with exogenous DNA that has the correct complementary sequence to form a renatured duplex or hybrid molecule. The exogenous DNA is called the probe and it can come from a variety of sources. The simplest probe is made from total genomic DNA isolated from a standard organism. More elaborate probes can be prepared by cloning fragments of an organism's DNA into plasmid vectors. The most specific probes are produced by chemically synthesizing oligonucleotides of defined sequence. Probes are labelled with a radioisotope (usually ^{32}P) or with a fluorescent tag or by incorporating biotinylated bases that can subsequently be recognized by alkaline phosphatase linked to streptavidin. The labelling of the probe allows detection of the hybrid DNA (probe bound to its target DNA on the membrane). A review of these basic molecular biology methods can be found in Sambrook et al. (1989).

Wintero et al. (1990) used total genomic pig DNA as a probe to detect increasing amounts of raw pork admixtured to beef. The minimum amount of pork that could be detected by DNA hybridization was 0.5% compared to 0.4–1% by immunological techniques and 5.0% by IEF. The authors also used a more specific probe consisting of a cloned 2700 base pair repetitive element from the pig genome. This element does not occur in the cow genome and thus it was not surprising that it gave more consistent results than the total genomic DNA probe. In a similar fashion, Chikuni et al. (1990) applied the DNA hybridization assay to the detection of species-specific DNA fragments in the cooked meats of chicken, pig, goat, sheep and beef. The probes were total genomic DNA from the species in question and they found that the probes for pig and chicken only hybridized to DNA isolated from their respective meats. However, the probes from the three ruminant species cross-reacted with one another but not with the DNA from pig or chicken. These results reflect the close evolutionary relationships of sheep and goat and of these to cow. This report also showed for the first time that canned meat products such as smoked chicken, Vienna sausage, and corned beef could yield sufficient DNA of a high enough quality for analysis. These observations were followed up by Ebbehoj and Thomsen (1991) who investigated the effect of time from slaughter on the degradation of DNA and also the effect of heating meat on the stability of DNA. Even after heating meat for 30 min at 120 °C, DNA in the size range of 300 to 1100 base pairs could be recovered. This was quite sufficient for quantitative hybridization to be performed and the lower limit on detection of pork in beef was estimated as 0.5%. This method was applied to commercial samples and the results confirmed the manufacturer's declarations.

These papers from the early 1990s provided evidence for the proof of concept that DNA hybridization could be used for identifying the species origin of cooked and processed meats. However, they also pointed out the problems of using total genomic DNA as the probe and that results on mixtures of closely related species would not be accurate with this type of probe and a slot blot approach. Blackett and Keim (1992) tried to overcome the problem of probe specificity by using highly repetitive regions of genomes to identify tissue samples from big game species. To prepare the probe, high molecular weight DNA was isolated from a particular species (e.g. antelope) and then digested with a series of restriction endonucleases. The digested DNA was separated by electrophoresis through low melting point agarose gels and stained with ethidium bromide. Repetitive elements are seen as bright bands above the background smear of genomic DNA when the gel is subjected to ultraviolet light. The bands were excised from the gel and the DNA in the band was radioactively labelled for hybridization studies. Probes from antelope, elk and mule deer were tested on DNA prepared from a variety of big game species that are commonly involved in poaching cases. Rather than use the probe directly on the total DNA isolated from the test animals, Blackett and Keim (1992) digested the DNA samples with the same restriction endonuclease that had been used to prepare the probe DNA, and then subjected the DNA to Southern blot analysis (Southern, 1975). Briefly, the fragments of DNA are separated by agarose gel electrophoresis, then denatured *in situ* using sodium hydroxide in a high salt solution. The single-stranded DNA is transferred to a membrane by capillary action and covalently attached to the membrane by exposure to ultraviolet light or by heating. At this point the hybridization is the same as described above for slot blots. An advantage of a Southern blot over a slot blot (when the DNA is not digested and separated before being bound to the membrane) is that even if the probe hybridizes to the DNA from more than one species, the banding patterns are likely to be different thus allowing species identification to be made. This is a useful procedure for wildlife forensic cases in which specimens may come from a variety of tissues. It is not appropriate for the food industry if cooked or canned meat is to be examined as the DNA will be denatured and partially degraded by the processing and thus refractory to restriction enzyme analysis.

7.3.2 *PCR direct sequence analysis (FINS)*

PCR has been called a 'surprisingly simple method for making unlimited copies of DNA fragments' (Mullis, 1990). There can be no doubt that the discovery and development of PCR revolutionized molecular biology and in many regards made DNA hybridization obsolete as an analytical tool. Yet the fundamental part of PCR relies on the specific hybridization of oligonucleotides of defined sequence to denatured, single-stranded template or target DNA. In order to carry out PCR one has to have some a priori knowledge of the nucleotide sequence of the DNA that is to be amplified. In particular, the sequences flanking the target must be known. This allows two oligonucleotide primers, approximately 20 to 30 nucleotides in length, that are complementary to the flanking regions of the target DNA to be chemically synthesized. After the DNA sample has been isolated, it is heated to approximately 95 °C to denature it and make it single-stranded. It is then allowed to cool to 45–60 °C in the presence of excess primers when these oligonucleotides hybridize to their complementary sequences such that they face each other on the two opposing strands. The reaction mixture also contains all four of the deoxyribonucleotide triphosphates, which are the building blocks of DNA, and a heat-stable DNA polymerase that catalyses the synthesis of new DNA by extending the primer and incorporating the appropriate base under the direction of the DNA template. The temperature of the reaction is raised to approximately 70 °C, which is the optimum temperature of the heat-stable polymerase, and the elongation step takes place. This process effectively doubles the segment of DNA that is targeted by the primers. At this point the cycle is repeated and the newly formed DNA, as well as the original sample, act as targets for the primers and templates for the synthesis of more DNA. As each cycle effectively doubles the amount of DNA specifically between the primers, ten cycles produce 2^{10} or a 1000-fold increase in the target DNA and 20 cycles will yield 10^6 copies. It is normal to carry out 30 to 40 cycles in a PCR reaction and this can provide enough DNA from a single human hair for forensic analysis (Higuchi et al., 1988).

Soon after the announcement of PCR (Saiki et al., 1988), methods became available for the determination of the nucleotide sequence of the amplified DNA (Gyllensten and Erlich, 1988). A comparison of known sequences of mitochondrial DNA from a variety of different animals revealed several highly conserved regions. This information enabled the construction of primers that could be used pairwise in combination with PCR to amplify homologous regions of the mitochondrial genome from representatives of all vertebrate classes (Kocher et al., 1989). This combination of PCR/direct sequence analysis permitted an examination of the dynamics of mitochondrial DNA evolution in animals and paved the way for the development of forensically informative nucleotide sequencing (FINS), a procedure for the identification of the species origin of biological samples (Bartlett and Davidson, 1992a, b).

The initial driving force behind the development of the FINS technique was the difficulty of enforcing the quotas and regulations governing the bluefin tuna fishery off the east coast of Canada. Four commercially important tuna species in the genus *Thunnus* are caught in the Atlantic near Nova Scotia. The harvest of bluefin (*T. thynnus*) is regulated whereas those of bigeye (*T. obesus*), yellowfin (*T. albacares*), and albacore (*T. alalunga*) are not. With the value of an individual bluefin tuna ranging from $1000 to $20 000 in 1990 on the Japanese sushi market, it was not surprising that a black market in illegally caught bluefin tuna arose, especially as it is very difficult to tell a small bluefin from the other three species (Mather, 1962). The usual characters used to differentiate among these species are skeletal and visceral in nature (Gibbs and Collette, 1966) and these identifying features are removed once the fish has been landed and gutted. Protein studies using IEF of water soluble muscle proteins revealed identical patterns for bluefin and yellowfin which differed from bigeye and albacore which were also identical (Bartlett and Davidson, 1991). Bartlett and Davidson (1991), therefore, turned to PCR/direct sequencing analysis of a segment of the mitochondrially encoded cytochrome b gene as a means of distinguishing these tuna species in the genus *Thunnus*. Although variation was observed within each of the four tuna species, distinct differences which characterize each of the species were found. These species-specific markers in the nucleotide sequence of the cytochrome b gene make it possible to use this procedure together with phylogenetic analysis to determine unambiguously the species identity of muscle, or other tissue, from an individual tuna. This method was adopted by Canada's Department of Fisheries and Oceans and, after a few test cases, fishermen were made aware of the procedure and the illegal fishing of bluefin tuna essentially stopped.

Bartlett and Davidson (1992a) quickly realized that FINS offered a novel approach that could rapidly, reliably, and reproducibly determine the species origin of a biological specimen when traditional protein-based methods are inadequate. FINS is composed of four steps (Fig. 7.1). First, DNA is extracted from the sample. Second, a specific segment of the isolated DNA is amplified using PCR and the universal cytochrome b primers of Kocher *et al.* (1989) or another pair designed specifically for the species of interest. The third step is the determination of the nucleotide sequence of the amplified DNA, and the fourth is a comparison of this sequence with standard sequences in a data base.

The isolation of DNA from processed meats is often not possible using standard procedures. These involve protease digestion in the presence of an anionic detergent followed by organic extraction and then ethanol precipitation

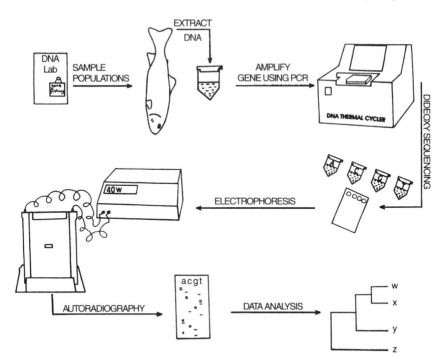

Figure 7.1 Schematic flow diagram of the FINS procedure. The four steps are: (1) isolate DNA; (2) amplify the desired region of DNA using PCR; (3) determine the nucleotide sequence of the segment that was amplified; and (4) compare the sequence with those in a database using a phylogenic approach.

of the DNA. The DNA in processed meats and other samples of interest to the food industry is usually degraded to less than 1000 nucleotides in length and because of the heat involved in the processing, it is likely to be denatured and in the single-stranded form. A method initially designed to isolate single-stranded RNA (Chomczynski and Sacchi, 1987) has been modified (Turner et al., 1989) and routinely yields amplifiable DNA from a wide variety of sources including smoked salmon and mackerel, pickled herring, salted cod, partially cooked chicken nuggets and battered cod, canned salmon and preserved skins (Bartlett and Davidson, 1992a).

Several factors have to be considered before a specific segment of DNA is selected for its ability to be used to identify the species origin of a sample. As described above, this region of the genome has to accept mutations quickly enough so that closely related organisms have different sequences but not so rapidly that the amount of variation within a species is substantial. The length of sequence must be long enough to permit the detection of sequence differences between congeneric species, but recognizing that the DNA in the sample will be degraded. A protein coding region of a gene is a good choice as amplification or sequencing errors are more likely to be detected by translating the nucleotide sequence and comparing the inferred amino acid sequence with known amino acid sequences of the gene. Highly conserved, invariant amino acids should be maintained. The creation of a database of sequences is a lengthy process for one lab and, therefore, it is desirable to take advantage of a gene whose sequence has been determined in many diverse organisms by many groups. The mitochondrially encoded cytochrome b satisfies these criteria admirably. Part of the nucleotide sequence of this gene has been determined for hundreds of vertebrates. (These sequences and those of other genes referred to in this chapter can be found by searching GenBank at http://www.ncbi.nlm.nih.gov.) The cytochrome b gene is very popular with population geneticists and molecular systematists and for this reason the amount of sequence information on this gene continues to expand. The pattern of nucleotide substitutions has been examined in mitochondrial genes encoding proteins and it has been shown that the cytochrome b gene follows the standard pattern with silent mutations predominating and transitions being more abundant than transversions (Thomas and Beckenbach, 1989; Irwin et al., 1991). Inter- and intraspecific studies have also shown that the amount of variation in cytochrome b within a species is less than between closely related species (McVeigh et al., 1991; Carr and Marshall, 1991). A mitochondrially encoded gene has the added advantage of behaving as if it were a haploid locus and therefore, the potential problem of heterozygosity is avoided.

Another major factor in favour of the cytochrome b gene is that primer design is not a problem. A 307 base pair segment of the cytochrome b gene corresponding to amino acids 33 to 134 can be amplified in all vertebrates tested to date using the primers described by Kocher et al. (1989). This latter point is very important as it allows one to amplify the DNA derived from a meat source without any prior knowledge of what the biological origin of the sample might be.

The final step involves comparing the sequence of the amplified DNA with sequences in a database to determine the closest relative (species). No single nucleotide sequence of the cytochrome b gene, or any other for that matter, is diagnostic of a particular species. This is because of the possibility of mutations and genetic variation within a species. This must be taken into account when techniques such as FINS are used because it is impossible to sample enough individuals from all the species that one may wish to have in a database to cover all the genetic variation within every species. A procedure that overcomes this potential problem involves phylogenetic analysis. This is a well-established field in molecular evolutionary genetics and several computer programs (e.g. PAUP, Swofford, 1989; PHYLIP, Felsenstein, 1989) are available. It is recommended that a search of the GenBank database using the BLASTN program (see www site listed above) be carried out first. If an exact match occurs between the sequence from the sample and one in the database, then there is no doubt about the identification of the species from which the sample came. If, on the other hand, the sequence of the species corresponding to the sample is not in the database, then a phylogenetic analysis will reveal the most closely related species. At this point it is advisable to expand the database by obtaining 'type specimens' for the closest relatives of these organisms. It has been our experience that the amount of intraspecific variation does not preclude the identification of even closely related species (Bartlett and Davidson, unpublished observations). As a rule of thumb, individuals whose cytochrome b sequences differ by less than 5% come from the same species.

FINS is a rapid, accurate and reproducible procedure that is based on established techniques. It is not subject to operator bias and can be performed independently in any lab equipped to carry out simple standard molecular biology. This means that there is no need for standards to be run alongside the sample. The method has been accepted by the court system in Canada and is the method of choice for Canada's Department of Fisheries and Oceans and the Canadian Wildlife Service if there is any question concerning the species identification of a biological sample.

Forrest and Carnegie (1994) used the FINS procedure to examine a supposed emu shish kebab purchased from a local restaurant. The cytochrome b sequence obtained from the 'emu' shish kebab bore little resemblance to that of authenticated emu (<70% identity) and a comparison of this sequence with cytochrome b sequences from the 14 species most commonly encountered in the meat trade in Australia indicated that the 'emu' shish kebab was most likely made from buffalo meat (>97% identity). The FINS process has also been used to determine the molecular genetic identification of whale meat and dolphin products from commercial markets in Korea and Japan (Baker and Palumbi, 1994; Baker et al., 1996). Instead of the cytochrome b gene, these researchers decided to use a 550 base pair fragment of the mitochondrial DNA control region. This is the most variable segment of the mitochondrial genome and the rationale for choosing this region rather than the cytochrome b gene was that it could potentially reveal the populations from which the samples were derived in addition to the species. This proved to be the case in some instances. A total of 34 sequences were obtained from 31 products and at least 8 species were represented. Some of the species identified in the products from the Japanese and Korean commercial markets were not consistent with recent catch records made available to the International Whaling Commission. The results of this survey raised questions concerning the legality of some of the products tested and point to the need for a standardized form of commercial labelling of whale products. It also suggested that the international and domestic management of whale products was not being practised as it should be. The FINS process is regarded as an important tool for monitoring the regulations covering the international trade in whale products and the domestic sale from unregulated local whaling or fisheries by-catch. As with the bluefin tuna problem that existed in Canada, the realization that there is now a reliable method for detecting cheating in the whaling industry might prove useful in conserving these mighty mammals of the sea.

The identification of the species in canned tuna presents a special problem. Unlike canned salmon which is usually smoked and then canned without much in the way of heat treatment, tuna is first cooked thoroughly and then autoclaved in the can. This leads to extensive degradation of the DNA in the sample and DNA isolated from commercial cans of tuna is no larger than 300 base pairs (Le Roux *et al.*, 1997) with the average size being about 100 base pairs (Unseld *et al.*, 1995). This precludes the use of the cytochrome b primers that flank the 307 base pair region of the cytochrome b gene for use in the FINS process with these samples. A comparison of cytochrome b sequences from a wide variety of tuna and tuna-like species (Block *et al.*, 1993) provided the information for the design of primers that amplified a 59 base pair fragment (123 base pairs including the primers) (Unseld *et al.*, 1995). This allowed the FINS procedure to be carried out on 30 commercially distributed cans of tuna and it was possible to determine a reliable sequence in each case. However, although this region of the cytochrome b gene provides unambiguous molecular markers for the identification of nine of the eleven tuna species commonly used by the canning industry, it was not able to distinguish between *Thunnus thynnus* (bluefin) and *Thunnus albacares* (yellowfin) tuna. This turns out to be a major impediment for the use of these primers as 5 of the 30 cans yielded the sequence corresponding to the indistinguishable species. Le Roux *et al.* (1997) independently solved the problem of working with highly degraded tuna DNA from cans. They designed two pairs of primers that amplify a 125 base pair fragment and a 101 base pair fragment, respectively. The information from both of these sequences is sufficient to identify unambiguously all the species that one might expect to encounter in cans of tuna. These primers were successful when tested with skipjack tuna and yellowfin tuna that had been processed to extreme conditions of sterilization. These minor modifications of the original FINS process illustrate the widespread utility of this procedure.

Having to sequence the amplified product as part of the FINS process to identify the species from which a sample is derived is both time consuming and expensive. It is, however, necessary if one wishes to obtain an unambiguous identification. Often it is sufficient to show the absence of a species in a product (e.g. no pork present). In other cases it is desirable to have methods that can be used as a rapid screen for detecting possible violations of regulations. The following sections illustrate techniques that have been developed to satisfy these demands.

7.3.3 Species specific primers

As has been demonstrated in the FINS process, the specificity afforded by the hybridization of two oligonucleotide primers to total extractable DNA allows PCR to amplify a particular region of a genome. By a judicious choice of primers, it should therefore be possible to design primers that are specific for one species and only that species. This twist on the use of universal primers has been used by Meyer *et al.* (1994) to detect pork in heated meat products. A comparison of the nucleotide sequences of the growth hormone genes from pig, sheep, goat, and cow permitted the synthesis of a pair of primers that can be used to amplify a 108 base pair segment from pig DNA but which yield no product when tested on the DNA from other species. As the absence of a product is taken as a positive result for no contamination of a sample by pork, it is necessary to have a control that can assess the potential of the extracted DNA from the sample to amplify in a PCR assay. A highly conserved 137 base pair fragment corresponding to part of the 18S ribosomal RNA gene provided this important control and it is worth noting that this fragment is larger than the diagnostic segment. This approach provides a rapid screening method for testing for the presence or absence of a specific DNA. The assay consists of carrying out the DNA extraction, the two PCR reactions, and then agarose gel electrophoresis to determine the presence or absence of any product and the size where appropriate. If a positive result is observed, the fragment can then be sequenced to confirm that it is indeed pork and not some other spurious contaminant. This method was able to detect pork in beef at levels below 2% and was substantially superior to available commercial kits for the immunological analysis of pork in beef mixtures. These kits were not able to detect levels of pork less than 20% in cooked meats or less than 10% in fresh meat (Meyer *et al.*, 1994).

A PCR assay has also been developed for detecting sheep and goat meats (Chikuni *et al.*, 1994). The species specific primers in this instance were designed to amplify a 374 base pair fragment from sheep satellite I, a repetitive element that accounts for approximately 12% of the sheep genome. Both sheep and goat DNA amplified to give this product but DNA from beef, buffalo, deer, horse, pig, rabbit and chicken yielded no result. The sequences of the sheep and goat amplification products were determined and it was found that there was an *Apa I* restriction enzyme recognition sequence in the sheep sequence that was absent from goat. Discrimination between the sheep and goat amplification products could be achieved by challenging them with this enzyme and separating the products on a 4% NuSieve agarose gel. The sheep fragment is cleaved to produce two fragments (236 and 138 base pairs) while the goat fragment remains intact. This combination of PCR amplification followed by restriction enzyme digestion is described in more detail in section 7.3.5.

A similar but more extensive approach has been taken to determine the species designations of commercially available lots of caviar (DeSalle and Birstein, 1996). The increasing demand for caviar has had a serious impact on the population sizes of commercial sturgeon species. The high cost of the product (>$2 per g) has increased the likelihood of poaching or the misrepresentation of one type of caviar for another. The traditional method for identifying the species of a particular shipment of caviar has relied upon such factors as egg size, appearance, smell, texture, and colour. De Salle and Birstein (1996) compared the sequences of portions of the cytochrome b genes and 12S and 16S ribosomal RNA genes from all extant sturgeons and paddlefishes and designed pairs of primers specific for diagnostic nucleotide substitutions for each of the three main commercial Russian sturgeon species: beluga sturgeon (*Huso huso*); Russian sturgeon (*Acipenser guelenstaedt*); and sevruga (*Acipenser stellatus*). They then surveyed 23 commercially available lots of caviar purchased in New York and found that five of them were misrepresentations. Three of these samples came from species that are threatened with extinction. The development of this procedure provides wholesalers with an alternative to the crude identification methods presently in use and should assist international conservation efforts to preserve what is left of commercial sturgeon populations. In addition, it should help protect the other species that are likely to be substituted as the commercial species disappear.

7.3.4 Sex specific primers

Sex is an important factor in establishing the commercial value of many animal carcasses. For example, meat from steers is usually more valuable than meat from cows of a similar age. EC regulations demand that carcasses be labelled both with the sex of the animal and its age at death. However, once the sexual morphological characters are removed from the animal it is practically impossible to tell the sex of the meat by a conventional examination. In mammals, sex is determined by the *Sry* gene that normally resides on the Y chromosome (Sinclair *et al.*, 1990; Gubbay *et al.*, 1990). The Y chromosome also contains repetitive DNA that is unique to this chromosome. Primers to either the *Sry* gene or to Y chromosome specific repetitive DNA can therefore be used to determine the sex of the animal from which a biological sample came. Tagliavini *et al.* (1993) examined the ability of two pairs of primers specific for bovine Y chromosome sequences to determine the sex of beef. Both of these primer pairs (BOV97M and BRY.1) gave amplification products of the expected size from DNA extracted from male bovine muscle but produced nothing from DNA derived from the meat of female bovine muscle. The BRY.1 primer pair that had been developed by Schroeder *et al.* (1990) gave the more reproducible results and it was recommended that this be used if the EC regulations regarding sex labelling had to be investigated. Primers related to those that amplify the BRY.1 repetitive DNA of the bovine genome also amplify Y chromosome specific sequences from male samples of cattle, buffalo, goat and sheep (Appa Rao *et al.*, 1995). The ability to determine the sex of big game carcasses is vitally important for the enforcement of permits that stipulate a male or female. Several labs have developed sex specific primers based on conserved sequences in *Sry* genes (e.g. Richard *et al.*, 1994) and these are being used for ecological studies and to enforce hunting regulations.

7.3.5 *Amplification of restriction site polymorphisms*

It has been shown above that a 374 base pair amplified region of satellite DNA that occurs in both the sheep and the goat genome could be distinguished by the presence or absence of an *Apa I* restriction site (Chikuni *et al.*, 1994). Restriction digests of PCR amplified DNA can provide another rapid screening procedure for species identification or even population identification. Murray *et al.* (1994) examined the mitochondrial D-loop regions of 15 species of ungulates using six restriction endonucleases that have four base pair recognition sequences. On average these enzymes are expected to cleave DNA once every 256 base pairs. The D-loop of mammals varies in length from 1100 to 1500 base pairs. The combination of restriction enzyme digestion patterns for each of the six enzymes and the variation in the length of the D-loop, allowed species specific sets of haplotypes to be developed. There seems to be sufficient species specific variation in the D-loop region of these ungulate species to establish the species origin of a haplotype obtained from a test sample. This adds another simple test for the analysis of big game meat that may be seized during a raid in a poaching case.

7.3.6 *PCR-single strand conformational polymorphisms (PCR-SSCP)*

PCR-single strand conformational polymorphism (PCR-SSCP) is a variation on the theme of IEF that uses PCR products for the rapid species identification of meats. It is based on the observation of Orita *et al.* (1989) that single-stranded DNAs that differ by one or more nucleotide substitutions have different mobilities when separated by neutral polyacrylamide gel electrophoresis. Hara *et al.* (1994) developed this assay for many species of fish. In their protocol, DNA is extracted from the sample and a fragment of approximately 110 base pairs is amplified using a pair of primers that correspond to highly conserved regions of the 28S ribosomal RNA gene. The region between the primers was known to be highly variable between species, both in sequence and in length (Naito *et al.*, 1992), and this provides the resolving power of the system to discriminate between species. The amplified fragment is denatured by heating in a formamide solution or by using strong base and then applied to a 12% native polyacrylamide gel. After electrophoresis, the gel is silver-stained and the resulting banding patterns are diagnostic of different species. Twelve taxa of fish were examined by this technique and the indications were that this method could be used for species typing using small samples such as eggs or larvae.

PCR-SSCP has also been used for fish species identification of canned tuna (Rehbein *et al.*, 1995). Two different regions of the cytochrome b gene were selected for amplification, the 123 base pair region described by Unseld *et al.* (1995) and another region that gave a product of 148 base pairs. Separation of the denatured 123 base pair fragment on 7.5 or 10% polyacrylamide gels revealed species specific patterns for six different tunas and bonitos which were different from the pattern for bluefin and yellowfin tuna which, as expected, were identical. The universal application of PCR-SSCP has been demonstrated by Pharmacia Biotech (Application Note 18-1108-90). In this report it was shown that the 307 base pair segment of the cytochrome b gene can be used to resolve deer species, poultry, pork, beef and lamb, and fish species. PCR-SSCP provides a rapid screening method for the analysis of the species origin of a biological sample. The main disadvantage of this technique is that, like IEF, it is necessary to run reference standards and samples side by side on the same gel. The information content of a PCR-SSCP result is obviously less than that of the FINS procedure. However, it is possible to cut out bands from the PCR-SSCP gel and determine the nucleotide sequence of them after reamplification. By this means the two methods complement one another. PCR-SSCP can also indicate that a sample contains mixtures from different organisms. Here again species identification would be possible by examining the DNA sequences of the multiple bands.

7.3.7 *Arbitrary primers (AP-PCR and RAPD)*

In 1990 two groups independently discovered that PCR would work if only one primer was added to the amplification reaction. The rationale for this result is that the single primer is binding to identical sequences that happen to be facing one another on opposite strands of the DNA. The primers used in this procedure tend to be smaller (often only ten nucleotides) than those usually used in primer pair PCR reactions. This procedure was termed arbitrarily primed PCR (AP-PCR) by Welsh and McClelland (1990) and randomly amplified polymorphic DNA (RAPD) by Williams *et al.* (1990). It is the latter terminology that has generally been adopted and 'RAPDs' are now accepted as genetic markers. RAPD analysis requires no a priori knowledge of the sequence of DNA that is amplified. In fact, it is by chance alone that a segment of DNA is amplifiable in a RAPD reaction. This means that many primers can be synthesized and tested to see if they produce any products. This provides a very quick and inexpensive means of screening organisms for genetic variation. The technique is expected to sample evenly throughout the genome and variant alleles are identified by the presence or absence of a PCR product, or a change in its size, as determined by agarose gel electrophoresis followed by staining the gel with ethidium bromide. The absence of a band on a gel where one is expected is explained by a change in the sequence of one of the priming sites such that the primer no longer binds to it. Allelic bands of different size reflect insertions or deletions between the priming sites. RAPD markers are dominant and this leads to one of the drawbacks of this technique. The presence of a band on a gel is observed in samples from either a homozygote (present/present) or a heterozygote (present/absent) individual in diploid organisms. Therefore, this procedure is particularly useful for distinguishing species and strains of haploid micro-organisms or for discriminating between highly inbred varieties of plants.

RAPD analysis has been used to examine strains of micro-organisms differing in their pathogenicity. Welsh and McClelland (1990) first demonstrated the generality of the method by its ability to distinguish 24 strains from five species of *Staphylococcus*. They also showed that this method could discriminate between three varieties of rice. RAPDs were used to determine the parentage in maize hybrids and this suggested its application for constructing genetic maps (Welsh *et al.*, 1991). Williams *et al.* (1990) originally verified the usefulness of the RAPD method by adding genetic markers to the soybean genomic map. Since then, RAPD markers have become the preferred route for plant genome mapping. For example, 151 primers produced a total of 558 markers in *Eucalyptus* sp. and, using a 'two-way pseudo-testcross' mapping strategy, approximately 250 loci were mapped in *Eucalyptus grandis* and *Eucalyptus urphylla* (Grattapaglia and Sederoff, 1994).

Another application of RAPD analysis is the development of sex specific probes for non-mammalian species. It is often very difficult to determine the sex of young immature birds and some species are sexually monomorphic even as adults. RAPD markers specific to the W chromosome can be identified by employing a technique known as bulked segregant analysis. Separate pools of DNA are prepared from a group of males and a group of females and the pooled DNAs are screened for RAPDs that appear in the female pool but are absent from the pooled male DNA (females being the heterogametic sex in birds). This method was used successfully to isolate a RAPD that was linked to the sex of individual roseate terns (Sabo *et al.*, 1994). The RAPD fragment was subsequently cloned and its sequence was determined. This information allowed a pair of primers to be synthesized that were longer and more specific than the short RAPD primer. The combination of arbitrary primers and PCR has become one of the most common approaches for searching for novel genetic markers. RAPD analysis is extremely quick and easy and it should be considered as the initial step in such projects.

7.3.8 Microsatellites

No description of PCR methods and applications, especially their use for authentication purposes, would be complete without mention of microsatellites. It is beyond the scope of this chapter to do proper justice to the impact that microsatellite analysis has had on the mapping of the human genome (diagnosis of human diseases), forensic science (individual identification), and agriculture (selective breeding programmes). However, I shall endeavour to give a flavour of some of the remarkable progress that has been achieved in a very short period of time.

Microsatellites are tandemly repeated, short (two to five base pairs) simple sequences that are widely dispersed throughout the genomes of all eukaryotes (Weber and May, 1989). Their function, if indeed there is one, is still not fully understood although they have been associated with the regulation of gene expression in some instances. They are usually less than 300 base pairs in length. Variability in the length of a microsatellite is common and tends to be correlated with the number of repeats at a particular locus. Changes in length result from the loss or gain of one or more of the units in the repeated sequence. As genetic markers, microsatellites are both efficient and informative. Mutations in repeat number can be detected as size differences in the PCR products of loci which are amplified using oligonucleotide primers specific to the non-repetitive flanking regions of the microsatellite. Amplification products are typically separated according to size by electrophoresis through polyacrylamide gels under denaturing conditions and are detected by having one of the primers labelled with ^{32}P or with a fluorescent tag.

The discovery of microsatellites has led to the rapid refinement and development of genetic maps in many diverse organisms. Thousands of microsatellite loci have now been mapped in the human genome and they are used every day to find the location of genes that are associated with genetic diseases. Microsatellites are being used extensively to produce low resolution genetic maps of economically important animals such as sheep (Crawford *et al.*, 1995), cow (Barendse *et al.*, 1994), pig (Rohrer *et al.*, 1994), and chicken (Cheng *et al.*, 1995). The driving force behind the production of these genetic maps is to improve the effectiveness of selective breeding programmes through marker assisted selection of desired traits and the mapping of quantitative trait loci (QTLs) (Whittaker *et al.*, 1995). This involves identifying associations between variable genetic markers and complex phenotypic traits such as growth rate, disease resistance, fat content, etc. The results of this type of work will become evident in the next century when it will only be with racehorses that one breeds the best with best and hopes for the best. In other domesticated animals it should be possible to give reliable predictions concerning the presence of particular production traits.

Microsatellites have also become the markers of choice for forensic work for individual identification (Fregeau and Fourney, 1993). The amount of polymorphism observed at microsatellite loci and the small amount of genetic material that is required to carry out the analysis has allowed criminal investigations to proceed when even a few years ago there was no method for gathering the information or evidence required to solve the case. For example, it is possible to isolate enough DNA from the saliva left on the back of a stamp (e.g. on a ransom letter or a letter bomb) to carry out PCR to obtain a genetic profile of a suspect using microsatellite markers. As the PCR products of microsatellites are not large, DNA that has been degraded by exposure to the elements can be examined by microsatellite analysis whereas 'DNA fingerprinting' using minisatellite markers requires high molecular weight DNA and many orders of magnitude more material (Jeffreys et al., 1985). The resolving power of microsatellites for forensic work is not restricted to humans. Poaching cases often revolve around the ability to match samples taken from a suspect with those from a 'kill site' in the field. Microsatellites are available for a growing number of big game animals and population data bases are being developed in many labs (see for example Paetkau and Strobeck, 1994). Microsatellites are particularly useful in this respect as the DNA isolated from tissue at a 'kill site' is often degraded and samples from a suspect may consist of a few flakes of blood on the blade of a knife. We have entered a new era when enforcement officers are conversant with the term PCR and they appreciate what these techniques have to offer.

7.4 Summary and future trends

There are three major questions that are of interest to the food industry when it comes to the authentication of meat and meat products:

1. what is the species origin of the meat;
2. what was the sex of the animal; and
3. was it a particular variety or strain?

These can all be answered by variations on existing methods that use PCR. It is simply a matter of choosing the most appropriate type of genetic marker and finding it for the species in question. There are, however, some basic analytical tools that it would be very useful to develop and some codes of practice that could be adopted. For example, sex specific markers have been developed for many domesticated mammals but the same is not true to the same extent for poultry. Connoisseurs of paté de foie gras want to be assured that the paté was made with goose not pork liver and some even go so far as to have a preference for liver from the gander rather than from the female of the species. It would be relatively easy to develop a sex specific marker for every bird species on a case by case basis but a universal avian sex marker would be very desirable and far more cost effective for regulatory agencies. For food producers, it is important to be able to confirm that the semen purchased for artificially inseminating the herd is authentic. In the future the label on this product will undoubtedly reflect the genotype of the donor and this will make it easy to spot check with microsatellite markers. Transgenic fruit and vegetables are already in the market place and some jurisdictions require that they be identified whereas others do not. Some consumers wish this information and others have no preference. As transgenic animals (e.g. those containing growth hormone genes) make their appearance, it is likely that this will be of more concern to the consumer. Producers of transgenic organisms, whether they be animal, vegetable or fruit, should be required to have a marker that can easily be identified by a PCR method to verify their product. An interesting aspect of authentication is whether a product was prepared at a particular plant on a given day. It has been suggested that food processors incorporate different oligonucleotides into the ink that is used on their date and place stamps. These oligonucleotide markers could then be 'retrieved' by extracting nucleic acid from the label and applying a PCR type of method to 'read' the card. PCR/DNA techniques are relatively inexpensive, rapid and sensitive. Most government labs charged with the task of enforcing health and safety regulations use them already as do law enforcement agencies. Many of the techniques lend themselves to automation and this will increase the throughput without increasing the cost of testing significantly. PCR/DNA procedures will be the norm in the next century.

Acknowledgements

None of this would have been possible without the support and assistance of my colleague Sylvia Bartlett. She has been the driving force behind the development of the FINS procedure and is constantly examining novel applications of PCR for species identification and general application to the food industry and enforcement agencies. I also thank Colin McGowan for many valuable discussions, particularly concerning RAPDs and microsatellites.

References

AOAC (1984) *Official Methods of Analysis*, 14th edn, 18.108–18.113, p. 349, Association of Official Analytical Chemists, Arlington, VA.
Appa Rao, K.B.C., Kesava Rao, V., Kowale, B.N. and Totey, S.M. (1995) Sex-specific identification of raw meat from cattle, buffalo, sheep and goat. *Meat Sci.*, **39**, 133–26.
Ashoor, S.H., Monte, W.C. and Stiles, P.G. (1988) Liquid chromatographic identification of meats. *J. AOAC*, **71**, 396–401.
Ashoor, S.H. and Osman, M.A. (1988) Liquid chromatographic quantitation of chicken and turkey in unheated chicken-turkey mixtures. *J. AOAC*, **71**, 401–3.
Baker, C.S., Cipriano, F. and Palumbi, S.R. (1996) Molecular genetic identification of whale and dolphin products from commercial markets in Korea and Japan. *Mol. Ecol.*, **5**, 671–85.
Baker, C.S. and Palumbi, S.R. (1994) Which whales are hunted? A molecular genetic approach to monitoring whaling. *Science*, **265**, 1538–9.
Barendse, W., Armitage, S.M., Kossarek, I.M. *et al.* (1994) A genetic linkage map of the bovine genome. *Nature Genet.*, **6**, 227–35.

Bartlett, S.E. and Davidson, W.S. (1991) Identification of *Thunnus* tuna species by the polymerase chain reaction and direct sequence analysis of their mitochondrial cytochrome b genes. *Can. J. Fish. Aquat. Sci.*, **48**, 309–17.

Bartlett, S.E. and Davidson, W.S. (1992a) FINS (Forensically Informative Nucleotide Sequencing): A procedure for identifying the animal origin of biological specimens. *BioTechniques*, **12**, 408–11.

Bartlett, S.E. and Davidson, W.S. (1992b) Erratum. *BioTechniques*, **13**, 14.

Berger, R.G., Mageau, R.P., Schwab, B. and Johnston, R.W. (1988) Detection of poultry and pork in cooked and canned meat foods by enzyme-linked immunosorbent assays. *J. AOAC*, **71**, 406–9.

Blackett, R.S. and Keim, P. (1992) Big game species identification by deoxyribonucleic acid (DNA) probes. *J. Forensic Sci.*, **37**, 590–6.

Block, B.A., Finnerty, J.R., Steward, A.F.R. and Kidd, J. (1993) Evolution of endothermy in fish: mapping physiological traits on a molecular phylogeny. *Science*, **269**, 210–14.

Carr, S.M. and Marshall, D. (1991) Detection of intra-specific DNA sequence variation in the mitochondrial cytochrome b gene of Atlantic cod (*Gadus morhua*) by the polymerase chain reaction. *Can. J. Fish. Aquat. Sci.*, **48**, 48–52.

Cheng, H.H., Levin, I., Vallejo, R.L. *et al.* (1995) Development of a genetic map of the chicken with markers of high utility. *Poultry Sci.*, **74**, 1855–74.

Chikuni, K., Ozutsumi, K., Koishikawa, T. and Kato, S. (1990) Species identification of cooked meats by DNA hybridization assay. *Meat Sci.*, **27**, 119–128.

Chikuni, K., Tabata, T., Kosugiyama, M. and Monma, M. (1994) Polymerase chain reaction assay for detection of sheep and goat meats. *Meat Sci.*, **37**, 337–45.

Chomczynski, P. and Sacchi, N. (1987) Single step method of RNA isolation by acid guanidinium thiocyanate-phenol-chloroform extraction. *Anal. Biochem.*, **162**, 156–9.

Crawford, A.M., Dodds, K.G., Ede, A.J. *et al.* (1995) An autosomal genetic linkage map of the sheep genome. *Genetics*, **140**, 703–24.

DeSalle, R. and Birstein, V.J. (1996) PCR identification of black caviar. *Nature*, **381**, 197–8.

Dincer, B., Spencer, J.L., Cassens, R.G. and Greaser, M.L. (1987) The effects of curing and cooking on the detection of species origin of meat products by competitive and indirect ELISA techniques. *Meat Sci.*, **20**, 253–65.

Durand, P., Laudrein, A. and Quero, J-C. (1985) *Catalogue Electrophoretique des Poissons Commerciaux*, l'Institut Francais de la Recherche pour l'Exploitation de la Mer, Nantes, France.

Ebbehoj, K.F. and Thomsen, P.D. (1991) Species differentiation of heated meat products by DNA hybridization. *Meat Sci.*, **30**, 221–34.

Felsenstein, J. (1989) PHYLIP (Phylogeny Inference Package) Version 3.5c. © Copyright 1986–1993 by Joseph Felsenstein and the University of Washington.

Forrest, A.R.R. and Carnegie, P.R. (1994) Identification of gourmet meat using FINS (Forensically Informative Nucleotide Sequencing). *BioTechniques*, **17**, 24–5.

Fregeau, C.J. and Fourney, R.M. (1993) DNA typing with fluorescently tagged short tandem repeats: a sensitive and accurate approach to human identification. *BioTechniques*, **15**, 100–19.

Gibbs, R.H. and Collette, B.B. (1966) Comparative anatomy and systematics of the tunas, genus *Thunnus*. *Fish Wildl. Ser. Fish. Bull.*, **66**, 65–130.

Grattapaglia, D. and Sederoff, R. (1994) Genetic linkage maps of *Eucalyptus grandis* and *Eucalyptus urophylla* using a pseudo-testcross: mapping strategy and RAPD markers. *Genetics*, **137**, 1121–37.

Gubbay, J., Collignon, J., Koopman, P. *et al.* (1990) A gene mapping to the sex-determining region of the mouse Y chromosome is a member of a novel family of embryonically expressed genes. *Nature*, **346**, 245–50.

Gyllensten, U.B. and Erlich, H.A. (1988) Generation of single-stranded DNA by the polymerase chain reaction and its application to direct sequencing of the HLA-DQA locus. *Proc. Natl. Acad. Sci. USA*, **85**, 7652–6.

Hara, M., Noguchi, M., Naito, E. *et al.* (1994) Ribosomal RNA gene typing of fish genome using PCR-SSCP method. *Bull. Jap. Sea Natl. Fish. Res. Inst.*, **44**, 131–8.

Higuchi, R., Beroldingen, C.H., Sensabaugh, G.F. and Erlich, H.A. (1988) DNA typing from single hairs. *Nature*, **332**, 543–6.

Higuchi, R.G., Bowman, B., Freiberg, M. *et al.* (1984) DNA sequences from the quagga, an extinct member of the horse family. *Nature*, **312**, 282–4.

Hitchcock, C.H.S. and Crimes, A.A. (1985) Methodology for meat species identification: a review. *Meat Sci.*, **15**, 215–24.
Irwin, D.M., Kocher, T.D. and Wilson, A.C. (1991) Evolution of the cytochrome b gene of mammals. *J. Mol. Evol.*, **32**, 128–44.
Jeffreys, A.J., Wilson, V. and Thein, S.L. (1985) Individual specific fingerprints of human DNA. *Nature*, **316**, 76–9.
Kang'ethe, E.K. and Gathuma, J.M. (1987) Species identification of autoclaved meat samples using antisera to thermostable muscle antigens in enzyme immunoassay. *Meat Sci.*, **19**, 265–70.
Kim, H. and Shelaf, L.A. (1986) Characterization and identification of raw beef, pork, chicken and turkey meats by electrophoretic patterns of their sarcoplasmic proteins. *J. Food Sci.*, **51**, 731–5.
Kocher, T.D., Thomas, K.K., Meyer, A. *et al.* (1989) Dynamics of mitochondrial DNA evolution in animals: Amplification and sequencing with conserved primers. *Proc. Natl. Acad. Sci. USA*, **86**, 6196–200.
Le Roux, M-G., Pascal, O., Lostanlen, A. *et al.* (1997) Identification of tuna species in preserved canned samples. *J. Sci. Food Agric.* (In press).
Lundstrom, R.C. (1981) Rapid fish species identification by agarose gel isoelectric focusing of sarcoplasmic proteins. *J. AOAC*, **64**, 38–43.
Mather, F.J. (1962) Tunas (genus *Thunnus*) of the western North Atlantic, in Symposium on Scombroid Fishes. Part 1. Marine Biology Association of India, Mondapan Camp, pp. 395–410.
McCormick, R.J., Collins, D.A., Field, R.A. and Moore, T.D. (1992) Identification of meat from game and domestic species. *J. Food Sci.*, **57**, 516–20.
McVeigh, H.P., Bartlett, S.E. and Davidson, W.S. (1991) Polymerase chain reaction/direct sequence analysis of the cytochrome b gene in Salmo salar. *Aquaculture*, **95**, 225–33.
Meyer, R., Candrian, U. and Luthy, J. (1994) Detection of pork in heated meat products by the polymerase chain reaction. *J. AOAC Int.*, **77**, 617–22.
Mullis, K.B. (1990) The unusual origin of the polymerase chain reaction. *Sci. American*, April, 56–63.
Murray, B.W., McClymont, R.A. and Strobeck, C. (1995) Forensic identification of ungulate species using restriction digests of PCR-amplified mitochondrial DNA. *J. Forensic Sci.*, **40**, 943–51.
Naito, E., Dewa, K., Ymanouchi, M.D. and Kominami, R. (1992) Ribosomal ribonucleic acid (rRNA) gene typing for species identification. *J. Forensic Sci.*, **37**, 396–403.
Orita, M., Iwahana, H., Kanazawa, H. *et al.* (1989) Detection of polymorphisms of human DNA by gel electrophoresis as single-strand conformational polymorphisms. *Proc. Natl Acad. Sci. USA*, **86**, 2766–70.
Paabo, S., Gifford, J.A. and Wilson, A.C. (1988) Mitochondrial DNA sequences from a 7000 year old brain. *Nucleic Acids Res.*, **20**, 9775–87.
Paabo, S., Higuchi, R.G. and Wilson, A.C. (1989) Ancient DNA and the polymerase chain reaction. *J. Biol. Chem.*, **264**, 9709–12.
Paetkau, D. and Strobeck, C. (1994) Microsatellite analysis of genetic variation in black bear populations. *Mol. Ecol.*, **3**, 489–95.
Rehbein, H., Mackie, I.M., Pryde, S. *et al.* (1995) Fish species identification in canned tuna by DNA analysis (PCR-SSCP). *Inf. Fischwirtsch.*, **42**, 209–12.
Richard, K.R., McCarrey, S.W. and Wright, J.M. (1994) DNA sequence from the *SRY* gene of the sperm whale (*Physeter macrocephalus*) for use in molecular sexing. *Can. J. Zool.*, **72**, 873–7.
Rohrer, G.A.L., Alexander, L.J., Keele, J.W. *et al.* (1994) A microsatellite linkage map of the porcine genome. *Genetics*, **136**, 231–45.
Sabo, T.J., Kesseli, R., Halverson, J.L. *et al.* (1994) PCR-based method for sexing roseate terns (*Sterna dougallii*). *The Auk*, **111**, 1023–7.
Saiki, R.K., Gelfand, D.H., Stoffel, S. *et al.* (1988) Primer-directed enzymatic amplification of DNA with a thermostable DNA polymerase. *Science*, **239**, 487–91.
Sambrook, J., Fritsch, F.E. and Maniatis, T. (1989) *Molecular Cloning: a Lab Manual*, 2nd edn, Cold Spring Harbor Press, CSH.
Schroeder, A., Miller, J.R., Thomsen, P.D. *et al.* (1990) Sex determination of bovine embryos using polymerase chain reaction. *Animal Biotech.*, **1**, 121–8.
Sinclair, A.H., Berta, P., Palmer, M.S. *et al.* (1990) A gene from the human sex-determining region encodes a protein with homology to a conserved DNA-binding motif. *Nature*, **346**, 240–4.
Sotelo, C.G., Pineiro, C., Gallardo, J.M. and Perez-Martin, R.I. (1993) Fish species identification in seafood products. *Trends Food Sci. Tech.*, **4**, 388–421.

Southern, E.M. (1975) Detection of specific sequences among DNA fragments separated by gel electrophoresis. *J. Mol. Biol.*, **98**, 503–17.
Swofford, D.S. (1989) PAUP: phylogenetic analysis using parsimony, Version 3.0b. Illinois Natural History Survey, Champaign, IL.
Tagliavini, J., Bolchi, A., Bracchi, P.G. and Ottonello, S. (1993) Sex determination on samples of bovine meat by polymerase chain reaction. *J. Food Sci.*, **58**, 237–8.
Thomas, W.K. and Beckenbach, A.T. (1989) Variation in salmonid mitochondrial DNA: evolutionary constraints and mechanism of substitution. *J. Mol. Evol.*, **9**, 233–45.
Turner, B.J., Elder, J.E. and Laughlin, T.F. (1989) DNA fingerprinting of fishes: a general method using oligonucleotide probes. *Fingerprint News*, **4**, 15–16.
Unseld, M., Beyermann, B., Brandt, P. and Hiesel, R. (1995) Identification of the species origin of highly processed meat products by mitochondrial DNA sequences. *PCR Methods and Applications*, **4**, 241–3.
Weber, J.L. and May, P.F. (1989) Abundant class of human DNA polymorphisms which can be typed using the polymerase chain reaction. *Am. J. Hum. Genet.*, **44**, 388–96.
Welsh, J., Honeycutt, R.J., McClelland, M. and Sobral, B.W.S. (1991) Parentage determination in maize hybrids using the arbitrarily primed polymerase chain reaction (AP-PCR). *Theor. Appl. Genet.*, **82**, 473–6.
Welsh, J. and McClelland, M. (1990) Fingerprinting genomes using PCR with arbitrary primers. *Nucleic Acids Res.*, **18**, 7213–18.
Whittaker, J.C., Curnow, R.N., Haley, C.S. and Thompson, R. (1995) Using marker-maps in marker-assisted selection. *Genet. Res. Camb.*, **66**, 255–65.
Williams, J.G.K., Kubelik, A.R., Livak, K.J. *et al.* (1990) DNA polymorphisms amplified by arbitrary primers are useful as genetic markers. *Nucleic Acids Res.*, **18**, 6531–5.
Wintero, A.K., Thomsen, P.D. and Davies, W. (1990) A comparison of DNA-hybridization, immunodiffusion, countercurrent immunoelectrophoresis and isoelectric focusing for detecting the admixture of pork to beef. *Meat Sci.*, **27**, 75–85.

8 Electrophoretic methods
A.T. ANDREWS

8.1 Introduction

The term 'electrophoretic methods' covers a very wide range of techniques that can be used for the analysis of complex mixtures and the purification of small amounts of material. Under appropriate conditions electrophoresis can be used to study almost any class of molecules ranging in size from simple inorganic salts to nucleic acids, viruses and even quite large particles and aggregates. The word electrophoresis describes the movement of a charged particle under the influence of an electrical field, the motion being the product of the charge on the particle and the field strength. Under conditions of constant velocity this driving force is balanced by the frictional resistance of the medium.

There are in essence only four groups of parameters that can be used to separate molecules from one another for either analytical or preparative purposes; they are differences in molecular size, charge or hydrophobicity and specific interactions with other molecules. Clearly electrophoretic separations can exploit differences in charge and also in size since this will influence frictional resistance and in turn this influences mobility, but under appropriate conditions hydrophobicity differences are also revealed by a technique referred to as charge shift electrophoresis, while the various forms of affinity electrophoresis and immunoelectrophoresis are examples of specific molecular interactions. There are thus electrophoretic methods that can make use of any of the possible separation parameters, or indeed combinations of them, so it can be seen that there is a considerable range of techniques, much broader in scope than can be covered in the confines of a single chapter. It is probably no exaggeration to say that well over half of all published papers in the fields of biochemical research or analysis and related areas include some electrophoretic methodology. There are two international journals (*Electrophoresis* and *Theoretical and Applied Electrophoresis*) that are devoted exclusively to methodological development and applications, and of course many comprehensive chapters in books and complete books on the subject (e.g. Allen and Maurer, 1974; Smith, 1975; Gordon, 1975; Righetti, 1983; Andrews, 1986; Dunbar, 1987; Hames and Rickwood, 1990).

Nearly 200 years ago, the first form of electrophoresis involved observing through a microscope the motion of charged particles moving in an electric field, and in terms of technique little changed for the next 140 years or so, until the use of anticonvective media was introduced to prevent remixing of the separated zones under the influence of the heat inevitably generated by the passage of the electric current. Coupled with the development of better ways of detecting the separated zones, this opened the door to the wide range of applications that we have today. Early separations on a paper matrix, then cellulose acetate strips and in starch gels have given way to polyacrylamide as an anticonvective medium with excellent and reproducible properties that can be varied in a defined manner to optimize gel properties for a given separation problem. The molecules to be separated must pass through the pores of the gel, and if these are similar to or smaller than the dimensions of the sample molecules little progress will be made; for very large molecules this is the case with polyacrylamide gels. Fortunately agarose gels can then be used and extend the molecular size range that can be studied to include, for example, large DNA molecules. Although a natural polymer, agarose is now available commercially in a refined state that also is capable of giving gels with highly reproducible properties. The most recent development in electrophoretic methods is almost a case of reverse evolution in that there may be no anticonvective medium at all, but by performing the separation at very high voltages in only a few minutes in very narrow bore capillary tubes (which have excellent heat-dissipation properties) both convective mixing and diffusion spreading of separated zones are minimal.

Although there are many different electrophoretic methods geared to particular problems or apparatus, this chapter will cover only those methods currently most widely used, namely polyacrylamide gel electrophoresis in the absence or presence of detergents, agarose gel electrophoresis, isoelectric focusing, immunoelectrophoresis and capillary methods.

8.2 Gel-based methods

8.2.1 *Polyacrylamide gel electrophoresis (PAGE)*

Acrylamide is a low molecular weight organic compound that in solution readily forms long chain polymers via the free-radical initiated vinyl polymerization mechanism to give rather viscous solutions. If a few percent of the chemically very similar but bifunctional reagent N,N'-methylene-bis-acrylamide (for convenience known as Bis) is mixed with the acrylamide, then during the polymerization reaction this introduces a number of cross-links between the growing acrylamide chains and a gel structure results (Fig. 8.1). Clearly the value of n, effectively the average number of acrylamide units between the cross-links, determines the pore size of the gel. Not surprisingly this can be varied in a simple and reproducible manner by changing the ratio of Bis to acrylamide in the starting mixture. It is this ability to adjust the gel porosity to suit the samples being examined that makes PAGE such a useful and popular analytical method. It is common practice to describe gels in terms of % T, the total monomer concentration (w/v) of acrylamide plus Bis and % C, the weight percentage of the cross-linker in T. Thus a T = 10%, C = 5% gel would contain 9.5 g of acrylamide and 0.5 g of Bis per 100 ml. The porosity of the gel can in fact be varied in two ways, by varying values of either C or T. Much work has been done on the mechanical, physical and chemical properties of gels with different values of T and C and is described in books such as those referred to above, but in summary satisfactory gels are difficult to make when $T < 3.0\%$, and if more dilute gels with bigger pore sizes are required, more mechanically stable gels are obtained with agarose. Likewise although T can be as high as 40–50%, even with relatively small peptide samples for example there is seldom any advantage in going beyond $T = 20$–25%. The proportion of cross-linker C is usually in the range 2.5–10%.

Assuming a more typical value for C between about 3.5% and 5.0%, Table 8.1 shows values of T that are likely to give the most satisfactory separations. Thus for large proteins like immunoglobulins for example, or nucleic acids low

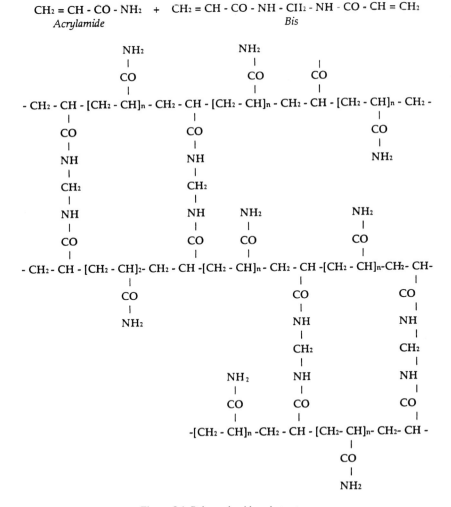

Figure 8.1 Polyacrylamide gel structure.

Table 8.1 Values of %T and best separation ranges

T (%)	Mol. wt. range
<5	Above 200 000
5–10	20 000–200 000
10–15	10 000–100 000
15–20	3000–30 000
20+	Below 5000

% T gels are required while for small peptides, histones, etc. more concentrated gels with small pore sizes will tend to accentuate small size differences in this range and give better separations. Once the gel needed is decided, the required amounts of acrylamide and Bis are weighed out, dissolved in buffer and the catalysts for the polymerization reaction added. A typical gel-making protocol is shown in Table 8.2. The free-radicals required to initiate (catalyse) the vinyl polymerization are generated chemically by the oxidizing agent ammonium persulphate acting on a base, N,N,N',N'-tetramethylethylenediamine (TEMED), or less frequently by photopolymerization with riboflavin acting on TEMED. Most of the components are relatively stable in solution and can be made up and mixed beforehand, but as soon as ammonium persulphate is added, the reaction begins and the solution must be added to the gel mould.

Both acrylamide and Bis monomers are toxic via dust inhalation or skin absorption, so gloves should be worn when handling solutions or preparing gels. It is common practice to avoid weighing the monomers by purchasing commercial mixtures, either as a solid mix to which a measured volume of water is added or as stabilized solutions (acrylamide is very slowly hydrolysed to acrylic acid but this is seldom a problem unless storage is prolonged). Once polymerized, acrylamide is non-toxic, but there is frequently one or two per cent of unpolymerized monomer present, so it is advisable always to wear gloves when handling gels. Some books refer to the desirability of recrystallizing the monomers before use, but so long as reasonably good quality materials are purchased this is totally unnecessary and should not even be contemplated because of the potential hazards involved. In many years of experimentation no poor results from any laboratory that could be attributed to poor quality monomers have been seen by the authors.

The precise technique for preparing and running PAGE gels differs according to the particular design of apparatus used and on whether a vertical or horizontal format is being employed. Most designs for PAGE use a vertical slab of gel typically measuring about 140 × 160 mm (e.g. Fig. 8.2) and for this it is usual for the gel mould to consist of two glass plates, separated by 1–2 mm, and sealed along three sides with a silicone rubber gasket. The gel mixture is poured between the plates (held in a vertical stand) and a plastic comb pushed into the top, so that when the gel has polymerized and the comb is withdrawn its teeth leave moulded

Table 8.2 Protocol for PAGE analysis

Buffers

Running gel buffer (pH 8.9) – Add 18.3 g Tris and 2.07 ml conc. HCl to 60–70 ml H_2O and when dissolved add 0.2 ml TEMED and make up to 100 ml.
Stacking gel buffer (pH 6.7) – Add 1.5 g Tris and 1.03 ml conc. HCl to 60–70 ml H_2O and when dissolved add 0.1 ml TEMED and H_2O to 100 ml.
Apparatus buffer (pH 8.3) – Dissolve 18.0 g Tris and 86.4 g glycine in H_2O to 6 l.

Procedure

1. Assemble 2 gel moulds according to manufacturer's instructions.
2. Prepare 60 ml of running gel mixture (sufficient for 2 gels in a typical vertical gel apparatus employing gels of approximately $140 \times 160 \times 1.0$ mm) as follows:
   Mix 15 ml running gel buffer + 30 ml H_2O + 15 ml of 40% stock acrylamide/bisacrylamide solution (prepared by adding the required volume of H_2O to commercially available pre-weighed material, e.g. Sigma-Aldrich; Bio Rad Laboratories; etc.).
3. Prepare a fresh solution of 25 mg/ml ammonium persulphate in H_2O and add 1.0 ml to the running gel mixture. (Note: this should be prepared fresh daily.)
4. Stir and add mixture to the gel moulds, filling them to about 50 mm short of the top.
5. With a micro-syringe or micro-pipette carefully overlay the gel mixture with 3–4 mm of H_2O or 10% ethanol or isopropanol.
6. As the polymerization reaction proceeds, due to refractive index changes the initially sharp water–gel mixture interface becomes indistinct but once the gel has formed a very sharp interface is apparent. Gelation time should be within the range 10–60 min. If it is not, discard the mixture and start again (too fast or too slow polymerization leads to gels with poor separation properties) altering the proportions of TEMED or ammonium persulphate. If stock buffer solutions have been stored (at 4 °C) for more than 1–2 weeks it is often beneficial to add a little more TEMED to them before use.
7. Once gels have polymerized remove the H_2O overlay.
8. Prepare 15 ml of stacking gel mixture by adding 2.5 ml 40% acrylamide/bisacrylamide to 7.5 ml stacking gel buffer + 5 ml H_2O.
9. Add 0.4 ml ammonium persulphate, mix and add to gel moulds.
10. Place plastic well-forming combs in the top and then completely fill moulds with stacking gel mixture, taking care not to trap any air bubbles.
11. When polymerized, remove sample comb and place the gels in the apparatus.
12. Fill apparatus with apparatus buffer ensuring electrodes are well covered.
13. Apply sample with micro-syringe or micro-pipette to sample wells in the top of the gels.
14. Connect to power supply with the +ve electrode at the bottom.
15. Turn on cooling water.
16. Switch on and adjust voltage to 300–500 V (do not exceed 200 W).
17. When bromophenol blue tracking dye has reached a position close to the end of the gel (about 90 min at 500 V; 2.5–3.0 h at 300 V) switch off, disassemble and place gel in staining bath.

Notes:
1. This protocol is suitable for typical large vertical slab PAGE gels, but individual designs of apparatus may require small modifications and clearly for the popular small (e.g. 80×80 mm) vertical gel or horizontal format, both solution volumes and voltage will need to be altered.
2. The above protocol is for making T = 10% separation gel with a T = 6.4% stacking gel (the proportion of cross-linker, %C, will depend on the proportion in the preweighed mixture purchased, but should be in the range 2.5–7.0% for most purposes). Gels with different %T are prepared by altering the volumes of 40% stock acrylamide/bisacrylamide and water in steps 2 and 8 above.
3. For PAGE under dissociating conditions, addition of 12.0 g urea at step 2 and 3.0 g urea at step 8 (and correspondingly less H_2O) will give gels containing 3.3 M urea.

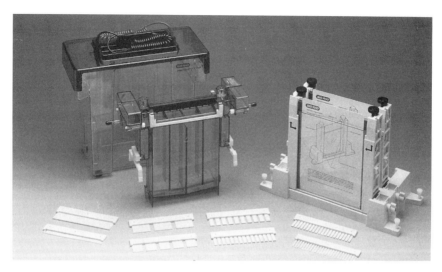

Figure 8.2 A typical vertical gel electrophoresis apparatus. The BioRad Protean II xi cell. (Reproduced courtesy of BioRad Laboratories Ltd.)

into the top of the gel a number of slots to which the samples are applied. In practice the gel mixture is best applied in two parts. First, most of the gel mould is filled with the separation gel mixture, which is carefully overlaid with a small volume of water or 5–10% ethanol (to avoid any mixing or disturbance of the acrylamide mixture). This protects the surface of the separation gel solution from atmospheric oxygen which inhibits the polymerization and will lead to an irregular surface. Once the gel has polymerized, the water is removed and persulphate added to the stacking gel mixture which is then used to fil the mould and the sample comb inserted. Ideally the relative amounts of separation and stacking gels should be such that samples have to pass through 1–2 cm of stacking gel first before reaching the top of the separation gel. The time taken for polymerization to occur after addition of persulphate should be 10–60 min as if polymerization is either too slow or too fast gels with poor mechanical and separation properties result. The amounts of TEMED and persulphate should be adjusted to achieve a satisfactory gelling time. Some published protocols recommend de-aerating gel-making solutions before persulphate addition as by removing inhibitory oxygen this enables less persulphate and TEMED to be used and this can be beneficial for samples that are peculiarly sensitive to oxidizing agents. If this is done the quantities of persulphate and TEMED in Table 8.2 should be reduced 5–10-fold, but for most samples de-aeration is unnecessary and it avoids another potentially hazardous handling step.

Samples can be made up in half-strength stacking gel buffer but other buffers can be used as long as they are of relatively low ionic strength because sample volumes are usually only a few microlitres. Buffer and other salts in the sample do not then interfere with the stacking process. Stacking is the reason for making the gel slabs up in two stages with two different buffers. Basically stacking is a device for improving resolution (the ability to separate closely similar components). It arises because on starting the run, Tris ions throughout the system migrate to the cathode while in the gel the chloride ions move to the anode. The chloride ion is fully ionized and has a high mobility while the glycinate ion is poorly ionized at pH 6.7 and has a low mobility. The chloride would like to accelerate away from the glycinate but this would leave a zone of very low conductivity and since the electric current is constant throughout this would generate a very high voltage gradient across this zone, which would accelerate the glycinate back up to the chloride ion boundary. At the pH (6.7) of the stacking gel proteins have a mobility between those of chloride and glycinate so become concentrated as very thin zones at this boundary. This is termed stacking, and the sample 'sandwich' moves rapidly through the stacking gel to the junction with the separation gel phase at which there is a change in both pH and % T, the consequence of which is that the glycinate becomes more ionized, increases in mobility and accelerates through the protein zones which thereafter move on through the gel slab in a tris-glycine medium.

Because there are discontinuities between the stacking and separation gel phases and buffer this technique is often referred to as disc gel electrophoresis, to distinguish it from more simple approaches with a gel of constant composition and a homogeneous buffer system. As there is no stacking effect such gels have lower separation capabilities than disc gels. The term 'disc' is sometimes misused – it refers to gel and buffer composition not to the earlier technique of making and running gels in a series of glass tubes, a method which no longer finds favour as it is much more difficult to compare accurately a number of samples than with a slab gel format.

A little sucrose or sorbitol and a few microlitres of an ethanolic 0.25% solution of bromophenol blue tracking dye are also added to the sample. This is because with most designs of vertical gel electrophoresis equipment, both anodic and cathodic compartments are filled with apparatus buffer before the samples are applied. They then have to be injected through the upper chamber buffer into the gel slots. Increasing sample density with sucrose or sorbitol prevents the samples from floating out of the slots, while the blue dye enables the samples to be clearly seen. Thus correct application of samples can be confirmed and the dye also gives an indication of when the separation is complete. Once the samples are applied, power is switched on and electrophoresis allowed to proceed until the tracking dye reaches the bottom of the gel slab. Power is then switched off, the glass plates separated and the gel immersed in the detection mixture. The electrical conditions for the separation are not critical, but ideally the run time should be about 1–3 h. Faster than this will usually require a high voltage with a relatively high current and as the amount of heat generated is proportional to the square of the current, adequate cooling could present difficulties because local temperature variations can distort the pattern of separated zones leading to a loss in resolution.

A popular alternative to 'standard size' vertical PAGE gels slabs as described above is the use of smaller (typically 60–80 mm square), often thinner (e.g. 0.05–0.1 mm), vertical gel slabs, which are prepared in essentially the same way but with reduced quantities of all reagents, smaller sample volumes (typically 1–10 µl) and shorter running and staining times. While very economical and convenient these smaller slabs generally do not give quite such high resolution separations. They usually accommodate fewer samples per gel slab but since most aspects of the analysis are quicker the overall throughput of samples would be similar. Small (about 50 mm square) gel slabs run in a horizontal format in an apparatus capable of automating the running and staining steps are also popular. A semi-automated system (e.g. the Pharmacia Phast system) will give even more rapid analysis times of perhaps 1.5–2 h from start to finish. The gels for this system are usually purchased ready-made from the manufacturer, and indeed if one wishes to avoid making up gels altogether a number of suppliers sell pre-packed gels for all types of apparatus. They tend to be relatively expensive but if PAGE or the related gel techniques are only being used occasionally they may be a viable alternative to preparing one's own.

8.2.2 SDS-PAGE

This is simply a variation on PAGE in which the separation is performed in the presence of the detergent, sodium dodecyl sulphate (SDS). If an excess of SDS is present one molecule of SDS will bind to each peptide bond along the protein polypeptide chain (its use is confined to protein or glycoprotein analysis). All proteins bind a very similar amount of SDS, about 1.4 g/g of protein. SDS binding is accomplished by a complete unfolding of the three-dimensional protein structure and a dissociation of multimeric proteins into their component subunits (i.e. loss of secondary, tertiary and quaternary structures). SDS molecules carry a negative charge over a wide pH range, so the effect of binding so much SDS is to 'swamp' any differences in native molecular charge contributed by the charged amino acid groups. All protein molecules will thus have a very similar overall charge to mass ratio. This means that during electrophoretic separation charge differences do not operate and only frictional resistance to passage through the gel pores operates to bring about the separation. This is proportional to the hydrodynamic radius of the SDS-protein complex, which is approximately proportional to polypeptide chain length, or to protein molecular weight. Thus SDS-PAGE complements iso-electric focusing (PAGIF, see below) which separates proteins purely on the basis of differences in molecular charge and PAGE in which both size and charge play a role.

There are few differences between PAGE and SDS-PAGE and exactly the same apparatus, gel and buffer compositions, running conditions and staining and destaining can be used. The main difference is that in addition to the other constituents, 0.1% (w/v) SDS is added to the apparatus buffer and both gel buffers. The only other difference is in the pre-treatment of samples. Whereas for PAGE the samples are made up or diluted in a sample buffer and then applied directly, for SDS-PAGE it is first necessary to ensure that all the protein components have been completely unfolded and the maximum amounts of SDS bound. This is achieved by adding a reducing agent, e.g. 0.1 M 2-mercaptoethanol or 0.05 M dithiothreitol to the sample buffer to break all disulphide bonds and facilitate complete randomization of the protein structure. All the samples are then heated in a boiling water bath for 2–3 min to ensure total denaturation; SDS can then bind to all available sites. Once this has been done and the samples cooled, they are applied to the gel and the experiment completed as for PAGE.

8.2.3 *Agarose gel electrophoresis*

Agarose is a purified form of agar, a naturally occurring polysaccharide. Agar itself used to be employed for electrophoresis but the relatively high content of sulphate and carboxyl groups gives rise to considerable electroendosmosis (EO), which is in effect the transport of solvent relative to the gel matrix. This results in 'flooding' around the electrodes, especially the cathode and can be avoided by using a purified agarose, commercially available in a variety of forms with differing levels of EO and differing melting points. Gels made with agarose have much larger pore sizes than polyacrylamide gels, and are preferable for separating nucleic acids and other large molecules. Since a single type of polymer with no cross-linking agent is used the gelation stage is a purely physical process relying on physical bonds and no new covalent bond formation. This means both that gelation is reversible and that agarose gels can be readily solubilized merely by warming or by adding chemical dissociating agents such as urea or guanidine hydrochloride. This property can be very helpful if it is desired to recover components separated by electrophoresis used as a small-scale preparative procedure. All that is needed is to detect the separated zones, cut out the relevant part of the gel and resolubilize it.

For agarose gel electrophoresis (Serwer, 1983) agarose concentrations in the range of 0.4–2.5% (by weight) are usual and since there is no cross-linking agent changing the agarose concentration is the only way that gel pore size can be altered. Gels with 0.4–0.5% agarose have the largest pore size and are best for the biggest molecules (e.g. genomic DNA etc.) but are mechanically very weak. While 1.5–2.5% gels are much stronger and easier to handle they are never as strong as polyacrylamide gels and are easily torn or damaged. For this reason agarose gels are nearly always run in a horizontal format (e.g. Fig. 8.3) because this provides much better support for the gel than there is in a vertical slab apparatus.

The preparation of agarose gels is straightforward. All that is needed is to suspend the required weighed amount of agarose in the desired buffer and to heat the mixture in a hot or boiling water bath with occasional stirring until the agarose has completely dissolved giving a clear solution. To avoid possible

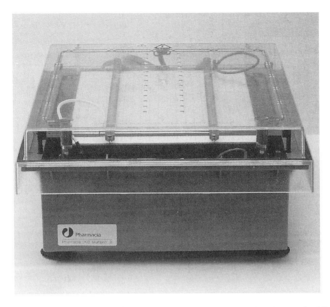

Figure 8.3 A typical horizontal slab gel electrophoresis apparatus. The Pharmacia Biotech Multiphor II unit. (Reproduced courtesy of Pharmacia Biotech.)

charring it is better to heat the mixture indirectly rather than to apply direct heat. Most buffers are satisfactory as long as the ionic strength is relatively low (e.g. 0.02–0.1 M). The hot agarose solution is cooled to 50–60 °C, poured into the gel mould, which should have also been pre-warmed, and then allowed to cool. The precise way in which gel moulds are assembled and used depends upon the particular design of equipment but because without support gels are difficult to handle, it is usual for gels to be mounted on a hydrophilic plastic support, such as Gel Bond, available from most suppliers of electrophoresis chemicals and equipment. A sheet of Gel Bond cut to the required size is placed on the bottom of the gel mould, hydrophilic side up. The warm agarose mixture is then poured in, covered with a plastic sheet or glass plate, and allowed to cool and gel. In some designs this covering plate will incorporate a number of 'teeth' so that when it is removed after gelling, the gel surface has a number of sample wells formed in it. Suitable sample wells can otherwise be cut at the required position with a scalpel or other cutting tool.

Depending upon the type of agarose used, its gelling temperature is usually in the range 35–45 °C, so cooling only takes a few minutes. Once gelled, all that is required is to dismantle the gel mould, lift out the gel on its Gel Bond backing, lay it flat on the cooling plate of the electrophoresis apparatus, apply the samples to the sample wells with a micro-syringe or micro-pipette, attach the electrodes (how this is done depends on apparatus design) and switch on. It is difficult to arrange for discontinuous buffer system when gels are run in the horizontal format. Some useful sharpening of separated zones can be achieved by ensuring that ionic strength of buffer in the sample is lower (preferably 5-fold or more) than that of the gel buffer (Hjerten *et al.*, 1965). Electrical conditions are similar to those for PAGE but it is easy to melt agarose gels, and high currents should be avoided; effective cooling is helpful.

8.2.4 *Isoelectric focusing (IEF)*

In this method a complex mixture of small molecules carrying a variety of charged ionized groups, called ampholytes, is used to set up a pH gradient. Under the influence of an applied electric field, larger molecules such as proteins and peptides (and it is a technique used almost exclusively for the analysis of protein or peptide samples) move through the pH gradient until they reach a position corresponding to their isoelectric points (pI) where they stop. Thus IEF can be regarded as electrophoresis within a pH gradient. All proteins have a molecular charge determined by the ionization of acidic and basic groups belonging to the side chains of the amino acids making up the protein primary sequence. The ionization of these groups is very pH dependent; a strongly acidic environment suppressing the ionization of acidic groups for example. The protein then has a net positive charge and will move towards the cathode, while in a basic environment the basic groups are suppressed and the protein moves towards the anode. By definition the pI is the pH where the number of positive charges balances the number of negative ones; the protein then has no net molecular charge and does not move towards either electrode. Because different proteins have different compositions, they also have different pI values and so when subjected to IEF in a pH gradient they stop at different points in the gradient.

When first developed into a popular viable method (e.g. Vesterberg and Svensson, 1966; Vesterberg, 1971), IEF was performed in free solution in a vertical column stabilized by a sucrose density gradient which prevented convection currents produced by the heat from the electric current from disturbing the pH gradient. This approach has now been superseded by IEF performed in gel matrices which is faster and needs smaller amounts of both sample and of expensive ampholytes. The same apparatus can be used for other electrophoretic techniques (e.g. Fig. 8.3), but IEF has a greater resolving power and most importantly can be arranged to analyse a large number of different samples (40–50 or more) on a single gel (the earlier column method could only handle one sample at a time). If protein samples are to move to their pI positions in the pH gradient, any sieving effect as they pass through the gel pores will slow them down and needlessly prolong the experiment. Thus, unlike PAGE or SDS-PAGE where size-fractionation due to this sieving effect is a useful or essential part of the separation process, in IEF it is a disadvantage and gels with a large pore size should be chosen to minimize this. For peptides and small or medium size proteins, IEF is usually performed in relatively dilute (e.g. T = 5%, C = 4%) polyacrylamide gels and the technique is often referred to as PAGIF. Low-electroendosmosis agarose gels are also widely used and are essential for the analysis of larger molecules. Whichever is used, the gel constituents are made up in water (buffers would prevent the formation of a good pH gradient), ampholytes added, followed by TEMED and persulphate for polyacrylamide gels, or by warming for agarose gels. IEF gels are nearly always run in the horizontal format so for ease of handling they are usually prepared on a supporting plastic sheet (e.g. Gel Bond) and once gelled, are placed on to the cooling plate of an apparatus identical or very similar to that used for agarose gel electrophoresis.

Although gel slabs of many different sizes and thicknesses can be used, they are typically 180–200 mm × 100–150 mm and about 1 mm thick. Designs of equipment differ according to the manufacturers, but the electrodes are usually placed along the longer dimension so separation distances are typically 80–120 mm. Much smaller gel slabs for use in the semi-automated Phast system (Pharmacia Biotechnology Ltd) are also available commercially.

One advantage of the horizontal gel format is the ease of reaching the surface of the gel in order to apply samples. In IEF, this is usually done by placing small pieces, 3–5 mm square, of filter paper (e.g. Whatman 3 MM) on to the surface of the gel in a line parallel to and close to one of the electrodes. A few µl of protein sample is then applied with a micro-syringe directly to these. Alternatively a plastic applicator strip is laid on the gel surface and again samples applied with a micro-syringe, while in the Phast system a special grooved applicator is used. Another difference between IEF and PAGE or agarose gel electrophoresis is that in IEF there are no buffer reservoirs. Instead, strips of thick filter paper or thin sponge soaked in electrode solution are laid along the edge of the gel and electrodes pressed into direct contact with them. The cathodic strip should be well moistened (but not too wet) with a strongly basic solution such as 1MNaOH; strong acid (e.g. $1MH_3PO_4$) should be used for the anode electrode solution. The electrical conditions are also very different to the other methods. Adequate cooling of the gel is vital (especially if agarose gels are used) and the gels must not be allowed to dry out, so once the samples and electrodes have been applied, the main area of the gel is covered with a thin plastic sheet to prevent evaporation and cooling water is passed through the base plate. During the separation, the wattage should probably not exceed 40–60 W at most, which with many systems will mean a starting voltage of 400–500 V. It takes only a few minutes for the ampholytes to move into a pH gradient and as this happens, the current falls so at constant wattage the voltage will rise. The sharpness of the separated protein zones in the final gel pattern is determined largely by the voltage gradient, which therefore should be as high as possible without causing overheating. Final voltages may be 2000–2500 V or more but should always be at least 1000–1200 V. If the power supply being used does not have a constant wattage setting, the gel should be run at constant voltage which can then be increased as the run progresses and the current falls. Typical running times are 1–2 h. With the small Phast system voltages are lower and runs take 30–45 min.

The final resolution obtained depends on other factors as well as voltage gradient, probably the most important being the pH range of the ampholytes used. Commercially available ampholyte mixtures vary from those giving a broad pH gradient (e.g. 3–10) to narrow range mixtures spanning one pH unit. The narrower the range, the more shallow the pH gradient and the greater the ability to separate two proteins with closely similar pI. In fact the resolution of IEF is excellent and two proteins differing in pI by only 0.005 of a pH unit should be separated.

One problem sometimes encountered is pH gradient drift, caused largely by electroendosmosis, which at its most extreme can result in flooding of the gel surface near the electrodes and dehydration and drying out elsewhere in the gel slab. To avoid this, runs should not be unnecessarily prolonged and adding 10% glycerol or sucrose to the gel-making solution to increase viscosity also helps. (Note: sucrose may contain traces of protein which will also focus in the pH gradient and can be confused with sample proteins.) As an aid to deciding how long the experiment should be run we have found it helpful to apply two samples of a coloured protein, such as haemoglobin, to each gel slab one being placed in a line with the unknown samples and the second opposite the first and close to the other electrode. During the run the two haemoglobin samples will move from opposite sides in the pH gradient towards the pI where they should meet and coalesce. Haemoglobin consists of several components of rather similar pI, so gives a group of bands in the region of pI = 6.8. Haemoglobin is a relatively small protein but in a non-restrictive gel matrix most other proteins will move towards their pI positions at a similar rate, so once the sharp bands of haemoglobin have met and focused they provide a very simple visible indication of when the run can be terminated. Other coloured standard proteins can be used in a similar way for IEF experiments in pH ranges that do not include the pI of haemoglobin (e.g. phycocyanine pI = 4.4–4.8; myoglobin pI about 7; cytochrome c pI = 9.6).

Once separated by IEF, as with other gel procedures it is necessary to visualize the protein zones in some way, and this is nearly always done by one of the gel staining methods (section 8.3.1). In many cases as well as merely analysing the proteins in a mixture for purity or for identification by comparison with samples of known proteins, an objective of the IEF experiment will be to obtain pI values for the sample components. This can only be done if the positions of the protein zones in the pH gradient are known. Placing a gel slab into a bath of staining mixture which usually contains both strong acid and an organic solvent completely destroys the ampholyte pH gradient in the gel. The pH gradient can be recorded by taking a series of measurements with a pH meter fitted with a surface electrode which is touched on to the gel surface before it is stained. Such electrodes are, however, expensive and taking a large number of measurements is time consuming. A better approach is to add one or more samples of standard proteins which are run and stained alongside the unknown samples on the same gel slab. The standards can be purified individual proteins or well-defined mixtures of proteins of known pI. Such mixtures are available commercially and focus into a 'ladder' of zones. The position of each component is plotted in a graph against its pI value to give a standard plot so that the pI of any unknown in the sample can be read off from the plot. As a further aid to identification of individual standard proteins in the standard mixture, it can be made up of coloured proteins. Mixtures where each standard protein has a different colour are available commercially (e.g. BioRad Laboratories). Such markers, being coloured initially, also give a useful visual means of checking the progress of the IEF separation during running.

The method referred to above utilizes a complex mixture of so-called carrier ampholytes to establish the pH gradient, so is sometimes called CA-IEF, but in 1982 Bjellqvist *et al.* introduced Immobilines which are commercially available from Pharmacia Biotechnology. Like conventional ampholytes these too are a mixture of small molecules with groups of varying pK_a but in this case they also possess carbon–carbon double bonds and are incorporated and fixed into the polyacrylamide gel as part of the vinyl polymerization reaction process. This means that the pH gradient is not formed by the migration of components in an electrical field (as in CA-IEF) but has to be preformed. This is done using two Immobiline solutions, one acidic and one basic and the pH gradient is prepared with the aid of a gradient-making device. In summary a dense acidic solution containing acrylamide, bis, glycerol and an appropriate (acidic) mixture of Immobilines is placed in one chamber of the gradient-making unit and a lighter alkaline mixture of acrylamide, bis and basic Immobilines (no glycerol) is placed in the other chamber. TEMED and ammonium persulphate are added to both chambers and the mixtures pumped (or allowed to flow) into the gel mould. Once polymerized, Immobilines-IEF gels (often called IPG, immobilized pH gradient, gels) are treated in the same way as CA-IEF gels in terms of sample application, running and staining. The precise mixture of Immobilines needed to give any desired pH range is quite complicated to work out, but useful tables are available in applications literature from Pharmacia. They have also been published by Bjellqvist *et al.* (1982), Cleve *et al.* (1982) and Righetti *et al.* (1983) and these sources also give other useful information on the preparation of IPG gels. The resolution of IPG gels is such that proteins differing by 0.001 pI unit can be separated. There tends to be less electroendosmosis with IPG gels than those with CA-IEF, so flooding and drying out of parts of the gel are less troublesome. The pH gradient is also much more stable and being immobilized does not slowly drift as can happen with CA-IEF gels. The protein load carrying capacity of the gradient is also greater without causing zone distortions, and for preparative work there is the advantage that there are no ampholytes in the purified samples. A further advantage of IPG gels is that unlike CA-IEF gels relatively high concentrations of salts in the samples cause little disturbance, so no sample pre-treatment is needed. Against these advantages, IPG gels are expensive, even more so than CA-IEF gels, and especially if one wishes to avoid the relatively demanding preparation process by buying (from Pharmacia

Biotechnology) prepacked gels. All aspects of IPG gel electrophoresis have been reviewed by Righetti (1984).

8.2.5 *Immunoelectrophoresis*

There are a number of varieties of immunoelectrophoresis methods suited for particular purposes (Axelsen *et al.*, 1973) but for all of them the protein sample is regarded as the antigen and the corresponding antibody is either prepared or purchased and used for detecting the antigen. Thus, while in a few techniques the antibodies move, most of them consist of an electrophoretic separation of the antigen protein components followed by an immunological detection step with antibodies. Antibodies belong to the γ-globulin group of blood proteins which all have pI values close to 8.0 so by choosing a buffer pH of 8.5 they would have a slight negative molecular charge and during electrophoresis move slowly towards the anode. Immunoelectrophoresis experiments are usually performed in 1–2% (w/v) agarose gels and these show some electroendosmosis which effectively moves solvent towards the cathode. At pH 8.5 this approximately balances the γ-globulin migration and the antibodies remain almost stationary in the gel. The various methods differ in how the antigen components are separated and how the antigen–antibody interaction is brought about, but the main advantage is that the great specificity of the interactions results in ideally only the selected antigenic protein being detected and other components, even in extremely complex mixtures, do not interfere. When an antibody interacts with its antigen an insoluble complex is formed and in a transparent gel the resulting zone of precipitation can be observed visually. Some further gain in sensitivity of detection can be achieved by staining the zone of precipitated protein (e.g. as in Table 8.3), but the increase is not great and visual assessment is often sufficient. More widely used, and giving much greater sensitivity in detection, is to use antibodies labelled with radioactive atoms (radioimmunoassay, RIA) or coupled to enzymes (ELISA) (section 8.3.2). The methodology and both the advantages and disadvantages of the various forms of immunoelectrophoresis have been discussed in detail elsewhere (e.g. Axelsen *et al.*, 1973; Andrews, 1986) so will only be summarized here.

For maximum efficiency in forming antigen–antibody complexes both antigen and antibodies must be able to move or diffuse freely through the gel, and since the γ-globulins are very large proteins, highly porous gels are best. In practice this nearly always means agarose gels are used and they are always set up in a horizontal gel slab arrangement. Gels are often quite small, seldom bigger than 10 cm square, and agarose-covered glass microscope slides are popular, inexpensive and easy to handle (Scheidegger, 1955). There is also the benefit that low-cost constant-voltage power supplies with outputs of 50–100 V are adequate.

In the original method (Grabar and Williams, 1953) the antigen proteins are subjected to electrophoresis in an agarose gel; when they have separated a narrow trough is cut in the gel parallel to the strip of separated antigens and filled with diluted antibody solution. The gel slab is incubated at 20–40 °C for several hours or days so that the antigen and antibody proteins both diffuse through the gel; where an antibody meets its corresponding antigen a line or arc of precipitated complex (often called a precipitin arc) is formed. By using an antibody of known specificity a particular antigen in the mixture of sample proteins can be detected. The method is rather slow and only gives qualitative data of whether a particular component is present in the antigen mixture, but it can be a very sensitive detection method.

Rocket immunoelectrophoresis (Laurell, 1966) consists of making an antibody-containing slab of gel and cutting in it a line of small sample wells which are filled with antigen solution. Then an electric voltage is applied to 'drive' the sample antigens through the gel. As antigen molecules meet appropriate antibody molecules a precipitate is formed and movement ceases, so a 'streak' or 'rocket' of precipitate is formed extending from the sample well to the point where all the antigen has been precipitated out. Thus the more antigen, the longer the rocket and the method therefore gives a quantitative measure (by comparison with known amounts of standard sample run on the same gel slab) of a particular antigen in each of the samples in the sample wells. There is no separation of antigen components from one another so highly specific antibodies (e.g. monoclonal antibodies) must be used if samples are at all complex.

Crossed immunoelectrophoresis is a two-dimensional technique (Laurell, 1965; Clarke and Freeman, 1968) in which antigen components are first separated by agarose gel electrophoresis and a strip of gel, containing the separated components, is then applied along one side of a slab of antibody-containing gel. In effect it forms a strip of samples components for the second-dimension at right angles to the first, in which antigen components are driven electrophoretically into the antibody-containing gel so that if polyclonal antibodies are used each antigen component forms its own individual 'rocket'. Thus many components in a complex antigen sample mixture can be analysed quantitatively in the same experiment. Only one sample can be examined per gel slab and quantification relies on comparison with standards run at the same time but on separate gels.

In 1964, Culliford introduced counter immunoelectrophoresis which differs from other methods in that agar gels with high electro-osmotic flow are used, so that the antibodies migrate in the electric field as well as the antigens. By cutting pairs of sample wells about 5 mm apart in the gel slab and applying antigen to the well on the cathodic side and antibody to the well on the anodic side of each pair, when the voltage is applied, antigen and antibody move towards each other and if they match they form a line of precipitate. Many pairs of wells can be cut in a single slab and the experiment takes only 10–15 min, so although only qualitative data are obtained, it is a very rapid method for scanning a large number of samples for a particular antigen protein.

All these (and other) immunoelectrophoretic methods are very sensitive and can detect as little as 1–10 µg/litre of antigen in a solution, but in recent years with the advent of monoclonal antibodies they have become less popular. They have been largely superseded by polyacrylamide gel separations, followed by electroblotting of the separated proteins on to immobilizing membranes and immunochemical detection and measurement (section 8.3.2). This tends to be as fast and as sensitive, but more convenient and versatile than earlier methods.

8.2.6 Two-dimensional methods

We have already discussed a number of two-dimensional (2D) variations of immunoelectrophoresis but 2D can also be applied to PAGE gels. In 2D-PAGE the best resolution is obtained if the basis of the electrophoretic separation is different in the two dimensions. Virtually all possible combinations have been tried at one time or another, but by far the most successful has been the use of PAGIF (or IPG) in the first dimension to separate components in the sample according to differences in molecular change (i.e. in pI), followed by SDS-PAGE separating according to size (molecular weight) differences in the second dimension. This was first introduced in 1975 by O'Farrell and whereas the single dimension separations such as PAGE, SDS-PAGE or PAGIF can separate 50–60 components in a mixture at most and IPG perhaps up to 80–100, the 2D approach can resolve well over 1000 on a single 200×200 mm gel slab (O'Farrell, 1975). In fact, it has been calculated that the maximum resolution on a similar or only slightly larger slab is close to 10 000 different proteins and this is probably almost the maximum number that from genetic information could be expressed in any single cell type. Because it is therefore the highest resolution technique available, 2D-electrotrophoresis is the pivotal analytical method underlying the Human Genome Project. In food analysis, however, this level of resolution is not often required, although it certainly does play a role, for example, in studies of microbial metabolism or when examining closely related microbial strains or their protein products. The ultra-high resolution is bought at the expense of complexity and accurate work, requiring comparison of protein patterns on a number of gels, requires precise control of gel preparation, running and detection with automated gel scanning and sophisticated data handling, so it is both time-consuming and labour-intensive. Being a two-dimensional method only one sample can be applied to each gel slab, so for most food analysis applications and especially for screening purposes the more simple one-dimensional methods are preferred. If necessary they can be used in combination with specific immunological or enzymic detection steps to differentiate particular components in a complex mixture. It is also worth noting that variants of a protein or those that have been modified in any way (e.g. glycosylated, etc.) show up as different zones and since the conditions are dissociating, any multi-subunit protein will be broken down into its component subunits. Thus even a pure protein may appear as several different zones so 2D-PAGE patterns can be very complex. If goals require it the 2D-PAGE approach is unbeatable.

It is usual in 2D-PAGE for the PAGIF (or IPG) separation to be the first dimension, and while long thin PAGIF gels run in narrow (e.g. 3–5 mm internal diameter) glass tubes can sometimes be used, it is normal for a number of samples to be run in the usual way on a single PAGIF gel slab as above, except that rather longer gels (e.g. 150–180 mm) may be more appropriate. When the PAGIF run is complete, the gel slab is not stained but cut into strips 1–3 mm wide with the separated components from one sample being spread along one strip. The strips must be frozen immediately at $-20\,°C$ for subsequent 2D analysis, or, to minimize losses in the resolution of the IEF separation, must immediately be placed into SDS-PAGE stacking gel buffer for a few minutes. This washes out the ampholytes and urea and allows the proteins to equilibrate with SDS, before being positioned across the top of an SDS-PAGE gel slab. This is done by preparing the SDS-PAGE separation gel in the usual way and when that has polymerized (it should be ready as soon as the IEF first-dimension run has finished) the gel strip is pushed into the gap between the plates, parallel to the top of the separation gel and 10–20 mm above it. Once in position, the SDS-PAGE stacking gel mixture is carefully injected below and around the sample strip and also covering it by 0.5–1.0 mm to seal it in place. Care is taken not to entrap any air bubbles. A single sample slot is often cast next to the end of the first-dimension gel strip in the top of the SDS-PAGE stacking gel so that a mixture of known standard proteins can be applied to provide molecular weight calibration for the SDS-PAGE second dimension. Alternatively a small piece of gel containing a mixture of protein standards can be polymerized into the stacking gel at an appropriate position level with the sample strip. Once the stacking gel has polymerized, the gel is then run like any other SDS-PAGE electrophoresis gel.

O'Farrell (1975) added urea to 9 M strength, 2% (w/v) of non-ionic detergent (e.g. Nonidet P-40 or Triton X-100), 5% (v/v) 2-mercaptoethanol and 2% (v/v) of ampholytes to the samples for the first dimension IEF run. The purpose of the first three is to ensure that the cellular proteins he was studying were completely disaggregated and dissolved. For many samples they will not be necessary although a number of other workers have incorporated urea. It is now known that 2-mercaptoethanol causes interference if the gel is subsequently silver stained so dithiothreitol at a level of 10–20 mM is often used as an alternative. Similarly, O'Farrell also used a polyacrylamide concentration gradient gel for his SDS-PAGE separation gel usually going from $T = 9\%$ at the top to $T = 15\%$ at the bottom. Many other workers have also copied this as it tends to give slightly sharper zones than using a gel of constant T, although they are of course a little more difficult to prepare reproducibly. There are a number of inexpensive gradient-making devices commercially available and the technique of preparing gradient gels has been well explained elsewhere (e.g. O'Farrell, 1975; Andrews, 1986; Dunbar, 1987; Hames and Rickwood, 1990). Likewise there is no scope to cover here the scanning and computerized data handling leading to the 2D-gel databases that are now available. A simple visual comparison of complex 2D patterns can be made by superimposing two autoradiographs or photographic negatives of the gels, but it is likely that much valuable information will be missed and this is where the more sophisticated methods yield advantages. For those who are interested, most current books on electrophoresis refer to this aspect but developments in automatic scanning and computerized data handling have been progressing at a particularly fast pace, and the reader is recommended to consult current copies of the journal *Electrophoresis* for the most up-to-date methodology or to search the Internet.

It will be apparent from the above that the apparatus required for 2D-PAGE is no different to that used for one-dimensional gel electrophoresis methods. The IEF gels are run in the horizontal format in the same way in non-restrictive low % T polyacrylamide or agarose gels. (Note: agarose gels cannot be used if high concentrations of urea are added as this disrupts the physical bonds that are responsible for gelling.) Vertical SDS-PAGE gel slabs are used in the second dimension. For highly specialized work where there is a requirement for running large numbers of 2D-gels, special apparatus for preparing large batches of gels (preparing many gels together aids reproducibility) and for running 10–12 gels at the same time are available. This is when the most elaborate scanning and data handling also become justified. Pharmacia have developed 2D protocols for use with the small gels (approx. 50 × 50 mm) of the Phast system, but although fast and convenient these have nothing like the resolving power of larger gels.

8.3 Detection of separated components

8.3.1 *Staining procedures*

After the separation is completed the components must be detected. If the components of interest are enzymes then there are often specific methods of detecting them, for example, by incubating the slabs of gel in a suitable substrate solution. Since methods are specific there are many hundreds of different procedures, so it is not possible to go into details here. As a generalization, substrates with a substituted naphthyl grouping are often useful as the naphthol or naphthylamine derivative released can be coupled to an azodye also added to the substrate solution, to give visible coloured bands corresponding to the enzyme activity. Likewise umbelliferyl substrates give fluorescent zones. One problem with specific methods is that they do not detect all the other constituents present, and while this can be very helpful for identifying components in complicated mixtures by greatly simplifying zone patterns, it does mean that very little information about purity can be gained. Nevertheless, isoenzyme analysis (detecting all zones exhibiting a particular enzyme activity) is a widely used and valuable method in food authentication work. For example, detection of amylase, esterase and peroxidase activities have all been used for species identification and cultivar determination in cereal grains, potatoes and many other fruits and vegetables (e.g. Fig. 8.4).

The other principal group of detection methods are general methods detecting for example, all protein, carbohydrate or nucleic acid zones. Most general protein detection methods make use of textile dyes such as Coomassie Blue G250 or R250 or similar. A typical procedure is outlined in Table 8.3. For greater sensitivity, by at least 10-fold and perhaps 100-fold, one of a variety of silver-staining methods is used (e.g. Table 8.4). It is often quite helpful to apply Coomassie Blue staining to measure the major protein components (as a rough approximation, the lower limit of detection is likely to be about 0.1–0.5 μg of protein per zone) followed by silver staining to detect minor components. If this is done using the protocols given in Tables 8.3 and 8.4, the first steps of the silver-staining method, designed merely to remove salts, small possibly interfering contaminants and to fix the proteins in the gel to prevent them from being washed out, can be omitted. These have in effect been achieved during the Coomassie Blue staining, so one can start with step 4 of the method of Morrissey (1981) in Table 8.4. Whether Coomassie Blue or silver staining (or both) has been used, gel electrophoresis methods should be regarded as only semi-quantitative. It is a simple task to compare a number of different samples separated on a single gel slab and to observe the presence of extra or missing bands, and often quantitative differences are visually apparent but more rigorous quantification is difficult and unreliable. Laser densitometer gel scanners have

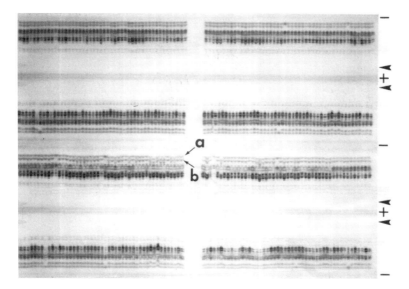

Figure 8.4 IEF-CA gel (pH 3–6), stained for PGM, showing the result of a brassica hybrid purity test. For this test the two most cathodic bands are of interest. The most basic one is the marker for the female parent (a-band), the lower one is the male parental marker (b-band). The F1 seeds have both bands. If the b-band is missing, it can be assumed that this was an inbred. Among the 384 seeds analysed, 12 inbreds were found. The anodes and cathodes are indicated with + and − symbols, respectively. The sample application points are indicated by horizontal arrows. (Reproduced with permission from Burg and van den Berg (1996).)

the necessary sensitivity and accuracy but there are always small variations in protein concentration even across a single protein zone in a gel. There will also be unevenness in both the staining and de-staining processes over different parts of a gel slab. Perhaps even more of a problem is that individual proteins have different dye binding capacities, so peak areas for identical amounts of two proteins can differ by a factor of 2 or more. For best results this necessitates the analysis and measurement of known standard proteins on the same gel at the same time. (With silver staining, quantitative response is even more variable and a small number of proteins do not stain at all.) Great care is needed to ensure maximum reproducibility of all the separation and staining steps and under ideal separation and measurement conditions quantification is possible with an accuracy of $\pm$ 5–10%. For accurate quantitative results other methods should be chosen (e.g. chromatographic methods, HPLC, FPLC, etc.). These objections are somewhat less valid if the proteins in the sample have been radioactively labelled before the electrophoretic separation and autoradiography or scintillation counting is used for their detection. This approach is very often used in 2D-PAGE work and has the advantage of a much wider response range than the dye-staining methods, i.e. minor components present at a thousand-fold lower concentration than the major constituents can often be detected. With staining methods this is often not possible without severely overloading the major

Table 8.3 Polyacrylamide gel staining for proteins

Staining mixture	Staining time (min)	Destaining mixture	Comments
Amido Black 10B Dissolve 2.5 g in 250 ml methanol, add 100 ml glacial acetic acid + 650 ml H_2O	60–120	10% acetic acid	Prefixing of gels by immersing in 10% trichloroacetic acid for 30 min may be advantageous
Coomassie Blue R250 Dissolve 1 g in 500 ml methanol, add 100 ml glacial acetic acid + 400 ml H_2O	20–60	methanol: acetic acid: H_2O = 5:1:4	Much more sensitive than Amido Black. Store gels in 7% aqueous acetic acid
Coomassie Blue G250 Dissolve 2.5 g in 500 ml methanol and add 500 ml 25% (w/v) trichloroacetic acid (prefixing step of 10–30 min in 10% trichloroacetic acid especially helpful for SDS-PAGE gels)	10–40	5% trichloroacetic acid for 15–30 min, then 7% aqueous acetic acid	Similar to R250 dye but greater fixing power so better for peptides and very soluble proteins/ glycoproteins which can be leached out of gel

Notes: 1. Staining solutions can be reused for up to about 10 gels, but fresh dye requires shorter staining times and fixing power deteriorates with reuse.
2. Large amounts of SDS can interfere with dye binding so for SDS-PAGE staining solutions should only be used 2 or 3 times and generally the methanol-acetic acid solvents appear to give better results with SDS-PAGE gels.
3. If band patterns are faint (as may happen particularly with SDS-PAGE), a restaining cycle often improves results.

components. Another problem area, particularly with any of the one-dimensional separation methods, is that a major zone can obscure a minor component running close to it.

There are detection methods specifically aimed at detecting glycoproteins. These do show up with the ordinary protein stains but the carbohydrate complexed to the polypeptide chains also enables them to be distinguished from non-glycosylated proteins. Traditionally this has been dome by periodic acid-Schiff staining (Table 8.5) which gives red zones on a colourless background (Zacharius et al., 1969). An alternative to staining, and offering greater sensitivity, is the exploitation of specific interactions between lectins and carbohydrates. The main difficulty is that a particular lectin binds only to one type of monosaccharide unit

Table 8.4 Silver staining protocols for proteins in polyacrylamide gels

Procedure	Time
Method of Morrissey (1981)	
1. Prefix gel in methanol: acetic acid: $H_2O = 5:1:4$	30 min
2. Wash in 5% methanol containing 7% acetic acid	7 min
3. Fix in 10% aqueous glutaraldehyde (use 6 ml for every ml of gel)	30 min
4. Soak overnight in H_2O then for 30 min in fresh H_2O or several changes of H_2O over at least 2 h	2 h+
5. Immerse in dithiothreitol solution (5 µg/ml in H_2O) (use 12 ml per ml of gel)	30 min
6. Decant off dithiothreitol solution and add 0.1% aqueous $AgNO_3$, using 12 ml for every ml of gel	30 min
7. Rinse rapidly with a small vol. of H_2O	10–20 sec
8. Rinse rapidly twice with a small vol. of developer solution (50 µl of 37% HCHO in 100 ml 3% Na_2CO_3)	30–40 sec
9. Immerse in developer solution (about 12 ml per ml of gel) until protein band pattern has reached the desired intensity	5–15 min
10. Add 5 ml of 2.3 M citric acid for every 100 ml developer solution and mix for 10 min (pH should be close to neutral or slightly acidic to stop development)	10 min
11. Wash gel in several changes of H_2O	
Method of Heukshoven and Dernick (1985)*	
1. Prefix gel in ethanol: acetic acid: $H_2O = 4:1:5$ (3 changes, 45 min each)	2.25 h
2. Immerse in a sensitizing solution containing: 5 ml glutaraldehyde + 2.5 g $K_2S_4O_6$ (potassium dithionite) + 68 g sodium acetate + 300 ml ethanol and made up to 1 l with H_2O	overnight
3. Wash with H_2O (4 changes, 30 min each)	2 h
4. Immerse in a solution of 2 g $AgNO_3$ + 250 µl HCHO made up to 1 l with H_2O	1–2 h
5. Rinse in H_2O	15 sec
6. Immerse in developer solution (30 g K_2CO_3 + 150 µl HCHO + 10 mg sodium thiosulphate pentahydrate made up to 1 l with H_2O)	10–25 min
7. Stop development with a Tris-acetate buffer (50 g Tris + 20 ml glacial acetic acid made up to 1 l with H_2O)	30 min
8. Wash gel in several changes of H_2O	1+ h

* As modified by Rabilloud (1992).
Notes: 1. The method of Morrissey is rapid but not quite so sensitive as that of Heukshoven and Dernick.
2. If gels have already been Coomassie Blue stained, start at step 4 of the Morrissey procedure or step 3 of the Heukshoven and Dernick method (acetic acid etc. must be removed before silver staining).

Table 8.5 Selective staining of glycoproteins by periodic acid-Schiff (PAS) procedure

Procedure	Time
1. Immerse gels in 12.5% (w/v) trichloroacetic acid	30 min
2. Rinse in H_2O	15 sec
3. Soak in 1% periodic acid in 3% acetic acid	50 min
4. Wash with at least six changes of H_2O over 1–2 h or overnight	2+ h
5. Place in fuchsin-sulphite (Schiff's reagent) in the dark*	50 min
6. Wash with three changes (10 min each) of 0.5% aqueous sodium metabisulphite	30 min
7. Wash in frequent changes of H_2O to remove background staining	2+ h
8. Store gels in 5% acetic acid	

* Schiff's reagent is available commercially (e.g. Sigma-Aldrich) but if necessary can be made by dissolving 2 g basic fuchsin (caution carcinogenic) in 400 ml H_2O with warming. Cool, filter and add 10 ml 2NHCl and 4 g $K_2S_2O_5$. Place in a stoppered bottle, keep cool and dark overnight, then add 1.0 g activated charcoal, stir, filter and add 2NHCl (at least 10 ml) until a drop dried on a glass plate does not turn red. Store stoppered in the dark and discard if it turns pink.

or one group of them, so the method is not a general one for detecting all glycoproteins at once. A range of lectins is needed to ensure a wide coverage. Nevertheless it is fast and sensitive and also gives information about the oligosaccharide groupings attached to the glycoproteins, so it has become popular in recent years. In the method of Furlan et al. (1979) gels are run and the proteins and glycoproteins fixed in acid in the usual way for 20–30 min, then washed in four changes of pH 7.0 buffer until they attain that value. Fluorescein isothiocyanate (FITC)-labelled lectins at 1 mg/ml in this same buffer are added and coupling allowed to continue overnight. The gels are then de-stained in this same (lectin-free) buffer for 2 days. Fluorescent glycoprotein bands can then be seen under UV illumination with a detection limit of about 100 ng of protein-bound carbohydrate. FITC-labelled lectins can be purchased commercially (e.g. Sigma-Aldrich Chemical Co.) or lectins labelled with FITC by the method of The and Feltkamp (1970). A similar approach but of even greater sensitivity is to use ^{125}I-labelled lectins followed by autoradiography (Koch and Smith, 1982).

Nucleic acids absorb UV light so if present in large amounts the zones can be observed by simple UV illumination (Clarke et al., 1982) but usually staining methods are used and are far more sensitive. Silver staining by methods identical, or only slightly modified, to those used for proteins, work well (Merril et al., 1982; Goldman and Merril, 1982), but much the most common approach is to stain the DNA or RNA with ethidium bromide. Great caution should be taken with this, however, as this reagent is a potent carcinogen and protective clothing must be worn at all times when handling solid, solutions and gels which have been so stained. It is also essential that all waste or any other contaminated material must be disposed of by approved procedures. Preferably a small room or area with limited access should be set aside for all operations involving ethidium bromide and gel viewing. Staining conditions are not very critical and there are many variations but Sharp et al. (1973) described an ethidium bromide reagent after immersion in which RNA or DNA zones containing down to about 0.05 µg of nucleic acid could be seen on short wavelength UV illumination.

8.3.2 *Blotting of gel patterns*

This is the procedure of transferring patterns of separated zones of proteins or nucleic acids from gels on to thin sheets of (usually) nitro-cellulose, nylon or PDVF where they are bound and immobilized (Gershoni and Palade, 1983). Although this adds a further stage to the analysis, advantages are that being on the surface of a thin membrane the separated macromolecules are far more accessible than when embedded in a gel slab. This is important in all the immunological detection methods, and processing times for staining, de-staining, incubating, washings etc. are much shorter than with gels. The transferred zone patterns can be dried easily and kept for months for record purposes or subsequent further analysis. Suitable membrane materials are readily available commercially (e.g. BioRad Laboratories).

Transfer by means of a solvent flow, sometimes called capillary blotting but more frequently referred to as Southern blotting, was introduced (Southern, 1975) for transferring nucleic acids from agarose gels on to a nitrocellulose membrane. It involves placing the gel on to buffer-wetted filter papers or thin sponge, positioning the membrane directly on to the surface of the gel, covering that with a few sheets of dry filter paper and paper towel, then a glass or plastic plate and placing a weight on top (e.g. 0.5 kg). Buffer is squeezed out of the wet papers through the gel and membrane (carrying nucleic acids from the gel on to the membrane) and is absorbed by the dry paper towel. After an overnight blotting, the 'sandwich' is dismantled and the separated DNA or RNA zones detected by hybridization with radioactively labelled probes (Southern, 1975).

The usual approach for proteins separated in polyacrylamide gels is electro-transfer, or 'Western Blotting' (Towbin *et al.*, 1979). The equipment for this is available from most suppliers of electrophoresis equipment (e.g. BioRad Laboratories etc.) who supply technical instructions for their particular apparatus, but basically this is usually performed in a vertical buffer-filled tank with an electrode array on each side. A 'sandwich' is prepared consisting of membrane next to the gel, filter papers on both sides and then sponge or foam outside. The whole assembly, mounted in a suitable frame, is positioned vertically between the electrodes and a voltage of typically 5–6 $V\ cm^{-1}$ applied for 1 h, so that proteins migrate out of the gel (at right angles to the plane of the gel) on to the membrane. Thus a relatively low voltage, high current capacity power supply is needed. The method works with all types of gel separation on polyacrylamide or agarose gels for proteins and nucleic acids and a variety of buffers have been used to fill the transfer apparatus. Care must be taken to ensure that the membrane is mounted on the correct side of the gel and, for example, if SDS is present, as in SDS-PAGE gels with neutral or weakly basic buffers, the proteins will migrate as anions and the membrane must be on the anode side of the gel.

As unusual feature, especially when nitrocellulose membranes are used, is that some protein will pass through the membrane. Thus if two or more sheets of membrane are placed next to each other one can obtain several replicate patterns from a single gel. This can be useful as one can be used for a general staining, for example, and a second for a more specific detection method. Unfortunately the transfer is variable with small proteins tending to pass quite rapidly through the first membrane and larger ones moving more slowly. Thus small molecules are underrepresented on the first membrane while larger ones may not even get to the second, so there are no good quantitative data.

Once transferred, the pattern of proteins on the membrane must be detected and Gershoni and Palade (1982) recommend a stain for proteins on nitrocellulose membranes based on Amido Black in aqueous isopropanol containing acetic acid with a 30 min destaining in the same solvent. The usual Coomassie Blue stain (Table 8.3) can also be used as long as staining and de-staining are kept short because nitrocellulose is not very stable in acidic methanol. Nylon and PVDF membranes have stronger binding properties than nitrocellulose and it is very difficult to remove background colour if this dye is used. These membranes are used with radiolabelled proteins or with immunological detection. The latter is done (Towbin et al., 1979) by soaking the membrane in 3% bovine serum albumen (BSA) or skim milk powder in buffer pH 7.4 containing NaCl to saturate any excess protein binding sites. This is followed by sequential rinsing in TBS, incubation in an appropriate antibody to the sample proteins on the blot, further washes in TBS and then a second incubation with TBS containing a second (indicator) antibody directed against the immunoglobins of the first antibody. This indicator antibody is either labelled with ^{125}I so that the reactive zones on the blot can then be detected by autoradiography or labelled with an enzyme such as alkaline phosphatase or horseradish peroxidase. Zones of peroxidase activity on the blot are revealed colorimetrically with a substrate solution containing 4-chloro-1-naphthol in buffer pH 7.4 containing NaCl and hydrogen peroxide. Peroxidase positive zones show up as blue areas in 2–15 min. Alternatively for greater sensitivity photochemical detection with luminol and luciferin can be used (Laing, 1986).

8.4 Capillary electrophoresis (CE)

Almost everything described so far relates to one or more of the gel-based electrophoretic methods and this prominence accurately reflects the overwhelming position of such methods. In the last few years, however, electrophoresis in free solution has made a comeback in the form of high speed, high voltage microanalysis in fine capillary tubes with internal diameters typically of 50–100 μm up to a metre long (but usually 50–70 cm) and separation voltages of up to 300 V cm^{-1}. By employing such narrow capillary tubes the surface area to volume ratio is large and heat generated by the passage of the electrical current is rapidly dissipated. This means that high voltage gradients can be used without causing undue heat-related problems, and mixing of separated zones by convection currents is relatively insignificant so that separation times of only a few minutes can easily be achieved. In terms of the overall analytical process, CE has been referred to as the electrophoretic analogue of HPLC, and like HPLC it is capable of giving rapid accurate quantitative results starting with only few μg or less of sample. Detection of separated zones is on-line, with the capillary passing through a suitable detector unit and the output of this being processed by computer to give the required display, calculation of retention times, quantification etc. The precise way this is all done varies from one manufacturer of equipment to another but all instruments incorporate UV detection of separated zones, with wavelengths down to about 190 nm. Many also incorporate fluorescence detection, although this usually requires some prederivatization of sample components. Conductivity, electrochemical refractive index and radiometric detectors are also available. Still largely at the research stage but with great future potential, the sample zones can be passed from the capillary directly into a mass spectrometer, which can give immediate identification of emerging components.

The first step of the technique simply involves introducing a small defined volume (typically 1–5 nl) of sample in an appropriate dilute buffer into one end of the capillary by gravity, vacuum or pressure application or by electrophoretic migration. The method of sample induction varies with the instrument, but most give a choice. For many separations electrophoretic induction which relies on a brief pre-electrophoresis step with the end of the capillary tube dipped into the sample vial, although convenient, is best avoided because there can be a selective enrichment of fast moving sample components leading to errors in quantification. Once the sample has been introduced, a high voltage of up to 30 000 V is applied to separate the sample components in just a few minutes. The method is extremely versatile and the same basic instrument, often with the same capillary, is capable of analysing a huge range of sample molecules from polymers, intact DNA, RNA and proteins to restriction fragments, peptides, amino acids and all kinds or organic and inorganic molecules. This can include chiral separations and even the analysis of metal actions and small anions. Very often a complete analysis of all types of molecules can be achieved in a single run. Non-UV absorbing species can be detected by incorporating a low level of UV absorbing background molecules in the separation buffer, so that the non-absorbing species show up as 'holes' or negative peaks in the background. Analysis by CE is not confined only to electrically charged sample molecules. This is because the capillary tubes used (often identical to those used for gas chromatography, so relatively inexpensive and readily available), are made of fused silica and the internal wall of the tube possesses a negative charge at most pH values. This causes a build up of positive ions in the layers of solution next to the wall and when a voltage is applied they are attracted towards the cathode, which results in a general movement of the bulk of the solution towards the cathode. The electrophoretic migration of ions is superimposed on this electroendosmotic flow and the net result is that at neutral or basic pH values cations are separated from each other according to differences in electrophoretic mobility. They are carried as a group rapidly towards the cathode by the electroendosmotic flow, followed by neutral species carried along by the flow only and followed lastly by the anionic species, again separated from each other electrophoretically but still with a net movement towards the cathode as a result of electroendosmosis.

The theory behind the various CE techniques and the design and operation of equipment has been well reviewed in a number of books, such as that by Grossman and Colburn (1992) and papers (e.g. Szökö, 1997), and many manufacturers of biochemical and electrophoretic equipment (e.g. BioRad Laboratories; Dionex; Perkin-Elmer; Beckman Instruments; Spectra-Physics (Thermo Separation Productions)) market CE equipment and supply useful applications literature.

8.4.1 *Capillary zone electrophoresis (CZE)*

This is CE in dilute aqueous buffers in the manner outlined above. The separation is very dependent upon the charge on the solute molecules and hence is greatly affected by pH. Although the mobility equations do include a size term, under these separation conditions it has little influence so size differences can be ignored, as can differences in molecular shape. Subtle shape differences can be exploited if the conditions are modified and, for example, chiral compounds can be separated in the presence of cyclodextrins. A frequent problem in CE is the adsorption of solutes on to the capillary wall. This can be avoided by performing the analysis at pH 2.0 which suppresses the charge on the walls or by adding a modifier which has a greater affinity for the wall than the solutes, but is most often combatted by using coated capillaries. A huge variety of coating agents have been successfully tried, including glycol, maltose, epoxydiol and polyethyleneimine but the most widely used are polyacrylamide (Dolnick *et al.*, 1989; Wainwright, 1990) polyethylene glycol (Terabe *et al.*, 1986), trimethylchlorosilane (Balchunas and Sepaniak, 1987) and polymethylsiloxane (Lux *et al.*, 1990). Coating of capillaries also reduces problems of gradually changing migration time. This can occur when a sample is repeatedly analysed with uncoated capillaries due to a gradual absorption of sample or accompanying contaminants on to the walls.

8.4.2 *Micellar electrokinetic capillary chromatography (MECC)*

This is a form of CE which was originally introduced (Terabe *et al.*, 1984; Sepaniak *et al.*, 1987) for separating neutral compounds that are washed along as a group but not separated from each other in CZE. It involves adding a surfactant such as 50 mM SDS to the buffer at a high enough concentration to form surfactant micelles and to form a two-phase system. The buffer represents the primary phase and the micelles the secondary phase which moves through the capillary by a combination of electroendosmotic flow and electrophoretic migration due to the electrical charge on the micelles. Solute molecules, both neutral and charged, partition into and out of the micelles and neutral species are separated in relation to the statistical proportion of time they spend associated with the charged micelles. In other words the retention characteristics of neutral molecules in MECC usually resemble the results of reverse-phase liquid chromatography. In MECC the attainment of reproducible retention times requires a precise control of experimental conditions and there is often a finite elution range which limits peak capacity and means that MECC is not so suitable for analysing extremely complex mixtures (Grossman and Colburn, 1992). Sample injection, running and detection are all similar to CZE and most frequently uncoated fused-silica capillaries have been employed. Although sample volumes are small they must be relatively concentrated for good detection. MECC has generally been used to analyse quite small, rather hydrophobic molecules, but it is a particularly useful technique for the analysis of enantiomers (chiral separations). Bile-salt surfactants such as sodium deoxycholate are particularly valuable in this context (e.g. Nishi *et al.*, 1990) but the use of chiral additives that form mixed micelles with a surfactant such as SDS has also been very successful (e.g. Dobashi *et al.*, 1989).

8.4.3 *CE in polymer solutions*

Because nucleic acids or SDS-denatured proteins of different sizes have constant charge to mass ratios it is not possible to separate these classes of compounds by CZE in ordinary buffers. If high molecular weight polymers are included they introduce a frictional resistance to the electrophoretic movement and solute molecular size then becomes a significant factor in the separation; a situation analogous to the sieving effect of gel pore size in SDS-PAGE. Because convection currents caused by ohmic heating are not a significant problem in CE, it is not necessary for the polymer molecules to be cross-linked into a gel and once the concentration of polymer reaches a certain threshold value the polymer chains interact and form a network of entangled molecules (Grossman and Soane, 1991). A wide range of uncharged, water soluble polymers can be added to the types of buffer used in CZE to give this size fractionation effect and for example Grossman and Soane (1991) achieved excellent separations of DNA restriction fragments in solutions with hydroxyethyl cellulose added at concentrations of 0.2–0.5% (w/v). Log of the mobility versus log of the inverse of molecular size (in base pairs) gave a straight line, the slope varying slightly with polymer concentration. Thus calibration with standards of known molecular size should enable that of an unknown sample run under identical conditions to be determined. Polymers of uncross-linked polyacrylamide (Heiger *et al.*, 1990) and polyethylene oxide (Chen *et al.*, 1989) have also proved popular. Using polymer solutions as opposed to gel matrices for size-based separations retains the advantages of speed and reproducibility of analysis, on-line detection and data handling, continuously renewable separation media and the ability to quickly and easily change the separation conditions merely by changing the buffer in the capillary that are all characteristic advantages of CE methods, and extends it to the range of applications and resolving power of gel-based systems.

8.4.4 CE in gel-filled capillaries

Using gel-filled capillaries, single-base separations of oligonucleotides (Karger et al., 1989), DNA sequencing runs (Drossman et al., 1990), chiral compounds separated in the presence of β-cyclodextrin (Guttman et al., 1988) and protein molecular weights determined in the presence of SDS (Cohen and Karger, 1987; Guttman, 1996) have all been successfully performed within a few minutes using only nanogram amounts of sample (Dubrow, 1992). These methods are essentially based on conventional gel electrophoresis methods as already described, but there is the advantage of speed of analysis and easier quantification. However, there are a number of serious drawbacks. Most importantly it is a difficult and skilled art to make good quality gel-filled capillaries, but they are available commercially. Samples have to be injected electrophoretically (electrokinetically) because with gel-filled capillaries the other methods cannot be used and while easy to perform it is difficult to determine the amount loaded. This can vary enormously depending upon the conductivity of both the sample and buffers. Gel composition and small differences in pH and ionic strength have a great effect so reproducibility of analysis is a major weakness, and gel-filled capillaries have a rather limited life. In summary while this technique is of considerable research interest, similar or better results can usually be obtained with polymer-filled capillaries or by the more conventional gel methods even though these may take rather longer to perform.

8.4.5 Capillary isoelectric focusing

This is the CE analogue of IEF in gels (section 8.2.4) and fused silica or even glass capillaries coated with hydrophilic, non-ionic polymer to virtually eliminate electroendosmosis must be used. This has the added benefit that the absorption of sample proteins to the walls is much reduced. The sample dissolved in a solution of 1% ampholytes in water is used to fill the capillary, the ends of which are dipped into acidic (anode) or alkaline (cathode) electrode solutions and the voltage applied. As with gels the ampholytes move rapidly to establish the pH gradient in which the proteins in the sample then migrate to focus as zones at their isoelectric points. Since sample proteins (which should be relatively free of salts) are mixed into the whole volume of the capillary and are focused into a sharp zone there is a substantial concentration effect. The technique can therefore handle much more dilute samples than the other CE methods. If proteins precipitate at their pI then addition of non-charged dissociating agents help, to minimize it. In a typical run at 4000–6000 V the ampholytes form the pH gradient in about 5–7 min and the steady state is reached when the electrical current has fallen to 10–20% of its initial value. In free solution the proteins also move quite quickly so the separation stage is essentially completed within this time and the voltage can be switched off. A major difference between this and other CE methods is that the protein components focus to almost stationary zones and will not pass the detector, which is at or close to one end of the capillary, so they must be mobilized in some way. Fortunately this is easily done by simply exchanging the two electrode solutions when the whole pH gradient migrates out of the capillary past the detector. Alternatively, addition of NaCl to the anode electrolyte moves the pH gradient and protein zones towards the anode. Addition to the catholyte causes movement towards the cathode (Hjertén, 1992). Like gel IEF, CE isoelectric focusing is essentially a technique for the analysis of proteins and peptides and this is reinforced in this case because ampholytes absorb short wavelength UV light so the separated zones have to be monitored at the protein-specific 280 nm. The only way the pH gradient can be measured is by analysing a series of standard proteins with known pI values. These are run separately to establish the peak pattern due to the standards and then a portion of standard solution is mixed with the sample to provide an internal standard mixture. Due to run to run variations a separate standard run is less satisfactory than this internal standard approach. A graph is plotted of the standard pI versus elution position and from this the pI values of sample proteins can be read off.

8.5 Some applications of electrophoretic analysis

Practical problems in the food analysis area generally fall into one of two broad categories; either screening studies for species or variety (cultivar) identification or compositional analysis for quality evaluation of food ingredients and products. The latter may involve the analysis of a small number of samples and comparison with known standards or may require the handling of a large number of samples as in a screening situation. In nearly all cases it will be comparison with known standards that is important and accurate absolute figures for the molecular weight, pI or mobility of sample constituents required less often. Likewise sample throughput, as reflected by speed of analysis, ease of interpretation of results and technical simplicity, all of which in turn influence labour costs and hence overall costs, are likely to be overriding considerations. A cost benefit analysis should be undertaken on the preparation of gels or capillaries in-house or the purchase of relatively expensive ready-made alternatives. The answer will depend on a number of factors including safety (avoidance of handling toxic solutions), reproducibility, convenience, availability of trained personnel and particularly upon how often a given technique is likely to be used.

Ampholytes for IEF are expensive and have a limited shelf life, so, for example, if this technique is only infrequently used it may well make economic sense to purchase ready-made gel plates.

For the first category of problems involving screening or analysing many samples this is the factor that largely determines the method to be used. Neither the complex 2D-PAGE approach, nor the two-dimensional immunoelectrophoretic methods are suitable, and in spite of the short analysis times in CE, realistically a minimum overall time of about 30 min between samples is needed. This rate of throughput is not adequate (i.e. 15–16 samples a day at maximum). This number of samples can easily be handled on a single gel slab and with the standard laboratory size gels (160–200 mm wide) and a conservative 20 samples per gel, 2 of which are run at the same time in standard vertical gel apparatus designs, it is possible to run 120 samples/day by PAGE or SDS-PAGE. If required there are commercial designs of equipment which can handle more samples per gel and more gels being run at the same time. When considering gels run in a horizontal format, small gel slabs, such as those used with the Pharmacia Phast system, will not handle enough samples for them to be the method of choice in screening applications, but larger gels such as those used in the Pharmacia Multiphor apparatus have similar capabilities to vertical gel designs. Gel IEF, however, has a rather greater capacity as 40–50 or more samples can be applied to a single gel slab. Indeed in a specially designed chamber fitted with multiple electrodes Burg and van den Berg (1996) were able to examine no less than 384 samples of tomato and brassica proteins on a single slab and identify hybrid variants (Fig. 8.4). In this case the identification of variants was simplified by using isoenzyme detection of phosphoglucomutase or alcohol dehydrogenase, rather than a general protein staining method which would have given extremely complex patterns of separated zones and would have been more difficult to interpret.

For this category of food analysis problems, the valuable methodology is confined essentially to PAGE, SDS-PAGE and PAGIF. Among the immunoelectrophoretic methods counter immunoelectrophoresis can deal very rapidly with a large number of samples and has been developed especially for screening applications in forensic work, but as far as I am aware has not yet been exploited in food analysis for which it would appear to have considerable potential. The potential use of these methods in food and agricultural analysis has been reviewed by Andrews (1991).

The other category (compositional analysis) provides greater scope for the application of other methods as well as those used for screening because generally greater in-depth information is required and often sample numbers are smaller. PAGE, SDS-PAGE and IEF have all found widespread use in the seed industry for plant testing and for monitoring plant breeding programmes (Cooke, 1984, 1989), in the baking industry for distinguishing hard from soft wheat flours (Günther et al., 1986), to examine malting quality in brewing (Smith and Lister, 1983), to study the basis of resistance to fungal diseases in barley and other cereals (Riggs et al., 1983) and for variety and cultivar identification in a wide range of cereals, legumes, brassicas, potatoes and many other vegetables and fruits. These same methods have also been widely used in quality control work relating to manufactured food products and for ingredient verification before manufacture. At present even here the more complex and sophisticated methods, such as 2D-PAGE, are primarily at the research and development stage although this has the ability to separate very closely related materials, such as individual durum wheat lines (Fig. 8.5) for example (Picard et al., 1997) or proteins in the different tissues of rice plants (Tsugita et al., 1996).

Rather more contentious applications in food analysis lie in the areas of food deterioration and adulteration and again electrophoretic methods of compositional analysis are among the most versatile and powerful approaches available. Most plant tissues and proteins derived from them are relatively free of proteolytic and lipolytic enzymes, but this is not the case with samples of animal origin. Lipases can hydrolyse lipids producing fatty acids which impart off-flavours while proteinases cleave protein polypeptide chains sometimes causing

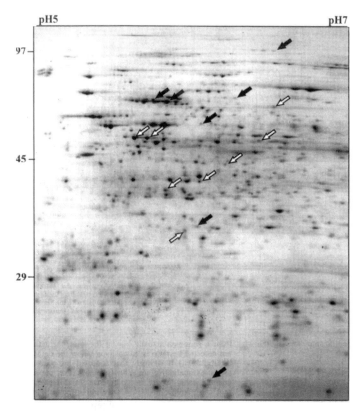

Figure 8.5 2-Dimensional electrophoresis to identify closely related durum wheat lines. (Reproduced with permission from Picard et al. (1997).)

textural, flavour or stability changes in foods, but always causing changes in the size or charge or both of the protein molecules which can be detected easily by electrophoresis. Similarly lipases and proteinases arising from microbial contamination cause the same types of changes and can be both detected and measured by most of the electrophorectic methods described. Not only does the formation of peptide fragments from proteins reveal the presence of proteinases but as individual enzymes cleave different bonds in the polypeptide chain they often give characteristic fingerprints of peptides which can be a useful pointer as to their source. This will establish if the problem is due to an indigenous proteinase enzyme or to one from a contaminating micro-organism, and if so sometimes even which one. For example, the breakdown of proteins in cheese during maturation is due to proteinases from starter micro-organisms, from any adventitious microbial contamination and from residual rennet action as well as the indigenous milk proteinase (mainly plasmin) (Fox, 1992). Thus electrophoretic methods play a crucial role in studying the deterioration of food products during both manufacture and storage.

A major application of electrophoretic technology is in the authentication and detection of fraud in food and drink samples. Although the analysis of DNA restriction fragments has been used for identifying pathogenic strains of bacteria such as *Listeria monocytogenese* (Casolari *et al.*, 1990) and the use of DNA probes for such work may have even greater advantages (Jones, 1991), most food authentication work to date has involved the analysis of proteins and peptides. Again the principal methods used have been PAGE, SDS-PAGE (Fig. 8.6) and PAGIF (Fig. 8.7) and the problem usually turns out to be one of species

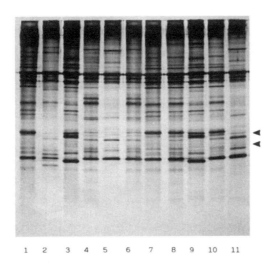

Figure 8.6 SDS electrophoretic separation profiles: 1, scampi (*nephrops norvegicus*); 2, sample C; 3, Pacific scampi (*Metanephrops andamanicus*); 4, sample D; 5, tropical shrimp (*Penaeus indicus*); 6, sample E; 7, scampi; 8, sample F; 9, Pacific scampi; 10, sample G; 11, tropical shrimp. (Reproduced with permission from Craig *et al.* (1995).)

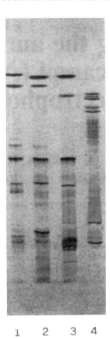

Figure 8.7 Isoelectric focusing profiles of aqueous extracts of : 1, cod (*Gadus morhua*); 2, haddock (*Melanogrammus aeglefinus*); 3, whiting (*Merlangus merlangus*); 4, scampi (*Nephrops norvegicus*). (Reproduced with permission from Craig *et al.* (1995).)

identification. For example the addition of as little as 1–2% of cow's milk to more valuable sheep or goat's milk is a common fraud that can easily be detected either in milk (Addeo *et al.*, 1990) or in cheese made from it. Similarly the fraudulent addition of cheap fish or meat species to products such as fish fingers, fish cakes, meat pies, sausages, etc. nominally made with more expensive ingredients is readily seen (Lundstrom, 1979) by protein analysis or more recently by DNA analysis. It is in such authentication work that immunoelectrophoretic methods with their ability to detect trace amounts of antigen protein in the presence of large amounts of other accompanying proteins find their greatest application.

The advent of CE methods has opened new doors in food analysis (e.g. Fig. 8.8), such as that in dairy products (van Riel and Olieman, 1995; Recio *et al.*, 1995), because it is now possible to rapidly identify not only proteins and peptides quantitatively but to extend this to the analysis of small molecules which are otherwise difficult to study. For example, Cancalon and co-workers (Cancalon and Bryan, 1993; Cancalon, 1994) have analysed in a single CZE experiment amines, amino acids, flavonoids, polyphenols, vitamins and preservatives in orange juice. Trace metal ions in beverages (Baechmann *et al.*, 1992) and organic acids in wine and coffee (Karovicova *et al.*, 1991; Cotter *et al.*, 1995) and fruit juices (Klockow *et al.*, 1994) are analysed with indirect UV

238 ANALYTICAL METHODS OF FOOD AUTHENTICATION

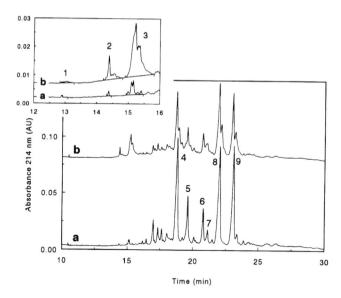

Figure 8.8 CZE of (a) casein of commercial Dutch half-skimmed pasteurized milk and (b) half-skimmed UHT milk. The inset shows the electropherogram of 12–16 min at a 5× higher sensitivity. Peaks: 1, BSA; 2, α-LA; 3, β-LG; 4, α_{s1}-casein; 5, α_{s0}-casein; 6, κ-casein; 7, β-casein B; 8, β-casein A1; 9, β-casein A2. (Reproduced with permission from Recio and Olieman (1995).)

detection or by conductivity, and similar methods can be used for sugar analyses, although a good alternative approach is to derivatize the sugar with a UV absorbing or fluorescent label (Schwaiger et al., 1994). Applications of CE methods in food analysis have recently been reviewed by Cancalon (1995).

Thus it can be seen that electrophoretic methods already hold a very prominent position in food authentication and analysis and new methods and applications are being developed all the time, so there is undoubtedly considerable further potential still to be tapped.

References

Addeo, F., Moio, L., Chianese, L. et al. (1990) *Milchwissenschaft*, **45**, 708.
Allen, R.C. and Maurer, H.R. (1974) *Electrophoresis and Isoelectric Focusing in Polyacrylamide Gel*, Walter de Gruyter, Berlin.
Andrews, A.T. (1986) *Electrophoresis; Theory, Techniques and Biochemical and Clinical Applications*, 2nd edn, Oxford University Press, Oxford.
Andrews, A.T. (1991) *Eurofood Trends and Technology*, 101.
Axelsen, N.H., Kroll, J. and Weeke, B. (1973) *A Manual of Quantitative Immunoelectrophoresis. Methods and Applications*, Blackwell Scientific Publications, Oxford.
Baechmann, K., Boden, J. and Haumann, I. (1992) *J. Chromatogr.*, **626**, 259.
Balchunas, A.T. and Sepaniak, M.J. (1987) *Anal. Chem.*, **59**, 1466.
Bjellqvist, B., Ek, K., Righetti, P.G. et al. (1982) *J. Biochem. Biophys. Meth.*, **6**, 317.
Burg, H.C.J. and van den Berg, B.M. (1996) *Electrophoresis*, **17**, 502.
Cancalon, P.F. (1994) *Am. Lab.*, **26**, 48 F.
Cancalon, P.F. (1995) *Food Technology*, 52.

Cancalon, P.F. and Bryan, C.R. (1993) *J. Chromatogr.*, **652**, 555.
Casolari, C., Facinelli, B., Fabio, U. *et al.* (1990) *Eur. J. Epidemiol.*, **6**, 319.
Chen, A.J.C., Zhu, M., Hansen, D. and Burd, S. (1989) *Bulletin* 1479, Bio-Rad Laboratories, Richmond, California, USA.
Clarke, H.M.G. and Freeman, T. (1968) *Chin. Sci.*, **35**, 403.
Clarke, P., Lin, H.-C. and Wilcox, G. (1982) *Anal. Biochem.*, **124**, 88.
Cleve, H., Patutschnick, W., Postel, W. *et al.* (1982) *Electrophoresis*, **3**, 342.
Cohen, A.S. and Karger, B.L. (1987) *J. Chromotogr.*, **397**, 409.
Cole, R.O., Sepaniak, M.J. and Hinze, W.L. (1980) *J. High Res. Chromatogr.*, **13**, 579.
Cooke, R.J. (1984) *Electrophoresis*, **5**, 59.
Cooke, R.J. (1989) *Advances in Electrophoresis* vol. 2, (eds A. Chrambach, M.J. Dunn and B.J. Radola), pp. 171–261, VCH Publishers, Weinheim.
Cotter, R.L., Benvenuti, M. and Krol, J. (1995) Pittsburgh Conference on *Capillary Electrophoresis*, Abstract 1365.
Culliford, B.J. (1964) *Nature*, **201**, 1092.
Dobashi, A., Ono, T., Hara, S. and Yamaguchi, J. (1989) *Anal. Chem.*, **61**, 1986.
Dolnick, V., Cobb, K.A. and Novotny, M. (1990) *J. Microcol. Sep.*, **2**, 127.
Drossman, H., Luckley, J.A., Kostichka, A. *et al.* (1990) *Anal. Chem.*, **62**, 900.
Dubrow, R.S. (1992) in *Capillary Electrophoresis Theory and Practice* (eds P.D. Grossmann and J.C. Colburn), pp. 133–57, Academic Press, San Diego.
Dunbar, B.S. (1987) *Two-Dimensional Gel electrophoresis and Immunological Techniques*, Plenum Press, New York.
Fox, P.F. (1992) *Food Enzymology*, Elsevier Applied Science, London.
Furlan, M., Perret, B.A. and Beck, E.A. (1979) *Anal. Biochem.*, **96**, 208.
Gershoni, J.M. and Palade, G. (1982) *Anal. Biochem.*, **124**, 396.
Gershoni, J.M. and Palade, G. (1983) *Anal. Biochem.*, **131**, 1.
Goldman, D. and Merril, C.R. (1982) *Electrophoresis*, **3**, 24.
Gordon, A.H. (1975) *Laboratory Techniques in Biochemistry and Molecular Biology*, Vol. 1, Electrophoresis of proteins in polyacrylamide and starch gels, (eds T.S. Work and E. Work), 2nd edn, North Holland Press, Amsterdam.
Grabar, P. and Williams, C.A. (1953) *Biochim., Biophys., Acta*, **10**, 194.
Grossmann, P.D. and Colburn, J.C. (1992) *Capillary Electrophoresis, Theory and Practice*, Academic Press, San Diego.
Grossmann, P.D. and Soane, D.S. (1991) *Biopolymers*, **31**, 1221.
Günther, S., Postel, W., Weser, J. and Görg, A. (1986) in *Electrophoresis '86* (ed. M.J. Dunn), pp. 485–8, VCH Publishers, Weinheim.
Guttman, A. (1996) *Electrophoresis*, **17,** 1333.
Guttman, A., Paulus, A., Cohen, A.S. *et al.* (1988) *J. Chromatogr.*, **448**, 41.
Hames, B.D. and Rickwood, D. (1990) *Gel Electrophoresis of Proteins. A Practical Approach*, IRL Press, Oxford.
Heiger, D.N., Cohen, A.S. and Karger, B.L. (1990) *J. Chromatogr.*, **516**, 33.
Heukeshoven, J. and Dernick, R. (1985) *Electrophoresis*, **6**, 103.
Hjertén, S. (1992) in *Capillary Electrophoresis, Theory and Practice* (eds P.D. Grossman and J.C. Colburn), pp. 191–214, Academic Press, San Diego.
Hjertén, S., Jerstedt, S. and Tiselius, A. (1965) *Anal. Biochem.*, **11**, 219.
Jones, J.L. (1991) *Trends in Food Science & Technology*, **2**, 28.
Karger, B.L., Cohen, A.S. and Guttmann, A. (1989) *J. Chromatogr.*, **492**, 585.
Karovicova, J., Polonsky, J. and Simko, P. (1991) *Nahrung*, **35**, 543.
Klockow, A., Paulus, A., Figueiredo, V. *et al.* (1994) *J. Chromatogr.*, **680**, 187.
Koch, G.L.E. and Smith, M.J. (1982) *Eur. J. Biochem.*, **128**, 107.
Laing, P. (1986) *J. Immunol. Methods*, **92**, 161.
Laurell, C.B. (1965) *Anal. Biochem.*, **10**, 358.
Laurell, C.B. (1966) *Anal. Biochem.*, **15**, 45.
Lundstrom, R.C. (1979) *J. Assoc. Off. Anal. Chem.*, **62**, 624.
Lux, J.A., Yin, H.F. and Schomburg, G. (1990) *J. High Res. Chromatogr.*, **13**, 436.
Merril, C.R., Goldman, D. and Van Keuren, M.L. (1982) *Electrophoresis*, **3**, 17.
Morrissey, J.H. (1981) *Anal. Biochem.*, **117**, 307.
Nishi, H., Fukuyana, T., Matsuo, M. and Terabe, S. (1990) *J. Chromatogr.*, **498**, 313.

O'Farrell, P.H. (1975) *J. Biol. Chem.*, **250,** 4007.
Picard, P., Bourgoin-Grenèche, M. and Zivy, M. (1997) *Electrophoresis*, **18,** 174.
Rabilloud, T. (1992) *Electrophoresis*, **13,** 429.
Recio, I., Molina, E., Ramos, M. and de Frutos, M. (1995) *Electrophoresis*, **16,** 654.
Riggs, T.J., Sanada, M., Morgan, A.G. and Smith, D.B. (1983) *J. Sci. Food Agr.*, **34,** 576.
Righetti, P.G. (1983) *Isoelectric Focussing: Theory, Methodology and Applications*, Elsevier, Amsterdam.
Righetti, P.G. (1984) *J. Chromatogr.*, **300,** 165.
Righetti, P.G., Delpech, M., Moisand, F. *et al.* (1983) *Electrophoresis*, 4, 393.
Scheidegger, J.J. (1955) *Int. Arch. Allergy*, **7,** 103.
Schwaiger, H., Oefner, P.J., Huber, C. *et al.* (1994) *Electrophoresis*, **15,** 941.
Sepaniak, M.J., Burton, D.E. and Maskarinec, M.P. (1987) in *Micellar Electrokinetic Capillary Chromatography* (ed. M.J. Comstock) Vol. 342, pp. 142–51, American Chemical Society, Washington, D.C.
Serwer, P. (1983) *Electrophoresis*, **4,** 375.
Sharp, P.A., Sugden, B. and Sambrook, J. (1973) *Biochemistry*, **12,** 3055.
Smith, D.B. and Lister, P.R. (1983) *J. Cereal Sci.*, **1,** 229.
Smith, I. (1975) *Chromatographic and Electrophoretic Techniques*, Vol 2, Zone electrophoreses, Heinemann Medical Books, London.
Southern, E.M. (1975) *J. Mol. Biol.*, **98,** 503.
Szökö, E. (1997) *Electrophoresis*, **18,** 74.
Terabe, S., Otsuka, K., Ichikawa, K. *et al.* (1984) *Anal. Chem.* **56,** 113.
Terabe, S., Utsumi, H., Otsuka, K. *et al.* (1986) *J. High Res. Chromatogr.*, **9,** 666.
The, T.H. and Feltkamp, T.E.W. (1970) *Immunology*, **18,** 865.
Towbin, H., Staehelin, T. and Gordon, J. (1979) *Proc. Nat. Acad. Sci.*, **76,** 4350.
Tsugita, A., Kamo, M., Kawakami, T. and Ohki, Y. (1996) *Electrophoresis*, **17,** 855.
Van Riel, J. and Olieman, C. (1995) *Electrophoresis*, **16,** 529.
Vesterberg, O. (1971) *Meth. Enzymol.*, **22,** 289.
Vesterberg, O. and Svensson, H. (1966) *Acta Chem. Scand.*, **20,** 820.
Wainwright, A. (1990) *J. Microcol. Sep.*, **2,** 166.
Zacharius, R.M., Zell, T.E., Morrison, J.H. and Woodlock, J.J. (1969) *Anal. Biochem.*, **30,** 148.

9 Antibody techniques
E. MÄRTLBAUER

9.1 Introduction

Immunochemical methods are based on the ability of antibodies (immunoglobulins) to recognize three-dimensional structures. They play a major role in biochemical research, clinical chemistry and food analysis. Being primarily a part of the immune system in most classes of vertebrates (Stanworth and Turner, 1979), immunoglobulins have been utilized as the key substances in any immunoassay for more than 40 years. Molecules capable of inducing the production of immunoglobins in a certain species are called immunogens or antigens. An immunogen must have a minimum size (mol wt. about 5000 daltons) and contain structures which are 'foreign' to the challenged animal species. Haptens, on the other hand, are substances which do not elicit an immunological response but are able to react with antibodies. Antibodies against haptens are induced by using artificial immunogens, e.g. haptens coupled to a carrier protein (macromolecule). Table 9.1 shows a list of substances and microorganisms, which are of importance in food analysis, and for which specific antibodies have been described.

Antibodies represent a group of glycoproteins possessing two distinct types of polypeptide chains linked by both covalent and non-covalent bonds (Fig. 9.1). Both the light chain (L, mol wt. 25 000 daltons) and the heavy chain (H, mol wt. 50–77 000 daltons) show a variable region (V) at their amino terminal end of about 110 amino residues, whereas the remaining part of the polypeptide chain is referred to as the constant region (C). The variable regions of both chains contain a hypervariable part which represents the antigen binding site (antibody

Table 9.1 Potential analytes for food immunoassay (Hitchcock, 1988, modified)

Category	Example
Bacterial toxins	*Staphylococcus aureus* toxins
Mycotoxins	Aflatoxins
Anabolics	Nortestosterone
Antimicrobial drugs	Chloramphenicol, sulfonamides
Pesticides	DDT, Alachlor
Contaminants	PCBs, PCDDs
Vitamins	B_{12}, D
Enzymes	Plasmin, papain
Hormones	Somatotropin
Food proteins	Soja, bovine IgG, casein
Bacteria	*Salmonella*
Moulds	*Penicillium*

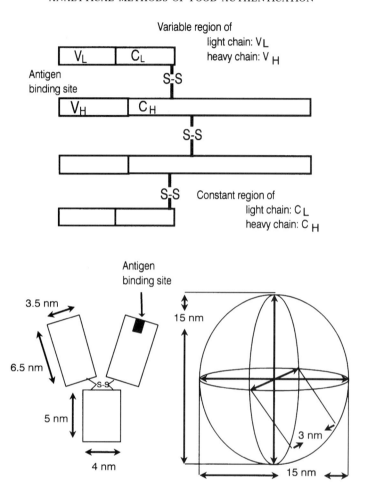

Figure 9.1 Schematic representation of an antibody molecule of the IgG-type with indication of the approximate size of the individual fragments. Under practical conditions (e.g. immunodiffusion) the antibody behaves like a spheroid molecule with a diameter of about 15 nm and a height of approximately 3 nm.

combining site) or 'paratope'. After folding and combining of the L- and H-chains, this hypervariable region of the immunoglobulin shows a structure complementary to the corresponding part of the antigen molecule, which is referred to as antigenic determinant or 'epitope'. The class and subclass of an immunoglobulin molecule is determined by its heavy chain type. Thus γ-chains are characteristic for IgG, μ-chains for IgM, α-chains for IgA, δ-chains for IgD and ε-chains for IgE. The light chains of most vertebrates have been shown to exist as κ- and λ-types. For analytical purposes, immunoglobulins of the IgG-type are mostly used. The IgG class is divided into subclasses, which differ from each other by different numbers and arrangements of the interchain disulphide

Table 9.2 Immunoglobulin G subclasses in mammalian species used for antibody production (Stanworth and Turner, 1979; Brock, 1980)

Species	Subclass	Electrophoretic mobility
Mouse	IgG1	Fast
	IgG2a	Slow
	IgG2b	Slow
	IgG3	Slow
Rabbit	IgG	Variable (different fractions described)
Cow	IgG1	Fast
	IgG2	Slow
Sheep	IgG1	Fast
	IgG2	Slow
Goat	IgG1	Medium
	IgG2	Medium

bonds resulting from different amino acid sequences influencing their electrophoretic mobility characteristics. Table 9.2 gives an overview of immunoglobulin G subclasses in some mammalian species.

The animal species used most often to produce antisera for diagnostic purposes are rabbits, mice, sheep and goats. In view of the differentiation of species specific proteins, the choice of the animal species used for immunization is, however, critical. As a general rule, the closer the phylogenetic relationship between the immunized host and the species of which the target antigen originates, the more specific the immune response will be. If, for example, rabbits (or mice) are used for the production of antibodies against cow milk proteins, the resulting antibodies will show cross reactivity with milk proteins of closely related species, such as goats and sheep. This cross reactivity can be avoided if sheep or goats are used for the immunization (Levieux, 1977; Radford et al., 1981; Garcia et al., 1991). In other words, the best way to produce a specific antiserum against cow milk proteins (IgG, casein, etc.) for the detection of cow's milk in ewe's milk is to immunize sheep.

The specific antibodies produced in an animal species are **polyclonal** in nature because usually a number of B-lymphocytes is stimulated by the immunization process. The antibodies are then synthesized and secreted by plasma cells, derived from these lymphocytes. In 1975, Köhler and Milstein developed a technique, in which antibody secreting lymphocytes are fused with myeloma cells, a type of B-cell tumour. The myeloma cells provide the genes for continued cell division, whereas the lymphocytes provide the functional immunoglobulin genes. The fused cells, which possess the properties of both parent cell lines, are called hybrid cells or hybridomas, and produce **monoclonal antibodies** (for further detail see Goding, 1986; Peters and Baumgarten, 1992). Main advantages of monoclonal compared with polyclonal antibodies are unrestricted availability, once a cell line is established, homogeneity, and often improved specificity. In addition, a further advantage of this technique is that impure antigens may be used for antibody production. Monospecific antibodies can be obtained by using complex mixtures of antigens since one hybridoma is cloned from one single antibody-producing cell, which is programmed to make only one kind of antibody (clonal selection theory; Roitt et al., 1985).

9.2 Principles

Immunochemical methods are based on the ability of antibodies to bind different substances specifically. The reversible association between antibodies and their corresponding antigens is called the immunological reaction. The binding forces involved are weak molecular interactions like Coulomb and van der Waals forces, as well as hydrogen bonds and hydrophobic binding. The antigen-antibody reaction is based on the law of mass action, and the amount of antigen or antibody present in the reaction mixture may be inferred from the extent of the reaction.

According to the law of mass action the following reaction equation is valid for the unbound antigen (Ag), unbound antibody (Ab), and the antigen-antibody complex (AgAb):

$$Ag + Ab \underset{k_{-1}}{\overset{k_{+1}}{\rightleftharpoons}} AgAb \quad (9.1)$$

At equilibrium, the association constant (affinity constant) K_a and the dissociation constant K_d are defined as:

$$K_a = \frac{[AgAb]}{[Ag][Ab]} = \frac{k_{+1}}{k_{-1}}$$

and

$$K_d = \frac{[Ag][Ab]}{[AgAb]} = \frac{k_{-1}}{k_{+1}}$$

where K_d is the reciprocal value of K_a having the dimension mol/l, whereas K_a is usually given as 1/mol.

The most important reaction parameter is the dissociation constant K_d. If K_d is known, theoretical binding curves may be calculated according to the equation described by Halfman and Schneider (1981). The 'quality' of any immunoassay is a function of the immunochemical principle of the method, the properties of the reagents, the assay design, and, of course, the experimental errors. These basic principles determine the sensitivity, specificity, precision and accuracy of the assay.

9.2.1 Marker-free methods

Concerning the detection principle, a distinction is to be made between marker-free immunological methods and techniques using labelled reagents. Marker-free methods use the primary reaction between antigen (Ag) and antibody (Ab), and the antigen-antibody complex formed acts itself as the 'marker system'. The main analytical techniques employing this type of detection are immunodiffusion, nephelometry and turbidimetry. This contribution, however, will give only a short introduction to immunodiffusion and immunoelectrophoresis which are mainly used for the detection of food adulteration. For detailed descriptions the reader is referred to Walker (1984), or Clausen (1988). Schematically, the general immunochemical principle of marker-free methods may be presented as:

$$Ag + Ab \rightarrow AbAb \rightarrow precipitate$$

Usually the immunoprecipitates formed between antigen and antibody are of a molecular size greater than the pore size of the gel and cannot diffuse further or be washed out of the gel. Unreacted reagents, however, may easily be removed by using a 'washing step'. This is of particular importance, if the formed immunoprecipitates are visualized using protein staining techniques.

Immunodiffusion. Immunodiffusion covers a variety of techniques that are useful for the analysis of species-specific antigens. The immunochemical principle is exactly the same as described above for antigen-antibody interactions in the liquid state. As increasing amounts of a multivalent antigen (i.e. an antigen containing two or more antibody binding sites) are allowed to react with a fixed amount of antibody, precipitation occurs, in part because of extensive cross-linking between the reactant molecules. Initially, the antibody is in excess and all of the added antigen is present in the form of an insoluble antigen-antibody aggregate. Addition of more antigen leads to the formation of more immune precipitate. However, a point is reached beyond which further addition of antigen produces an excess of antigen and leads to a reduction in the amount of the precipitate because of the formation of soluble antigen-antibody complexes.

The methods employing immunodiffusion in gels are often classified as single (simple) diffusion or double diffusion. In simple diffusion, one of the reactants (usually the antigen) is allowed to diffuse from solution into a gel containing the corresponding reactant, whereas in double diffusion both the antigen and the antibody diffuse into the gel.

If **single radial diffusion** (Mancini *et al.*, 1965) is used for species differentiation, the antigen (sample extract) is applied in a hole and allowed to diffuse radically into gel containing the antibody. It is a main prerequisite that the antiserum used is monospecific, otherwise multiple precipitin lines will be formed. The diameter or the area of the precipitation band is proportional to the concentration of the antigen. Usually the immunoprecipitates are not evaluated directly but after staining with protein stains (e.g. Amido Black).

In **double diffusion** (Ouchterlony, 1968), both the antigen and the antibody are applied in separate wells and allowed to diffuse into the gel. The position where the precipitin line is formed mainly depends on the diffusion rate of the two reactants. If different antigens, each capable of reacting with the antiserum used, are allowed to diffuse from separate wells, an indication of similarity of the antigens is given by the geometrical pattern of the precipitin lines produced. Figure 9.2 shows the basic precipitin patterns that can be produced in a balanced system between antigens and a particular test antibody.

Immunoelectrophoresis. In principle this method combines gel immunodiffusion and electrophoresis. Antigens are separated on the basis of their charge and then visualized by precipitation using specific antibodies. First, the sample containing an antigen mixture is separated by electrophoresis. In the second step, antiserum against one or more of the sample proteins is applied to the gel in parallel to the electrophoretic moving direction. After double diffusion of the antigen and the antiserum, precipitates are formed in a characteristic system of arcs alongside the electrophoretic zones. In most applications of this method described so far, a modification called **crossed immunoelectrophoresis** is used for either qualitative or quantitative measurements.

Rocket immunoelectrophoresis was originally introduced by Laurell (1966) and involves a comparison of a sample of unknown protein concentration with a series of calibration solutions of that protein. This technique requires a monospecific antiserum against the protein to be analysed. Samples and solutions are loaded on the antibody-containing gel, and are driven towards the anode after applying electrical power. Depending on the antigen concentration at a certain position on the gel, a point of equivalence will be reached, where the antigen-

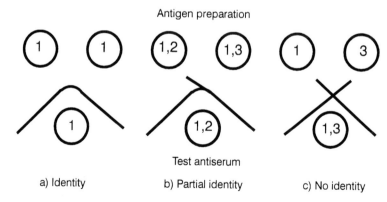

Figure 9.2 Schematic comparison of possible results obtained in double diffusion assays: (a) 'pattern of identity' – the precipitin lines formed fuse, indicating that the antibodies in the antiserum employed (all against epitope no. 1) cannot distinguish between the two antigens. (b) 'Pattern of partial identity' – the precipitin lines fuse, but show the addition of a 'spur' indicating partial identity between the antigen preparations (antiserum contains antibodies against epitope no. 1 and 2). The 'spur' is the result of an additional epitope present in one antigen preparation, but lacking in the other, which can also be recognized by the antibodies. (c) 'Pattern of non-identity' – independent precipitin lines are found, indicating that the antibodies in the antiserum react with different antigenic determinants (no. 1 and 3).

antibody complex is insoluble. The resulting precipitation line represents a 'rocket' spreading out from the loading well. Since peak height depends on the relative excess of antigen over antibody, a comparison of the areas of the unknown and standard samples allows the unknown protein concentration to be determined.

In **countercurrent electrophoresis** the antigen and the antibody move towards each other and precipitate after voltage is applied across the gel. This method is usually performed under conditions where the antibody is positively and the antigen negatively charged. Therefore the technique is applicable to any protein with an electrophoretic mobility at pH 8.6 greater than γ-globulin. In principle the method is similar to double diffusion, but it is quicker and provides better sensitivity.

9.2.2 Enzyme immunoassay

General remarks. Although marker-free methods are still useful for many purposes, this chapter will place most emphasis on so-called labelled reagent methods, particularly on enzyme immunoassays. The use of labelled reagent methods has improved the sensitivity of immunoassays by several orders of magnitude compared to marker-free methods such as simple agglutination techniques. The enzyme immunoassay (EIA) has become the most important immunochemical method in food analysis.

Labelled reagent methods need either labelled antigen or labelled antibody to permit observation of the antigen-antibody reaction. EIA is therefore a combination of two principles, the antigen-antibody reaction and the enzyme substrate reaction. Enzymes commonly used for labelling procedures are given in Table 9.3, together with the respective substrate (and chromogen) and the principle of measurement. Concerning the principle of the method, a distinction is to be made between competitive and non-competitive methods.

Competitive methods are based on the competition of free (Ag) and labelled or solid phase bound antigen (Ag*) for a limited number of antibody combining sites (AB). Schematically the reaction equation may be presented as:

$$Ag + Ag^* + AB \rightarrow AgAB + Ag^*AB \qquad (9.2)$$

Table 9.3 Common enzymes used for EIA

Enzyme	Analytical principle	Substrate
Alk. phosphatase	Photometry	4-Nitrophenol
	Fluorimetry	4-Methylumbelliferone
Peroxidase	Photometry	H_2O_2(+Chromogen)
	Fluorimetry	H_2O_2 (+Hydroxyphenylpropionic acid)
	Luminometry	H_2O_2 (+Luminol)
β-D-Galactosidase	Photometry	2-Nitrophenylgalactopyranoside
	Fluorimetry	Methylumbeliferyl-β-D-galactopyranoside
Glucose oxidase	Fluorimetry	Homovanillic acid
Luciferase	Luminometry	Luminol

In most cases the assay response represents the bound labelled antigen, but any other measure of the distribution of the labelled antigen is possible in principle.

In a typical **competitive EIA**, free and solid phase bound antigen compete for a limited number of antibody combining sites. Using a solid support (microtitre plate) provides the possibility to remove all reagents, which are not bound to the solid phase antigen, by a 'washing' step. The assay response represents the solid phase bound antigen, and is therefore inversely proportional to the concentration of the free antigen. The main steps of the assay procedure are shown in Fig. 9.3.

In non-competitive methods a limited number of antigen molecules is bound by the specific (partly) labelled antibodies (AB*). Schematically, the reaction equations may be presented as

$$Ag + AB + AB^* \rightarrow ABAgAB^* \qquad (9.3)$$

or

$$Ag + AB^* \rightarrow AgAB^* \qquad (9.4)$$

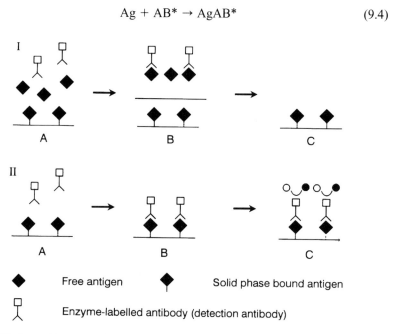

Figure 9.3 Competitive enzyme immunoassay: In the first step the microtitre plate is coated with a known amount of target antigen. Then the unknown sample and the specific antibody are added (A). After competition of free and solid-phase bound antigen for the antibody combining sites, unbound reagents are removed by a washing step. In the last step, enzyme substrate and chromogen are added, and the colour formation is estimated visually or by using a photometer. I. If the sample contains a sufficiently high amount of the target antigen, all the antibodies are absorbed by the free antigen and will be removed by the washing step (B). Since no enzyme-labelled antibody is bound, no colour formation will occur (C), and a positive result is obtained. II. If the sample contains no or low amounts of target antigen, the enzyme-labelled antibody will be bound to the solid phase antigen (B) and colour formation occurs (C).

The assay response is usually represented by the bound labelled antibody in order to obtain highest sensitivity.

Two typical variants of this type of assay are mainly used for species identification (Fig. 9.4). The **sandwich EIA** can only be used for the detection of macromolecules, such as species specific proteins like IgG, having at least two antigenic determinants in suitable steric positions, enabling two antibodies to bind to the antigen. In both variants, the target antigen is bound by the specific antibodies. In the first type of assay, a part of the antibodies (capture antibody) is bound to a solid-phase (microtitre plate) as the immunosorbent, so all the reagents, which are not bound by the antibody can be easily removed by 'washing' the solid phase. The other part of the specific antibodies (enzyme-labelled) is subsequently added, in order to monitor the extent of the immunological reaction. In the second variant the solid phase is coated directly with the antigen and the amount of antigen bound is determined using the specific enzyme-labelled antibodies. In both cases the assay response is directly

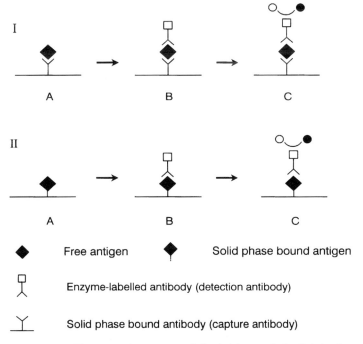

Figure 9.4 Non-competitive enzyme immunoassay: I. Sandwich assay. In the first step the sample is added to the microtitre plate coated with specific antibodies (A). Then the enzyme-labelled antibody is added and allowed to bind to the antigen (B). After a washing step, enzyme substrate and chromogen are added and the colour formation is observed for positive samples containing the target antigen (C). II. The microtitre plate is coated directly with the sample extract solution (A). If the sample contains the target antigen, the specific enzyme-labelled antibody is bound to the solid phase via the target antigen (B), and – after a washing step – colour formation occurs after adding enzyme substrate/chromogen solution (C).

proportional to the concentration of the target antigen. The antibody used for the detection of the bound antigen may be directly labelled with a suitable enzyme. Alternatively, a second antibody-enzyme conjugate, enzyme-conjugated protein A or an antibody-biotin conjugate detected with an avidin-enzyme conjugate could be used (Figure 9.5).

For the detection of macromolecules such as most target proteins for species identification, both the competitive and the non-competitive assay format may be used. For detection of low molecular weight compounds, which possess only one antibody binding site (epitope), the competitive assay format is mandatory. To provide distinction between unreacted and complexed reactants most assays use as immunosorbent either antibody or antigen bound to a solid phase. So all the reagents, which are not specifically bound by the antibody can be easily removed by 'washing' the solid phase.

Depending on the individual needs of the assay user, immunochemical test systems can be designed as quantitative routine tests (microtitre plate format) or rapid qualitative tests (membrane-based assay formats).

Microtitre plate assays. The most prevalent test format of enzyme immunoassays is still the microtitre plate assay, which is usually performed employing (semi)-automated absorbance measurement and calculation of the results. Depending on the specificity of the individual tests these assays are either quantitative or qualitative methods which can easily be performed in routine laboratories. Compared to physicochemical methods of analysis, microtitre tests – as well as immunochemical techniques in general – have advantages in aspects of the sample treatment necessary prior to analysis. In particular sample extract

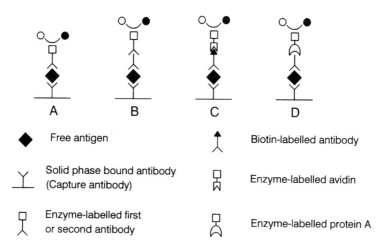

Figure 9.5 Labelling procedures (Section 9.3): A, antibody conjugated directly to an enzyme (direct label); B, antibody detected via second antibody-enzyme conjugate (second antibody label); C, antibody labelled with biotin and detected with an avidin-enzyme conjugate (avidin-biotin label); D, antibody detected via a protein A-enzyme conjugate (protein A label).

clean-up can be simplified or even totally omitted. This is mainly due to the fact that immunochemical methods are highly sensitive and specific for their target molecule. Table 9.4 shows a general scheme of working steps involved in this kind of assay.

Liquid materials such as milk present an ideal sample matrix for immunochemical assays and milk samples may be analysed without any extraction step necessary. For analysing meat and meat products in most cases the preparation of an aqueous extract including a defatting step is usually sufficient.

Rapid immunochemical test formats. During the last decade, a range of rapid on-site immunochemical tests in various test formats has been developed in clinical chemistry. In food analysis, however, only few such tests have been described. Some of the most promising approaches use membrane-based test devices either in a dipstick enzyme immunoassay or in a flow-through 'immunofiltration' system. These tests have been basically designed as visual tests requiring only low-cost instrumentation, some are even self-contained.

Dipstick and immunodot tests. Dipstick tests use either membranes (nitrocellulose, nylon, etc.) or plastic materials as the solid phase. Depending on the pretreatment of the solid support, antigens or antibodies are bound covalently or

Table 9.4 Scheme of working steps involved in a typical microtitre plate test (sandwich EIA)

Step	Example
Coating of the microtitre plate with specific antibodies	Add solution of antibodies against bovine IgG, diluted in carbonate-bicarbonate buffer (0.05 mol/l, pH 9.6) 100 µl per well; incubate 20 h at 4 °C
Blocking of free protein binding sites	Add 200 µl of a 1% solution of gelatine in phosphate buffered saline (PBS) (0.01 mol/l, pH 7.3; phosphate buffer containing 0.1 mol of NaCl per l); incubate 30 min at room temperature
Washing step	Add 3 times 200 µl of PBS containing 0.025% Tween 20 and make plate semi-dry
Incubation with antigen	Add 100 µl of sample extract or standard solution; incubate 1 h at room temperature
Washing step	see above
Incubation with labelled antibody	Add 100 µl of horseradish peroxidase labelled antibody solution diluted in PBS containing 1% Tween 20; incubate 1 h at room temperature
Washing step	see above
Colour reaction	Add 100 µl enzyme chromogen solution (potassium citrate buffer (0.2 mol/l, pH 3.9) containing 3 mmol of H_2O_2 per l and 1 mmol of 3,3′,5,5′-tetramethylbenzidine per l); incubate 20 min at room temperature
Stopping the enzymatic reaction	Add 100 µl of 1 mol/l sulphuric acid
Absorbance measurement	Read microtitre plate at 450 nm

just adsorbed by multiple non-covalent bonds on the surface of the membrane as dots or lines. The tests give qualitative or semiquantitative results in a relatively short time (<1 h). The working steps involved in this procedure are essentially the same as for the microtitre plate assay, except that usually the antibody-coated membrane is incubated in the individual test solutions rather than pipetting the solutions into the wells of a plate. Compared to the corresponding microtitre plate tests, most visual tests such as the dipstick assay show a loss of sensitivity. Several factors contribute to the reduced sensitivity, probably the most important and self-explanatory being the fact that visual evaluation is less precise and sensitive than measuring absorbance with a photometer. This is particularly important in competitive enzyme immunoassays, where the negative control has the most intense colour and a relatively large reduction in colour intensity (high analyte concentration) is required between the negative control and positive samples to avoid misinterpretations by an untrained user. The competitive format also requires that the negative control dot has a high colour intensity to facilitate the scoring of positives. Therefore antibody density, i.e. the amount of antibody for coating the membrane, as well as the concentration of the enzyme-labelled antigen, has to be relatively high. This reduces the sensitivity of a competitive enzyme immunoassay. Although the sensitivity of these tests for standard solutions of the respective analyte is in general less than that of the corresponding microtitre plate immunoassay, many dipstick immunoassays are on the other hand considerably less subject to matrix interferences and solvent composition of the extract.

Immunofiltration assay. A further reduction of assay time can be achieved by employing immunochemical, membrane-based flow-through systems, in which unbound reagents and sample matrices are removed by absorbance through a cellulose pad. Usually, the tests are performed in a plastic device (Fig. 9.6), in which antibody-coated nylon membranes are pressed tightly to an absorbent cellulose layer. Sample extract solution, antibody-enzyme conjugate, and enzyme substrate/chromogen solution are sequentially added to the membrane. The test can be evaluated visually by comparing the intensity of the resulting coloured (blue) dot with that of a negative control (Fig. 9.6). Qualitative results can be obtained within 10 min. These assays find their ideal application as on-site tests to detect the presence of the target molecule at a defined threshold level.

Immunoblot techniques. Immunoblot techniques combine the analytical potential of electrophoretical separation methods with the specificity of the antigen-antibody reaction, thus enabling the sensitive and reliable detection of single proteins in complex mixtures. First, proteins are separated by an appropriate electrophoretical technique, e.g. PAGE (polyacrylamide gel electrophoresis), SDS-PAGE (sodium dodecyl sulphate-PAGE), IEF (isoelectric focusing), and then transferred ('blotted') from the gel on to the surface of an immobilizing

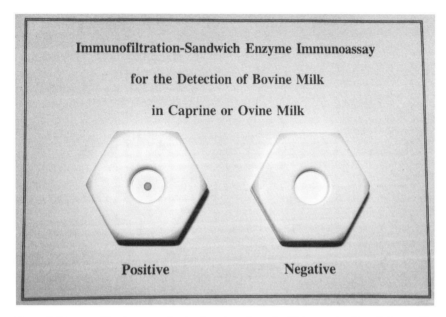

Figure 9.6 Immunofiltration assay for the detection of cows' milk in ewes' milk and cheese. The test principle is a non-competitive (sandwich) assay. Further details are outlined in the text.

membrane. For this purpose nitrocellulose, nylon and polyvinylidenedifluoride (PVDF) sheets are most commonly used. Before the blot is incubated with the specific polyclonal or monoclonal antibodies, free binding sites on the membrane have to be blocked. Then the membrane is incubated with a second, enzyme labelled antibody, which reacts with the primary, specific antibody. Finally, the antigen-antibody reaction is visualized by the addition of an enzyme substrate/chromogen solution, which yields a coloured, water-insoluble product. The detection of the analyte on a membrane rather than in a gel offers certain advantages: proteins are more accessible, membranes are easier to handle than gels, smaller amounts of reagents are needed, and processing times are shorter.

9.2.3 *Sensitivity*

The potential sensitivity of any immunoassay is directly related to the affinity of the antibody and may be calculated, if the equilibrium constant is known (Fig. 9.7). Since the antigen-antibody reaction may be described using reaction kinetics as well as thermodynamic equations, reaction time and temperature also influence assay sensitivity. Relying on antibody affinity there is a significant difference between competitive and non-competitive methods. In practice, non-competitive methods may show sensitivities which are one or two orders of magnitude greater than comparable competitive assays (Jackson and Ekins, 1986). The dominant factor causing the difference in sensitivity between both

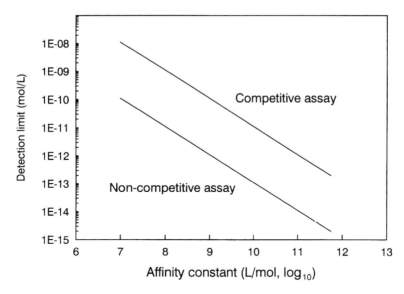

Figure 9.7 Potential sensitivity of competitive versus non-competitive immunoassays according to Jackson and Ekins (1986). By using excess concentrations of immunoreagents the non-competitive assay shows about 100 times higher sensitivity compared to a competitive assay using antibodies with identical affinity.

assay formats is the use of excess reagents in non-competitive assays. This theoretical advantage of non-competitive over competitive assays, however, can only be utilized if markers with high specific activity, such as horseradish peroxidase or alkaline phosphatase, are used for labelling.

9.2.4 Specificity

Next to sensitivity, the specificity of an immunochemical method is important for the performance of the assay. In principle, a 'specific reaction' in immunology may be defined as follows. In the presence of different molecules the specific antibodies must complex only one kind of molecules. The probability of forming a 'wrong' complex determines the specificity of the reaction. Specificity is determined by the steric (three-dimensional) match of antigen and antibody, as well as by the number of interactions taking place between both molecules. Discussion of specificity requires that both the structure of the antigen and the homogeneity or heterogeneity of the antibodies be considered. An antibody preparation is homogeneous if all the antibodies bind only to one and the same epitope, although possibly with different affinity. This definition applies both to monoclonal antibodies, and also to antisera against compounds of low molecular weight (haptens). On the other hand, an antibody preparation is heterogeneous if it contains different antibody populations specific for different epitopes. A typical example of the latter is an immune serum obtained after immunization with a high molecular weight protein such as an immunoglobulin. A mixture of different monoclonal antibodies may also represent a heterogeneous antibody preparation. From a practical point of view the term 'monospecific' is frequently used in publications dealing with species identification. In the context of this chapter, a **monospecific antiserum** is defined as an antiserum reacting only with one (or more) proteins from a single animal species. As mentioned in the introduction, rabbits immunized with bovine IgG will produce antisera that contain antibodies against bovine, ovine, and caprine IgG, because the IgG molecules of these phylogenetically closely related species share common epitopes. In order to make such an antiserum monospecific, it may be either absorbed with serum of sheep and goats, or purified by immunoaffinity chromatography. In its widest sense a monospecific antiserum is therefore an heterogeneous antibody preparation containing active antibodies only against epitopes specific for one animal species. Therefore, even monoclonal antibodies need not necessarily be monospecific because they may be directed against an epitope common to several animal species.

On a molecular level, consideration of antibody heterogeneity leads to the distinction of two types of cross reactivity, e.g. the 'true' cross reactivity and the shared reactivity (Berzofsky and Schechter, 1981). The first concept (true cross reactivity) describes the case in which at least two different antigens compete for the same antibody binding site (Fig. 9.8A). The second concept describes the competition of different antigens for different antibody combining sites (Fig. 9.8B). Almost exclusively the first type of cross reactivity is observed in competitive assays using either monoclonal antibodies or antisera against low

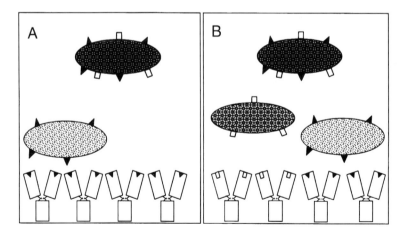

Figure 9.8 Principles of cross reactivity according to Berzofsky and Schechter (1981). A represents 'true' cross reactivity, where different antigens compete for the same antibody binding site, B shows the competition of different antigens for different antibody binding sites (shared reactivity). A typical example would be the reaction of a rabbit polyclonal antiserum against bovine IgG incubated in a mixture of cows' and ewes' milk. Only the bovine IgG contained in cows' milk would be able to saturate all antibody combining sites, whereas ovine IgG shares only some epitopes, and therefore occupies only a part of the antibody binding sites.

molecular weight compounds. The second type may be observed both in competitive as well as non-competitive systems, but only if an heterogeneous antibody preparation is used. In practice this means that a sufficiently high concentration of a truly cross reacting substance gives the same result as the 'right' substance in a competitive assay. A substance with shared reactivity, however, only influences assay response until the shared antibody binding sites are saturated.

It is important to keep in mind that non-specific influence on the assay, caused by matrix effects, for example, often cannot be distinguished from the specific influence of cross reacting substances. In both cases the assay response may be the same and a false positive or a false negative result is obtained. Due to the immunoassay principle, however, false negatives are unlikely in competitive assays but may occur in non-competitive assays. False positive results may be observed in both assay types.

9.3 Specific applications

9.3.1 *Milk and milk products*

The detection of cows' milk in ewes' and goats' milk and cheeses has been the main field for applying antibody techniques. Until the end of the last decade immunoelectrophoretic methods were predominant (Table 9.5), although a few immunodiffusion techniques were described. A double diffusion assay was described in 1974 by Durand *et al.* for the detection of 1% cows' milk in goats' milk. Later (1981) Gombocz *et al.* described an Ouchterlony-technique able to detect an admixture of about 5% of cows' milk in ewes' milk cheese. By using bovine whey proteins for the immunization of goats, Garcia *et al.* (1989) developed a double diffusion test (COMIT, cows' milk identification test) to detect about 3% of cows' milk in ewes' milk and cheese. A single radial diffusion assay (CV Test II), developed by Levieux (1977, 1978), is still in use and commercially available (CERB-Baugy, France). This assay was evaluated by several authors (Amigo *et al.*, 1989; Calvo *et al.*, 1989; Kaiser, 1989) and proved to allow the detection of about 1% cows' milk in ewes' and goats' milk, and about 5% in cheese.

To avoid the cross reactivity of the antisera against cows' milk proteins with proteins in ewes' or goats' milk, the authors immunized either sheep or goats (Levieux, 1977; Radford *et al.*, 1981; Garcia *et al.*, 1991), or absorbed the antisera against bovine casein by adding ovine or caprine casein (Rodriguez *et al.*, 1990) or more complex mixtures (Gombocz *et al.*, 1981). Immunoaffinity chromatography was used to purify polyclonal antisera by Aranda *et al.* (1988), Sargeant *et al.* (1989) and Garcia *et al.* (1991). Antibody preparations obtained after these purification steps could also be used to establish non-competitive (sandwich) EIAs for the detection of about 1% or less of cows' milk in ewes' milk (Table 9.6).

Table 9.5 Immunoelectrophoretic methods for the detection of cows' milk in ewes' and goats' milk or cheese, respectively

Method	Antigen	Samples tested	Detection limit (%)	Author(s)
Crossed immuno-electrophoresis	β-casein	ewe milk cheese goat milk cheese	0.01	Berner (1990)
Countercurrent electrophoresis	IgG	ewes' milk, goats' milk ewe milk cheese, goat milk cheese	0.1–1 0.1–2	Bernhauer et al. (1983)
Qualitative immuno-electrophoresis	bovine serum proteins (immunoglobulin, serumalbumin)	goats' milk		Durand et al. (1974)
Crossed immuno-electrophoresis	β-casein	ewe milk cheese goat milk cheese	0.1–0.2	Elbertzhagen (1987)
Crossed immuno-electrophoresis	α-lactalbumin β-lactoglobulin	ewe milk cheese whey		Elbertzhagen and Wenzel (1982)
Crossed immuno-electrophoresis	α_{s1}-casein	ewes' milk ewe milk cheese		Gombocz et al. (1981)
Qualitative immuno-electrophoresis	β-lactoglobulin			Kluge-Wilm (1988)
Rocket immuno-electrophoresis	cows' milk	goats' milk goats' milk	5 1	Radford et al. (1981)

The milk proteins used to produce specific antisera are either caseins or whey proteins. Caseins have the advantage that they represent the main part of the protein fraction in milk and are more or less heat-stable. This means that caseins may successfully be used as target antigens in heat treated (pasteurized or UHT) milk and milk products. The main disadvantage of caseins is their poor immunogenicity (Perez et al., 1992). In addition, caseins are susceptible to proteolytic degradation. In contrast, whey proteins are less susceptible to proteolysis and mostly good immunogens, but are less heat-stable than the caseins. Nevertheless, a very sensitive assay was established in our laboratory by immunizing sheep with bovine IgG (Sauer, 1992). The resulting antiserum was very specific for bovine IgG and cross-reacted with neither homologous immunoglobulins of closely related species (caprine or ovine IgG) nor with other bovine whey proteins. Utilizing a conventional microtitre-plate technique and instrumental test evaluation, the qualitative and quantitative determination of bovine IgG could be achieved. Milk samples were analysed directly after deflating by centrifugation, cheese samples were homogenized with 0.5% Tween 20/PBS and filtered before assay. The detection limit of the microtitre EIA for cows' milk and goats' milk was at 0.001%. In order to enable quantitative analysis of cows' milk in ewes' milk and goats' milk cheeses, bovine IgG was determined in ewes' milk cheeses (Feta, Pecorino, Roquefort) in different stages of ripeness

Table 9.6 EIA-systems for the differentiation of milk from cows, sheep, and goats in the microtitre plate

Antigen	Samples tested	Detection limit	EIA principle (label)	Author(s)
β-Casein	milk		non-competitive EIA (direct)	Anguita et al. (1995)
β-Lactoglobulin	cheese	0.5% cows' milk	competitive EIA (direct)	Beer et al. (1995)
κ-Casein fragment (139–152)	ewes' milk	0.25% cows' milk	competitive EIA	Bitri et al. (1993)
	goats' milk	0.25% cows' milk		
Bovine IgG	buffaloes' milk	0.0008% raw cows' milk	sandwich-EIA (direct)	Çan (1996)
		0.0015% past. cows' milk		
Bovine whey proteins	goats' milk	1% cows' milk	non-competitive EIA (biotin-avidin)	Castro et al. (1992)
Bovine whey proteins	ewes' milk	1% cows' milk	non-competitive EIA (biotin-avidin)	Garcia et al. (1990)
Bovine whey proteins	ewes' milk	1% cows' milk	non-competitive EIA (biotin-avidin)	Garcia et al. (1991)
Caprine whey proteins	ewes' milk	0.5% cows' milk	sandwich-EIA (biotin-avidin)	Garcia et al. (1993)
Caprine whey proteins	ewes' milk	1% goats' milk	non-competitive EIA (biotin-avidin)	Garcia et al. (1994)
Bovine β-lactoglobulin	ewes' milk	0.0001% cows' milk	sandwich-EIA (direct)	Levieux and Venien (1994)
	goats' milk			
γ_3-Casein	cheese	0.2% cows' milk	competitive EIA	Richter et al. (1995)
Bovine caseins	ewes' milk	1% cows' milk	non-competitive EIA (biotin-avidin)	Rodriguez et al. (1990)
Caprine caseins	ewes' milk	1% goats' milk	non-competitive EIA (biotin-avidin)	Rodriguez et al. (1991)
Bovine caseins	ewes' milk	0.5% cows' milk	sandwich-EIA (biotin-avidin)	Rodriguez et al. (1993)
Caprine caseins	ewes' milk	1% goats' milk	sandwich-EIA (biotin-avidin)	Rodriguez et al. (1994)
Bovine α_{s1}-casein fragment (140–149)	ewes' milk	0.125% cows' milk	competitive EIA	Rolland et al. (1993)
Bovine whey proteins	goats' milk		sandwich-EIA (direct)	Sargeant et al. (1989)
Bovine IgG	ewes' milk	0.001% cows' milk	sandwich-EIA (direct)	Sauer et al. (1991)
	goats' milk			
Bovine plasmin	ewes' milk	0.0002% raw cows' milk	sandwich-EIA (biotin-avidin)	Schilk (1995)
		0.0004% past. cows' milk		
		1.3% UHT cows' milk		
Caprine IgG	ewes' milk	0.002% raw goats' milk	sandwich-EIA (direct)	Schilk (1995)
	goats' milk	1% cows' milk	sandwich-EIA (direct)	Spencer and Patterson (1986)

produced with cows' milk added in known concentrations. These experimental data obtained for different types of cheeses were used as conversion factors to calculate the amount of cows' milk in cheese samples of unknown composition.

Using the same reagents, an immunofiltration assay was developed. The test indicates the presence of ≥1% cows' milk in ewes' or goats' milk via a visual discernible colour reaction. Using this technique to screen defatted milk, results were obtained within a total test time of 5 min. A third format of enzyme immunoassay, a dipstick EIA, was used to screen cheeses from ewes' milk or goats' milk. The test can also detect ≥1% cows' milk. Results are obtained by visual evaluation after a test time of 90 min. The microtitre plate and immunofiltration assay described above have become commercially available (RIDASCREEN® CIS and RIDA®QUICK CIS, R-Biopharm GmbH, Germany). A comparison of the working steps involved in both assays is given in Table 9.7.

During the last years, several other membrane-based assays for the detection of cows' and goats' milk have been described (Table 9.8). The detection limit provided by these tests is sufficient under practical aspects, although compared with microtitre-plate assay they show a reduced sensitivity. On the other hand rapid immunochemical methods are excellent tools to check milk delivered to dairy plants before processing, thus preventing financial losses for the cheese producing industry.

9.3.2 *Meat and meat products*

Immunoassay procedures for providing the authenticity of meat and meat products have been reviewed recently by Lumley (1996). The intention of this chapter is, therefore, to give a concise overview about selected properties of

Table 9.7 Comparison of the total assay time and the working steps involved in a microtitre plate and immunofiltration EIA

Microtitre plate assay	Immunofiltration assay
Addition of standard and sample solution (100 µl)	Addition of sample solution (400 µl) (2–3 min)
Incubation (30 min)	
Addition of washing solution, repeat 3 times (5 min)	Addition of 8 drops of washing solution (1 min)
Addition of antibody-enzyme-conjugate (100 µl)	Addition of 3 drops of antibody-enzyme-conjugate (1 min)
Incubation (30 min)	
Addition of washing solution, repeat 3 times (5 min)	Addition of 8 drops of washing solution (1 min)
Addition substrate/chromogen	Addition of 3 drops of substrate/chromogen
Colour formation (30 min)	Colour formation (2–5 min)
Addition of stop reagent	
Measurement of absorbance using microtitre plate reader	Visual evaluation of colour formation
Total assay time: approx. 100 min	Total assay time: approx. 10 min

Table 9.8 Membrane based immunoassay systems for the detection of cows' and goats' milk

Antigen	Samples tested	Detection limit	EIA principle	Author(s)
β-Casein	cheese	0.5% cows' milk	immunoblot	Addeo et al. (1995)
β-Casein	milk	1% cows' milk	non-competitive EIA (dipstick)	Anguita et al. (1996)
	cheese	0.5% cows' milk		
Bovine caseins	ewes' milk casein	0.1% bovine casein	non-competitive EIA immunodot	Aranda et al. (1988)
Caprine IgG	ewes' milk	0.5% cows' and goats' milk	sandwich-EIA immunodot	Aranda et al. (1993)
β-Lactoglobulin	cheese	1% cows' milk	immunoblot	Molina et al. (1996)
Bovine caseins Caprine IgG	ewes' milk	0.1% cows' milk	sandwich-EIA immunodot	Perez et al. (1992)
γ$_3$-casein	cheese	0.2% cows' milk	immunoblot	Richter et al. (1995)
Bovine IgG	ewes' milk and goats' milk	1% cows' milk	sandwich-EIA immunofiltration/ dipstick	Sauer (1992)

immunodiffusion (Table 9.9) and enzyme immunoassay procedures (Table 9.10). The target antigens used for both types of methods in most studies are blood or serum proteins, such as albumin. Some authors also used muscle proteins, preparations of adrenals, and sacroplasmatic extracts to produce specific polyclonal antisera in rabbits, goats, and sheep. In a few studies mice were immunized for the production of monoclonal antibodies. The proteins used for immunization cover a wide range of both domestic animal species, such as horse, cattle, pig, sheep, and also exotic species like impala and topi (Table 9.9). Immunodiffusion methods still represent an ideal tool for routine species identification of raw meats (see Table 9.9 for commercially available products). Usually reliability and sensitivity of these assays decreases with increased heating of the samples, even when relatively stable proteins, such as myoglobulin (Hayden, 1979), adrenal preparations (Hayden, 1981) or troponin (Schweiger et al., 1983) are used as the target antigens.

Polyclonal antisera have to be treated by absorption or immunoaffinity procedures to provide sufficient specificity for use in EIA. The main advantages of EIA over immunodiffusion or immunoelectrophoresis procedures are reduced assay time (usually less than 4 h), requirement of only small amounts of antisera, and the possibility of obtaining quantitative results. In addition the EIA microtitre-plate assay may be automated, thus allowing a large number of samples to be processed, whereas rapid tests like dipsticks (immunosticks) may be used as field tests to screen suspicious samples. A water or buffer solution extract of meat or meat products is sufficient for most EIA variants described. As described for the immunodiffusion methods, EIA is best suited to analyse raw meat or mildly heated meat products. A selection of the variety of assay formats described so far is presented in Table 9.10, using the definitions given in section 9.2.2. The most sensitive assays have been established using the sandwich EIA format.

Table 9.9 Agargel immunodiffusion and immunoelectrophoresis methods for species identification in meat and meat products

Method	Protein (species)	Antiserum (properties)	Extraction procedure	Detection limit (%)	Author(s)
Double diffusion (PROFIT)*	serum albumin (chicken and turkey)	goat	water (ground/formulated meat products) paper discs	7.5–10 (chicken) 3–5 (turkey)	Cutrufelli et al. (1986, 1987)
Double diffusion (PRIME)*	serum albumin (pig)	goat	paper discs	3–5	Cutrufelli et al. (1988)
Double diffusion (SOFT)*	serum proteins (sheep)	calf	paper discs	3	Cutrufelli et al. (1989)
Double diffusion, immunoelectrophoresis	serum proteins (pig, horse, rabbit)	goat, rabbit	PBS	1–3	Hayden (1978)
Double diffusion immunoelectrophoresis	myoglobulin (horse, pig, sheep)	rabbit, goat (absorbed)	water (heated, 90 °C, 30 min)	3–10	Hayden (1979)
Double diffusion	adrenal preparation (horse, cattle, sheep, pig)	rabbit	NaCl (heated, 98 °C, 1 h)	5–10	Hayden (1981)
Double diffusion	thermostable muscle (buffalo, impala, eland, waterbuck, wildebeest, oryx, cattle, Grant's gazelle, Thomson's gazelle, topi)	goat, sheep (absorbed)	NaCl (heated, 98 °C, 1 h)		Kang'ethe et al. (1986)
Double diffusion (ORBIT)*	serum and serum albumin (cattle)	rabbit, goat, sheep	paper discs	10	Mageau et al. (1984) Cutrufelli et al. (1987)

* Commercially available from Rhone-Poulenc Diagnostics Ltd.

Table 9.10 EIA systems for species identification in meat and meat products

Antigen (species)	Antiserum (properties)	Extraction solvent	Detection limit (%)	EIA principle (label)	Author(s)
Serum, albumin (pig)	rabbit (IAC)*	PBS/Tween	1–10	competitive EIA (second antibody)	Ayob et al. (1989)
Skeletal muscle extract (pig, chicken)	rabbit	water	250 ppm (pork) 126 ppm (poultry)	competitive EIA (direct) Sandwich EIA (biotin-avidin)	Berger et al. (1988)
Desmin (chicken)	mouse (monoclonal)	8 M urea	10	non-competitive EIA (second antibody)	Billet et al. (1996)
Serum (cattle)	rabbit	NaCl	<10	non-competitive EIA	Griffiths and Billington (1984)
Serum albumin (pig)	sheep, rabbit (IAC)*	water	1–3	competitive EIA (second antibody) sandwich-EIA (second antibody)	Jones and Patterson (1985)
Serum albumin (cattle, horse, pig)	sheep (absorbed)	water, NaCl	c. 3	non-competitive EIA (second antibody)	Jones and Patterson (1986)
Serum albumin (cattle, horse, pig)	rabbit (IAC)*	NaCl	c. 3	non-competitive EIA (second antibody)	Kang'ethe et al. (1982)
Sarcoplasmatic extracts (pig)	rabbit (IAC)*	NaCl	1	sandwich-EIA (direct)	Martin et al. (1988)
Soluble muscle proteins (chicken)	mouse (monoclonal), rabbit (IAC)*	NaCl	c. 1	sandwich-EIA (second antibody)	Martin et al. (1991)
Muscle protein (pig)	mouse (monoclonal)	NaCl	1	non-competitive EIA (second antibody)	Morales et al. (1994)
Serum proteins (buffalo, goat, donkey)	cattle, sheep, horse (IAC)*	water	0.1–1	sandwich-EIA (direct)	Patterson and Spencer (1985)
Serum proteins (sheep, goat, cattle, buffalo, kangaroo, horse, camel, pig)	sheep, cattle, rabbit (IAC)*	PBS/Tween	<1	sandwich-EIA (direct, protein A)	Patterson et al. (1984)
Autoclaved muscle extract (pig)	sheep (absorbed)	hot NaCl	0.5–1	non-competitive EIA (second antibody)	Sawaya et al. (1990)
Heated adrenals and muscles (buffalo, cattle, sheep, goat, pig)	rabbit (absorbed)	water	1	non-competitive EIA (second antibody)	Sherikar et al. (1993)
Serum protein (horse, cattle, sheep, kangaroo, pig, camel)	rabbit (IAC)*	citrate buffer	<10	non-competitive EIA (protein A)	Whittaker et al. (1983)

* Purified by immunoaffinity chromatography (IAC).

Compared to the non-competitive assay format, utilizing direct coating of the antigen to the solid phase, the sandwich EIA avoids variation of the assay response due to variations in coating efficiency if different types of samples are analysed. In addition, binding of antigen to the solid phase via a capturing antibody results in better presentation of the antigen for binding by the second (labelled) antibody. A disadvantage is that antibodies against two epitopes in suitable steric positions must be present in the test antiserum in order to be used as capturing and detecting antiserum. If the detecting antibody is not directly labelled with an enzyme, it must be considered that this antibody originates from an animal species different to that of the species used to elicit the capturing antibody. If these prerequisites cannot be matched, other assay principles, such as competitive tests, or non-competitive tests involving direct coating of the antigen, should be used.

EIAs utilizing blood or serum proteins, such as albumin, as the target antigen show limited suitability in testing heat-treated sample materials. A progressive loss in activity is observed with increased heat treatment due to denaturation of the antigen (Goodwin, 1992). Another drawback is that the amount of blood may be different in various types of beef, and that the antigenic properties of blood and serum proteins may change on storage (Griffiths and Billington, 1984). On the other hand, a number of assays has been described based on antibodies against heat stable or heat treated antigens (Berger *et al.*, 1988; Patterson and Jones, 1989; Sawaya *et al.*, 1990; Sherikar *et al.*, 1993). The main drawback of these methods is the need for complex extraction steps. Furthermore, the assays show reduced specificity and provide no quantitative results.

A general disadvantage of antibody techniques in this particular area is the limited availability of commercial test kits. Within Europe only one company is selling EIA systems for species identification of meat and meat products. Cortecs Diagnostics (UK) is providing dipstick (F.A.S.T., immunostick) assays for a range of animal species, including beef, horse, pork, sheep/goat, poultry, rabbit, kangaroo, goat, chicken, turkey and buffalo. The F.A.S.T. system can be used for testing **raw** meat and meat products, milk, animal tissue and serum. To analyse cooked samples, a microtitre-plate assay system (Biokit) is marketed for beef, pork, poultry and sheep by the same manufacturer. The Biokit system is also available for raw sample specimen originating from cattle, horse, pork, sheep, poultry, rabbit and kangaroo. The detection limit for all assays is in a range of 1–3%, quantification is, however, not possible.

9.4 Statistical parameters for quantitative evaluation of EIA

For the discussion of statistical parameters influencing the 'quality' of immunoassays it should be noted that general statistical methods to define the limit of detection, the limit of quantification, reproducibility etc., must also be applied to immunoassays. There are, however, some parameters influencing within assay precision (repeatability) and accuracy of immunoassay results, which are typical for enzyme immunoassays set up on 96-well microtitre plates (for details see Bunch et al., 1990). For routine analysis such methods usually specify a number of wells for preparation of the standard curve and a fixed number of replicates for standards and samples as well as a number of dilutions for each sample. In this process there are two factors that mainly affect the precision of the estimate. The first is the location of the observed absorbance values on the calibration curve. The second is the number of replicates used for each sample. From Fig. 9.9 it is clear that due to the non-linear shape of the calibration curve, absorbance readings near 50% relative binding give more precise results than those near 100% and 0% relative binding, respectively. As with any other method, the measurement becomes more precise, if the number of replicates is increased.

Accuracy of the estimates is mainly affected by the accuracy of the calibration curve. The calibration curve, however, is not known exactly, but estimated from absorbance readings of the standard concentrations on the plate. So the method of curve fitting as well as the number of calibration values used and the true concentrations of the standards determine the accuracy of the calibration curve. Among the mathematical methods used to describe immunoassay standard curves only the four-parameter-logistic model and approximation of a cubic spline should be used for enzyme immunoassays. Both methods give comparable results and at least one of them is implemented in most of the available programmes for immunoassay data processing.

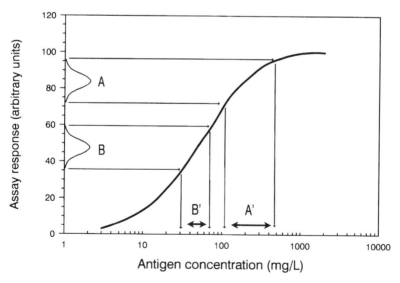

Figure 9.9 Effect of location of the observed assay response (absorbance, fluorescence, etc.) on the calibration curve on the precision of the assay. Assuming the same variation in recording the assay response over the whole measuring range (A, B), the variation of the calculated antigen concentration (A', B') will be different due to the non-linear shape of the standard curve and the semi-logarithmic scale of the plot.

Table 9.11 Factors with direct influence on major assay quality parameters of enzyme immunoassays (Märtlbauer, 1993)

Factor	Assay quality parameter			
	Sensitivity	Specificity	Precision	Accuracy
Immunochemical principle	yes	yes		
Antibody				
Affinity	yes	no		no
Specificity	no	yes		
Labelled reagent				
Activity	yes	no		
Design	yes	yes		
Stat. design				
Standards				
– replicates (n)			no	yes
– dilutions (n)			no	yes
		no		
Samples				
– replicates (n)			yes	no
– dilutions (n)			yes	yes
Curve fitting			no	yes
Experimental error	yes	yes	yes	yes
Sample preparation	yes	yes	yes	yes

Considering these parameters leads to the problem of how to optimize number and allocation of standards and samples within the 96 wells of the plate. Working on immunoassays since 1982 an assay template using six calibration values and quadruplicates for standards and samples was set up empirically in our laboratory. This template is recommended if the costs for the immunochemical reagents are moderate. In routine analysis using commercial available test kit duplicates for standards and samples may be employed. The use of single wells, however, cannot be recommended because of the uncertainties mentioned above.

Although the overall quality of an immunoassay is a complex function, some factors influencing major assay quality parameters are summarized in Table 9.11 in a simplified manner.

9.5 Conclusions

New regulatory limits, together with new sampling strategies and the consumer's demand for safe and authentic food require more sensitive analytical methods as well as an increase in the number of analyses. These requirements can only be fulfilled by reliable and rapid methods suitable for routine analyses. In principle, antibody techniques, particularly enzyme immunoassays, have the advantage of high sensitivity and speed together with simplicity of operation. Such methods are therefore easily automated and show good cost-effectiveness.

A main drawback of antibody techniques in this particular area of food analysis is the limited availability of commercial reagents and test kits. In the near future this situation is unlikely to change, because the market for immunochemical methods for food analysis is characterized by a broad range of potential products with a high degree of innovation but relatively low sales per product.

References

Addeo, F., Nicolai, M.A., Chianese, L. et al. (1995) A control method to detect bovine milk in ewe and water buffalo cheese using immunoblotting. *Milchwiss.*, **50**, 83–5.
Amigo, L., Ramos, M., Martin-Alvarez, J. et al. (1989) Effect of technological parameters on electrophoretic detection of cow's milk in ewe's milk cheeses. *J. Dairy Sci.*, **74**, 1482–90.
Anguita, G., Martin, R., Garcia, T. et al. (1995) Indirect ELISA for detection of cows' milk in ewes' and goats' milk using a monoclonal antibody against bovine β-casein. *J. Dairy Res.*, **62**, 655–9.
Anguita, G., Martin, R., Garcia, T. et al. (1996) Immunostick ELISA for detection of cow's milk in ewe's milk using a monoclonal antibody against bovine β-casein. *J. Food Prot.*, **59**, 436–7.
Aranda, P., Oria, R. and Calvo, M. (1988) Detection of cows' milk in ewes' milk and cheese by an immunodotting method. *J. Dairy Res.*, **55**, 121–4.
Aranda, P., Sanchez, L., Perez, M.D. et al. (1993) Rapid immunoenzymatic method for detecting adulteration in ewes' milk. *Food Control*, 4, 101–104.
Ayob, M.K., Ragab, A.A., Allen, J.C. et al. (1989) An improved, rapid, ELISA technique for detection of pork in meat products. *J. Sci. Food Agric.*, **49**, 103–16.
Beer, M., Krause, I., Rumpel, J. et al. (1995) Epitopspezifität von avinen Antikörpern gegen bovines ß-Lactoglobulin und deren Eignung zum Nachweis von hitzedenaturiertem Rinder-Molkenprotein in Schaf- und Ziegenkäse. *Milchwirtsch. Berichte*, **125**, 201.
Berger, R.G., Mageau, R.P., Schwab, B. et al. (1988) Detection of poultry and pork in cooked and canned meat foods by enzyme-linked immunosorbent assays. *J. Assoc. Off. Anal. Chem.*, **71**, 406–409.
Berner, I. (1990) Bestimmung von Kuhmilch in Schaf-, Ziegen- und Büffelmilchkäsen unter Berücksichtigung der Käsereifung. Diss. agr., München.
Bernhauer, H., Baudner, S. and Günther, H.O. (1983) Immunologischer Nachweis von Kuhmilchproteinen in Schaf- und Ziegenmilch bzw. -käse über ein spezifisches Immunglobulin. *Z. Lebensm. Unters. Forsch.*, **177**, 8–10.
Berzofsky, J.A. and Schechter, A.N. (1981) The concepts of crossreactivity and specificity in immunology. *Mol. Immunol.*, **18**, 751.
Billet, E.E., Bevan, R., Scanlon, B. et al. (1996) The use of a poultry-specific murine monoclonal antibody directed to the insoluble muscle protein desmin in meat speciation. *J. Sci. Food Agric.*, **70**, 396–404.
Bitri, L., Rolland, M.P. and Besançon, P. (1993) Immunological detection of bovine casinomacropeptide in ovine and caprine dairy products. *Milchwiss.*, **48**, 367–70.
Brock, J. (1980) Präparation von Immunglobulin G (IgG), in *Immunologische Arbeitsmethoden* (ed. H. Friemel), Gustav Fischer Verlag, Stuttgart, pp. 323–32.
Bunch, D.S., Rocke, D.M. and Harrison, R.O. (1990) Statistical design of ELISA protocols. *J. Immunol. Meth.*, **132**, 247.
Calvo, M.M., Amigo, L., Olano, A. et al. (1989) Effect of thermal treatments on the determination of bovine milk added to ovine or caprine milk. *Food Chem.*, **32**, 99–108.
Çan, E. (1996) Entwicklung und Anwendung von enzymimmunologischen Verfahren zum Nachweis von Kuhmilch in Büffelmilch bzw. -käse. Diss. med. vet., München.
Castro, C., Martin, R., Garcia, T. et al. (1992) Indirect enzyme-linked immunosorbent assay for detection of cow's milk in goat's milk. *Food Agric. Immunol.*, 4, 11–18.
Clausen, J. (1988) Immunochemical techniques for the identification and estimation of marcomolecules, in *Laboratory Techniques in Biochemistry and Molecular Biology* (eds R.H. Burdon and P.H. van Knippenberg), Elsevier, Amsterdam, New York, Oxford.
Cutrufelli, M.E., Mageau, R.P. and Schwab, B. (1986) Development of poultry rapid overnight field identification test (PROFIT). *J. Assoc. Off. Anal. Chem.*, **69**, 483–7.

Cutrufelli, M.E., Mageau, R.P., Schwab, B. et al. (1987) Detection of beef and poultry by serological field screening tests (ORBIT and PROFIT): Collaborative study. *J. Assoc. Off. Anal. Chem.*, **70**, 230–33.
Cutrufelli, M.E., Mageau, R.P., Schwab, B. et al. (1988) Development of porcine rapid identification method (PRIME) by modified agar-gel immunodiffusion. *J. Assoc. Off. Anal. Chem.*, **71**, 444–5.
Cutrufelli, M.E., Mageau, R.P., Schwab, B. et al. (1989) Development of serological ovine field test (SOFT) by modified agar-gel immunodiffusion. *J. Assoc. Off. Anal. Chem.*, **72**, 60–61.
Durand, M., Meusnier, M., Delahaye, J. et al. (1974) Détection de l'addition frauduleuse de lait de vache dans les laits de chèvre et de brebis par la méthode de l'immunodiffusion en gélose. *Bull. Acad. Vét.*, **47**, 247–58.
Elbertzhagen, H. (1987) Bestimmung von Kuhmilch-Casein in Schaf- und Ziegenkäse mittels Immunelektrophorese. *Z. Lebensm. Unters. Forsch.*, **185**, 357–61.
Elbertzhagen, H. and Wenzel, E. (1982) Nachweis von Kuhmilch in Schafkäse mittels Immunelektrophorese. *Z. Lebensm. Unters. Forsch.*, **175**, 15–16.
Garcia, T., Martin, R., Rodriguez, E. et al. (1990) Detection of bovine milk in ovine milk by an indirect enzyme-linked immunosorbent assay. *J. Dairy Sci.*, **73**, 1489–93.
Garcia, T., Martin, R., Rodriguez, E. et al. (1991) Detection of bovine milk in ovine milk by a sandwich enzyme-linked immunosorbent assay (ELISA). *J. Food Prot.*, **54**, 366–9.
Garcia, T., Martin, R., Morales P. et al. (1993) Sandwich ELISA for detection of caprine milk in ovine milk. *Milchwiss.*, **48**, 563–6.
Garcia, T., Martin, R., Rodriguez, E. et al. (1994) Detection of goat's milk in ewe's milk by an indirect ELISA. *Food Agric. Immunol.*, **6**, 113–18.
Goding, J.W. (1986) *Monoclonal Antibodies: Principles and Applications*. Academic Press, San Diego.
Gombocz, E., Hellwig, E. and Petuely, F. (1981) Immunologischer Nachweis von Kuhmilchcasein in Schafkäsen. *Z. Lebensm. Unters. Forsch.*, **172**, 178–81.
Goodwin, P. (1992) Immunoassay methods for animal speciation, in *Food Safety and Quality Assurance: Applications of Immunoassay Systems* (eds M.R.A., Morgan, C.J. Smith and O.A. Williams), Elsevier Science Publishers, London, pp. 33–40.
Griffiths, N.M. and Billington, M.J. (1984) Evaluation of an enzyme-linked immunosorbent assay for beef blood serum to determine indirectly the apparent beef content of beef joints and model mixtures. *J. Sci. Food Agric.*, **35**, 909–14.
Halfman, C.J. and Schneider, A.S. (1981) Optimization of reactant concentrations for maximizing sensitivities of competitive immunoassays. *Anal. Chem.*, **53**, 654–8.
Hayden, A.R. (1978) Determination of residual species serum albumin in adultered ground beef. *J. Food Sci.*, **43**, 476–92.
Hayden, A.R. (1979) Immunochemical detection of ovine, porcine and equine flesh in beef products with antisera to species myoglobin. *J. Food Sci.*, **44**, 494–500.
Hayden, A.R. (1981) Use of antisera to heat-stable antigens of adrenals for species identification in thoroughly cooked beef sausages. *J. Food Sci.*, **46**, 1810–13.
Hitchcock, C.H.S. (1988) *Immunoassays for Veterinary and Food Analysis – 1* (eds B.A. Morris, M.N. Clifford, and R. Jackman), Elsevier Applied Science Publishers, London, New York, p. 3.
Jackson, T.M. and Ekins, R.P. (1986) Theoretical limitations on immunoassay sensitivity. Current practice and potential advantages of fluorescent Eu^{3+} chelates as non-radioisotopic tracers. *J. Immunol. Meth.*, **87**, 13.
Jones, S.J. and Patterson, R.L.S. (1985) Double-antibody ELISA for detection of trace amounts of pig meat in raw meat mixtures. *Meat Science*, **15**, 1–13.
Jones, S.J. and Patterson, R.L.S. (1986) A modified indirect ELISA procedure for raw meat speciation using crude anti-species antisera and stabilised immunoreagents. *J. Sci. Food Agric.*, **37**, 767–5.
Kaiser, R. (1989) Zur Differenzierung von Kuh-, Schaf- und Ziegenmilch. *Dtsch. Molkerei-Ztg.*, **110**, 110–13.
Kang'ethe, E.K., Gathuma, J.M. and Lindqvist, K.L. (1986) Identification of the species of origin of fresh, cooked and canned meat and meat products using antisera to thermostable muscle antigens by Ouchterlony's double diffusion test. *J. Sci. Food Agric.*, **37**, 157–64.
Kang'ethe, E.K., Jones, S.J. and Patterson, R.L.S. (1982) Identification of the species origin of fresh meat using an enzyme-linked immunosorbent assay procedure. *Meat Science*, **7**, 229–40.

Kluge-Wilm, R. (1988) Untersuchungen zum Nachweis von Kuhmilch in Schafskäse. 41. *Arbeitstagung des Arbeitskreises Lebensmittelhygienischer Tierärztlicher Sachverständiger* (ALTS), Berlin, p. 65.
Köhler, G. and Milstein, C. (1975) Continuous cultures of fused cells secreting antibody of predefined specificity. *Nature*, **256**, 495–7.
Laurell, C.-B. (1966) Quantitative estimation of proteins by electrophoresis in agarose gel containing antibodies. *Anal. Biochem.*, **15**, 45–52.
Levieux, D. (1977) Une nouvelle technique de détection de l'adulteration des laits de chèvre et de brebis. *Dossiers de l'Elevage*, **2**, 37–46.
Levieux, D. (1978) CV Test II – An immunological method for the detection of cow's milk in supplies of goat's milk. C.E.R.B., 18800 Baugy, France.
Levieux, D. and Venien, A. (1994) Rapid, sensitive two-side ELISA for the detection of cows' milk in goats' or ewes' milk using monoclonal antibodies. *J. Dairy Res.*, **61**, 91–9.
Lumley, I.D. (1996) Authenticity of meat and meat products, in *Food Authentication* (eds P.R. Ashurst and M.J. Dennis), Blackie Academic & Professional, London, pp. 108–39.
Mageau, R.P., Cutrufelli, M.E., Schwab, B. *et al*. (1984) Development of an overnight rapid bovine identification test (ORBIT) for field use. *J. Assoc. Off. Anal. Chem.*, **67**, 949–54.
Mancini, G., Carbonara, A.O. and Heremans, J.F. (1965) Immunochemical quantitation of antigens by single radial immunodiffusion. *Immunochem.*, **2**, 235–54.
Martin, R., Azcona, J.I., Casas, C. *et al*. (1988) Sandwich ELISA for detection of pig meat in raw beef using antisera to muscle soluble proteins. *J. Food Prot.*, **51**, 790–94.
Martin, R., Wardale, R.J., Jones, S.J. *et al*. (1991) Monoclonal antibody sandwich ELISA for the potential detection of chicken meat in mixtures of raw beef and pork. *Meat Science*, **30**, 23–31.
Märtlbauer, E. (1993) Basis principles which determine the quality of immunoassays, in *Analytical quality assurance and good laboratory practice in dairy laboratories. IDF Special Issue No. 9302*, pp. 137–43.
Molina, E., Fernandez-Fournier, A., de Frutos, M. *et al*. (1996) Western blotting of native and denatured bovine β-lactoglobulin to detect addition of bovine milk in cheese. *J. Dairy Sci.*, **79**, 191–7.
Morales, P., Garcia, T., Gonzalez, I. *et al*. (1994) Monoclonal antibody detection of porcine meat. *J. Food Prot.*, **57**, 146–49.
Ouchterlony, O. (1968) *Handbook of Immunodiffusion and Immunoelectrophoresis*, Ann Arbor Science Publications, Michigan.
Patterson, R.M. and Jones, T.L. (1989) Species identification in heat processed meat products. Proceedings of 35th International (Copenhagen) Congress on Meat. *Sci. Technol.*, **2**, 529–36.
Patterson, R.M. and Spencer, T.L. (1985) Differentiation of raw meat from phylogenically related species by enzyme-linked immunosorbent assay. *Meat Science*, **15**, 119–23.
Patterson, R.M., Whittaker, R.G. and Spencer, T.L. (1984) Improved species identification of raw meat by double sandwich enzyme-linked immunosorbent assay. *J. Sci. Food Agric.*, **35**, 1018–23.
Perez, M.D., Sanchez, L., Aranda, P. *et al*. (1992) Use of an immunoassay method to detect adulteration of ewes' milk, in *Food Safety and Quality Assurance: Applications of Immunoassay System* (eds M.R.A. Morgan, C.J. Smith and O.A. Williams), Elsevier Science Publishers, London, pp. 41–8.
Peters, J.H. and Baumgarten, H. (eds) (1992) *Monoclonal Antibodies*, Springer Verlag, New York.
Radford, D.V., Tchan, Y.T. and McPhillips, J. (1981) Detection of cow's milk in goat's milk by immunoelectrophoresis. *Austr. J. Dairy Technol.*, **36**, 144–6.
Richter, W., Krause, I. and H. Klostermeyer, H. (1995) Verwendung polyklonaler Antikörper zum Nachweis von Kuhmilch in Schaf- und Ziegenkäse mittels ELISA und Immunoblotting. *Milchwirtsch. Berichte*, **125**, 201.
Rodriguez, E., Martin, R., Garcia, T. *et al*. (1990) Detection of cows' milk in ewes' milk and cheese by an indirect enzyme-linked immunosorbent assay (ELISA). *J. Dairy Res.*, **57**, 197–205.
Rodriguez, E., Martin R., Garcia, T. *et al*. (1991) Indirect ELISA for detection of goats' milk in ewes' milk and cheese. *Int. J. Food Sci. Tech.*, **26**, 457–65.
Rodriguez, E., Martin, R., Garcia, T. *et al*. (1993) Detection of cows' milk in ewes' milk and cheese by a sandwich enzyme-linked immunosorbent assay (ELISA). *J. Sci. Food Agric.*, **61**, 175–80.

Rodriguez, E., Martin, R., Garcia, T. *et al.* (1994) Sandwich ELISA for detection of goats' milk in ewes' milk and cheese. *Food Agric. Immunol.*, 6, 105–11.
Roitt, I., Brostoff, J. and Male, D. (1985) *Immunology*, Gower Medical Publishing, London, New York.
Rolland, M.P., Bitri, L., and Besançon, P. (1993) Polyclonal antibodies with predetermined specificity against bovine α_{s1}-casein: application to the detection of bovine milk in ovine milk and cheese. *J. Dairy Sci.*, **60**, 413–20.
Sargeant, J.G., Farnell, P.J. and Baker, A. (1989) Quantitative determination of cows' milk as an adulterant of goats' milk by ELISA. *J. Assoc. Public Analysts*, **27**, 131–6.
Sauer, S. (1992) Entwicklung und Anwendung enzymimmunologischer Verfahren zum Nachweis von Kuhmilch in Schaf- und Ziegenmilch bzw. -käse. Diss. med. vet., München.
Sauer, S., Dietrich, R., Schneider, E. *et al.* (1991) Ein enzymimmunologischer Schnelltest zum Nachweis von Kuhmilch in Schaf- und Ziegenmilch. *Arch. Lebensmittelhyg.*, **43**, 151–4.
Sawaya, W.N., Mameesh, M.S., El-Rayes, E. *et al.* (1990) Detection of pork in processed meat by an enzyme-linked immunosorbent assay using antiswine antisera. *J. Food Sci.*, **55**, 293–7.
Schilk, J. (1995) Enzymimmuntests für bovines Plasmin und caprines IgG - Nachweis von Kuh- und Ziegenmilch in Schafmilch. Diss. med. vet., München.
Schweiger, A., Baudner, S. and Gunther, H.O. (1983) Isolation by free-flow electrophoresis and immunological detection of troponin T from turkey muscle: an application in food chemistry. *Electrophoresis*, **4**, 158–63.
Sherikar, A.T., Karkare, U.D., Khot, J.B. *et al.* (1993) Studies on thermostable antigens, production of species-specific antiadrenal sera and comparison of immunological techniques in meat speciation. *Meat Science*, **33**, 121–36.
Spencer, T.L. and Patterson, R.M. (1986) Improvements in or relating to enzyme-linked immunosorbent assay. European Patent Application EP 0 177 352 A1 Ref. *Dairy Sci. Abstr.*, **48**, 6010.
Stanworth, D.R. and Turner, M.W. (1979) Immunochemical analysis of immunoglobulins and their sub-units, in *Handbook of Experimental Immunology* (ed. D.M. Weir), Blackwell Scientific Publications, Oxford, pp. 6.1–6.13.
Walker, J.M. (ed.) (1984) *Methods in Molecular Biology, Vol. 1: Proteins*, Humana Press, Clifton, New Jersey.
Whittaker, R.G., Spencer, T.L. and Copland, J.W. (1983) An enzyme-linked immunosorbent assay for species identification of raw meat. *J. Sci. Food Agric.*, **34**, 1143–8.

10 Trace element analysis for food authenticity studies
H.M. CREWS

10.1 Introduction

Interest in the use of analytical spectroscopic methods for agricultural products was first expressed in the 1930s and 1940s (McHard *et al.*, 1980). Henrik Lundegårdh was the first to establish the utility of flame and spark spectrographic methods for soil and leaf analysis (Lundegårdh, 1938, 1943). His results linked the health of plants and optimum fruit development to the levels of calcium, nitrogen, phosphorus, iron, manganese and copper in the soil. It is well established now that the mineral nutrition of crops of all types is a fundamental aspect of their growth and ability to flourish. The position of plants in the food chain means that their elemental composition affects in turn, that of fish, animals and of animal and plant products. It is not unreasonable, therefore, to postulate that the trace element and mineral content of foods and beverages should reflect their origins. However, the problem with this assumption is that, ironically, most trace metals and minerals are ubiquitous. The potential for adventitious contamination from soil, dust, metal utensils and instrumentation can be high. Thus, one can consider four possible options; firstly that the levels of one element are absolutely specific for an area of origin and that they can be measured accurately and precisely with no risk of the data being distorted by contamination or processing, for example; secondly, that single element data can be used in conjunction with information about other analytes (not trace elements or minerals) to determine likely geographic origin and/or authenticity; thirdly, that multi-element analyses can provide pattern recognition data sets which will permit the identification of likely geographic origin and/or authenticity and finally, that this latter option is used alongside information about other analytes. Examples of all of these approaches can be found in the literature and will be cited below. The fourth option is probably the way forward, offering robust data for different types of analyte from (ideally) primary methods such as isotope dilution mass spectrometry. The combination of good data sets with powerful multivariate data handling methods should enable an unbiased identification of authentic products.

The purpose of this chapter is to provide an overview of the methods used for elemental analysis for determining authenticity and/or geographical origin of foods and beverages, using published work to illustrate the success or usefulness of the approaches. The chapter will not be a detailed account of the principle behind each technique since there are numerous published texts (e.g. Willard *et al.*, 1988; McKenzie and Smythe, 1988; Sharp, 1988a, b; Jarvis *et al.*, 1992;

Alfassi, 1994; Crews, 1996) which provide this information and they will be referred to as necessary. Table 10.1 summarizes the energy source and measured quantity for the methods included in this chapter.

An indication of the general points to be considered when undertaking elemental analyses will be given in section 10.2. The statistical analysis of data

Table 10.1 Energy source and measured quantity (adapted from Willard et al., 1988), including the references quoted in the text, for some methods of elemental analysis which have been used to determine authenticity of geographical origin

	Energy source	Measured quantity	Reference
FAAS	Flame (1700–3200 °C)	Absorption of radiation	Brause and Raterman, 1982 Day et al., 1994, 1995a, b, Fernandez-Pereira et al., Fresno et al., Gonzales-Larraina et al., Latorre et al., Martin-Hernández et al., McHard et al., 1980, Moret et al., Ureña Pozo et al.
FES	Flame (1700–3200 °C)	Intensity of radiation	Fernandez-Pereira et al., Fresno et al., Gonzales-Larraina et al., Latorre et al., Moret et al.
AFS	Flame (1700–3200 °C)	Intensity of scattered radiation	McHard et al., 1980
ETAAS	Electric furnace (1200–3000 °C)	Absorption of radiation	Day et al., 1994, 1995a, b, Favretto et al., 1989a, b, Krivan et al., Morris and Greene
DCP-AES	Argon plasma produced by a dc arc (6000–10 500 °C)	Intensity of radiation	McHard et al., 1979, 1980
ICP-AES	Argon plasma produced by induction from high-frequency magnetic field (6000–8500 °C)	Intensity of radiation	Dettmar et al., McHard et al., 1980, Munilla et al., Nikdel, Nikdel and Attaway, Nikdel and Temelli, Nikdel et al., van der Schee et al.
NAA	Neutron bombardment of the sample	Gamma or beta radiation	Feller-Demalsy et al., Krivan et al., Teherani
TIMS	Heated surface ($\approx$0.2000 °C) e.g. tungsten filament	Evaporated ions	Eschnauer et al., Horn et al.
ICP-MS	Argon plasma produced by induction from high-frequency magnetic field (6000–8500 °C)	Transmitted ions	Augagneur et al., Baxter et al., Crews et al., Eschnauer et al.

will not be discussed but any statistics referred to will be considered in more detail in Chapter 12 of this book.

10.2 General considerations

For all trace analytical techniques, the prevention of possible contamination of samples, reagents and apparatus with the analytes of interest or potential interfering substances is paramount. When attempting to measure most elements in the Periodic Table this aspect becomes crucial. Many are present in dust, as trace contaminants in reagents and laboratory equipment as well as in human hair, sweat and clothing. Attention to laboratory cleanliness is therefore essential and the use of clean-room facilities is strongly recommended for ultra trace determinations. Wherever possible the purest available reagents should be used. In addition all glass and plastic ware should be appropriately cleaned in, for example, dilute nitric acid followed by rinsing with purified water. The instrument itself can contribute to contamination either by analyte build-up or retention in transfer tubes, nebulizers and spray chambers. Whenever new elements are to be measured, the efficiency of washout systems should be tested in advance of the analyses. Some very good recent accounts of methods for sampling and sample preparation (Jarvis, 1992; Woittiez and Sloof, 1994; Negretti de Bratter *et al.*, 1995), systematic errors in trace analysis (Tölg and Tschöpel, 1994), quality control (Vandecasteele and Block, 1993; Thompson and Wood, 1995), quantitative trace analysis of biological materials (McKenzie and Smythe, 1988) and analysis of foods (Gilbert, 1996), are available. In addition, annual reviews, for example, on the use of atomic spectrometry for the analysis of clinical and biological materials, foods and beverages (Taylor *et al.*, 1996, 1997) and for industrial analyses (Crighton *et al.*, 1996) are published. These reviews contain up-to-date information concerning the instrumentation and the methods employed for numerous combinations of elements and matrices, including the use of chemical modifiers for interference correction, preconcentration techniques and sample digestion methods. Detailed aspects of sample preparation will not be given here; however, the following sections will give examples of elemental anayses for authentication using particular techniques and any special precautions and approaches that were taken will be given.

The choice of technique will often depend upon the levels of the elements of interest and the nature of the sample. Where access to techniques is limited then appropriate preconcentration or dilution methods will need to be used to achieve the desired limit of detection (LOD). Figure 10.1 (adapted from Tölg, 1993) shows a comparison of estimated LODs and relative concentrations in solution for ICP-MS, NAA, FAAS, ETAAS, secondary ion MS (SIMS) and ICP-AES. It is apparent that with the exception of SIMS (absolute LOD 10^{-18} g), which cannot compete in terms of sample throughput, ICP-MS offers the best absolute

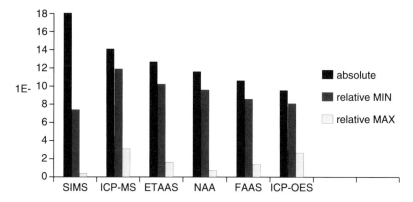

Figure 10.1 Rough estimates of absolute (1E-g) limits of detection and relative minimum and maximum concentrations (1E-g/ml). (Adapted from Tolg, 1993.)

LOD (estimated at 10^{-14} g). For ICP-MS, the estimated relative concentrations in solution which can be detected range from 10^{-3} to 10^{-12} g/ml. This compares with, for example, 10^{-2} to 10^{-10} g/ml for ETAAS and 10^{-3} to 10^{-8} g/ml for ICP-AES (Tölg, 1993). In practice, the LOD obtained is strongly dependent upon the sample being analysed, the dissolution procedure and the background levels of the analytes in reagents.

10.3 Flame emission and atomic absorption spectrometry (FES and AAS)

It was only in 1953 that the Australian physicist, Alan Walsh, first demonstrated that atomic absorption could be used as a quantitative tool (Walsh, 1955). By contrast, FES has been used since the early 1900s. This involves the energy from a flame transforming atoms from ground state to an excited electronic state (Willard *et al.*, 1988). The intensity of radiation emitted by the excited atoms returning to the ground state provides the basis for analytical determinations.

The term atomic absorption refers to the absorption of energy, by analytes in the flame, from an external source of radiation, with a consequent decrease in the radiant power transmitted through the flame. Measurement of this absorption corresponds to AAS (Christian, 1978). The most commonly used flames are air-acetylene and nitrous oxide-acetylene as they cause minimum interference. The latter flame provides a higher temperature and thus gives a better detection limit for refractory elements such as silicon, aluminium, scandium, titanium, vanadium, zirconium and the rare earth elements. In atomic fluorescence, the free analyte atoms formed in the flame absorb radiation from an external source, rise to excited electronic states and then return to ground state by fluorescence. For AFS, atomization efficiency, background emission and quenching characteristics

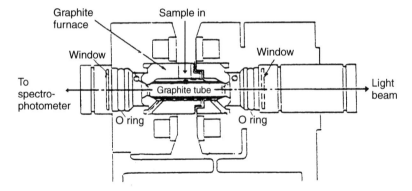

Figure 10.2 Cross-sectional view of a graphite furnace HGA500. (Adapted from *Modern Methods for Trace Element Determination*, John Wiley, 1993.)

must be considered when selecting a type of flame. The order for quenching efficiency is argon < hydrogen < water < nitrogen < carbon monoxide < oxygen < carbon dioxide. Air-acetylene and nitrous oxide-acetylene flames are most commonly used for AFS (Vandecasteele and Block, 1993).

However, a major development in AAS was the replacement of the flame by an electrothermal furnace (usually made of graphite, hence the term graphite furnace AAS), which resulted in improved LODs for many elements. Figure 10.2 shows a cross-sectional view of a furnace and its position in relation to gas flows and the light beam from the source lamp. Furnaces have been and are used in a number of different forms and configurations with samples introduced as liquids, slurries and solids. The ability to programme the furnace heating means that methods can be optimized for different elements and different matrices. The main disadvantage of ETAAS when compared with FAAS is the increased interferences caused by emission signals, light scattering, molecular absorption, chemical reactions within the furnace and contamination. This problem is becoming less severe as the reasons for interferences are becoming better understood. The use of matrix modification (chemical modifiers) and the development and use of background correction systems have been extensively described in text books (Willard *et al.*, 1988; Vandecasteele and Block, 1993) and regularly covered in reviews (Hill *et al.*, 1996). Figure 10.3 illustrates the use of temperature programming for ETAAS, with the dotted line showing the temperature profile, the upper line the overall transmission and the lower solid line the net absorption after background transmission (Vandecasteele and Block, 1993). The absorbance is only recorded during the atomization step.

Both FES and FAAS are commonly used world-wide and remain the workhorse technique for many laboratories. While they are inexpensive, the major drawback is their LODs for some elements (FES is still very good for alkali metals), which frequently require preconcentration methods to be used. For food

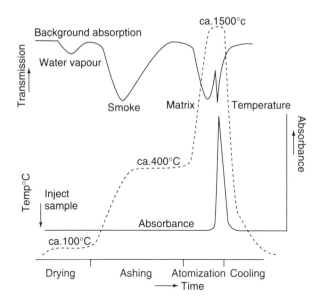

Figure 10.3 Temperature programming scheme with electrothermal atomization and resulting signal. (Adapted from *Modern Methods for Trace Element Determination*, John Wiley, 1993.)

authenticity purposes, the fact that only single analyte analyses are possible may also be a disadvantage in terms of the time taken if more than one analyte is required. LODs are not generally a limitation for ETAAS but the technique is more complex, has a higher background contribution and requires a skilled operator. In the future, however, the use of multi-element AAS is predicted with better LODs, and more predictable and extended calibration than most single element instruments currently available (Hill *et al.*, 1996).

10.3.1 *Applications using FAAS, FES, ETAAS and AFS*

Both FAAS and ETAAS have been used by Day and colleagues (Day *et al.*, 1994) to complement stable isotope methods for the characterization of grape musts. In this study, grapes which had been harvested from five regions of production in France, following a precisely defined protocol, were stored frozen ($-4\,^{\circ}C$) in sealed, colourless LDPE (low density polyethylene) bags until required. They were defrosted and digested using a microwave system. Whole grapes (300 g) were defrosted then pressed, using acid-washed polyamide tissue, into acid-washed HDPE (high density polyethylene) bottles and acidified to approximately 1% w/w nitric acid and stored at $-18\,^{\circ}C$ until required. The must (10 ml) was accurately weighed into a Teflon PFA bomb liner and nitric acid (3 ml) added. Samples were then digested in a microwave oven at a moderate pressure. The resulting solution was diluted to 20 ml with deionized water. It should be noted that all equipment was washed with acid (10% w/w hydrochloric acid) and then rinsed three times with deionized water. The major metals, potassium, calcium, magnesium, manganese, iron, zinc and copper were measured using FAAS after appropriate dilution and calibration with matrix matched standards. To reduce interferences, potassium measurements were carried out in an excess of caesium chloride (5 g/l as CsCl) and lanthanum (10 g/l) was added for the determination of calcium and magnesium. Aluminium and strontium were determined by ETAAS using a pyrolytic graphite tube and magnesium nitrate as chemical modifier for the measurement of aluminium. Data were subjected to analysis of variance having been standardized to a constant dry extract based on °Brix to remove differences in grape sugar richness. The four most significant elements for differentiating wines from different regions were calcium, magnesium, zinc and strontium, giving between 64.3 and 81.2% correctly classified wines depending upon region. When the trace element data were combined with deuterium/hydrogen ratio data, the range of correct classification improved to 82 to 100% depending upon region. However, the authors emphasized that the grapes were harvested under strictly controlled conditions and that further work needs to be done on 'real' samples. Similar results were obtained for 165 grape samples (grape juice and fermented products) from well-defined French vineyards following analysis of variance and principal component analysis of 16 parameters including minerals and trace elements (Day *et al.*, 1995a). Experimental plots (parcels), carefully described from the geological and pedological point of view were used in a study of the characterization of a region and the year of production of the wine (Day *et al.*, 1995b). Wines were again collected according to a defined protocol, ethanol was removed from the samples (10 ml) by evaporation in a microwave oven. Organic matter was destroyed by digestion with nitric acid in a Teflon bomb with the elements measured in variously diluted solutions by either FAAS or ETAAS as indicated above. In addition, rubidium was also measured by FAAS. Aluminium, rubidium and strontium permitted discrimination between wines produced in the same year from adjacent plots. The trace element content of wines was also found to correlate with the geological nature of the soil.

Spanish wines have been studied by several groups. In 1987 Fernández-Pereira *et al.* analysed 64 wines including most of the Certified Brand of Origin table wines. Sodium, potassium, calcium and magnesium were measured in wines following dilution with doubly distilled water. Other elements were measured after digestion with nitric acid and hydrogen peroxide. Lithium, sodium, potassium and rubidium were determined by FES and the remaining 14 elements were measured by FAAS. The range of concentrations for red, white and rosé wines was given. Using principal components analysis, the following elements were considered most likely to be of use in detecting the origin of the wine: calcium, magnesium, chromium, manganese, cobalt, zinc and rubidium. Similarly, Gonzales-Larrana *et al.* (1987) found that pattern analysis using lithium, manganese and rubidium allowed differentiation (93% correctly classified) of 50 red wines into the three Rioja regions: Alavesa, Baja and Alta. Lithium, sodium, potassium and rubidium were measured by FES and calcium, magnesium, manganese, copper and zinc by FAAS. Also in 1987, Ureña Pozo et al. determined iron, zinc, manganese, copper, lead, cadmium and cobalt in wines from Malaga using FAAS. The wines were evaporated to one-third of the original volume then further evaporated with nitric acid and hydrogen peroxide before analysis of the diluted residue. The data were compared with those for wines from two other regions (Valladolid and Montilla) but not used for assigning origin. Wines from Galicia in northwest Spain were classified according to Certified Brand of Origin following pattern recognition analysis using cluster analysis, PCA, discriminant analysis, K nearest neighbours and soft independent modelling of class analogy (Latorre et al., 1994). Seven metals, lithium, rubidium, sodium, potassium (these four determined by FES), and calcium, manganese and iron (the latter three determined by FAAS) were measured in all wines. Lithium, rubidium, iron and manganese were measured in wines without diluting them. The remaining elements required between 10-fold and 25-fold dilutions with lanthanum chloride added for calcium determinations. Using only lithium and rubidium, correct classifications of over 90% of wines to Certified Brand of Origin were achieved.

Flame emission spectrometry was used to measure lithium, sodium and potassium in Venetian wines (Moret et al., 1994). Calcium and magnesium were determined by FAAS. Wines were obtained directly from producers and stored in their bottles at 3–4 °C until required. Best results were obtained following canonical variate analysis with potassium being the third most important variable in a group of otherwise organic variables.

McHard et al. (1980) reported a comparison of AFS with FAAS, DC-AES and ICP-AES for the determination of barium, calcium, copper, iron, magnesium, manganese, potassium, rubidium, sodium and zinc in seven brands of Florida orange juice. Copper and zinc were not measured by ICP-AES, and rubidium was not measured by FAAS. The samples were dry ashed and dissolved in either nitric or hydrochloric acid with the acid concentration being 0.1 M for measurement. The data from FAAS were considered the most precise. However, in subsequent work to establish the origin of orange juice, DC-AES was used (section 10.4). Brause and Raterman (1982) used FAAS to measure potassium in apple juice; sugars and individual phenolics were also determined. The tests would detect fraudulent but not poorly processed apple juice.

The mineral content of 153 Spanish cheese varieties was measured by Fresno et al. (1995). Samples were ashed at 460 °C and then dissolved and diluted to 1 M nitric acid. Calibration solutions were also in 1 M nitric acid. Lanthanum was added to the solutions for sodium, potassium, calcium and magnesium to eliminate the interference from phosphorus. All the elements (calcium, magnesium, zinc, iron, copper and manganese) were measured by FAAS except for potassium and sodium which were measured by FES. Screening tests had been carried out to establish the likely range of concentrations and the precision of the method. Discriminant analysis was used to differentiate cheeses according to the animal milk species used in their manufacture and according to variety. It was possible to differentiate between the cheeses according to animal milk species using calcium, magnesium, potassium, zinc, copper and manganese (phosphorus was also included but was not measured by AAS), with success rates of about 76%. Similar results were obtained by Martin-Hernández *et al.* (1992) following stepwise discriminant analysis of data for the same elements obtained using AAS. They found 86.8% correct classification for pure-milk cheeses using iron, sodium, copper and potassium. However, Fresno *et al.* concluded that the mineral content of the cheese, the manufacturing technology (type of coagulation and intensity of salting and wheying) had more influence than the animal milk species from which it was made. Favretto *et al.* (1989a) used ETAAS to measure 11 elements in 45 cows' milk samples collected over a period of 3 years. Samples were ashed and measured using pyrolytically coated graphite tubes with instrumental conditions suggested by the manufacturer of the instrument. Principal components analysis was used to process the data. A tentative differentiation was made using aluminium, iron and zinc data for samples from two Italian regions.

Multivariate analysis was also used by Favretto *et al.* (1989b) to differentiate between mussels from two different sites. Ten trace elements were measured by ETAAS. Edible parts of the samples were lyophilized, transferred to a platinum crucible, charred under an infrared lamp and then ashed with nitric acid (1 ml) at 95 °C. The solution was then filtered into a calibrated flask and made to volume with 0.1 M nitric acid. Working standards were prepared in 0.1 M nitric acid. Using a two-dimensional plot of PC scores, 43 samples of unpolluted edible mussels from a hatchery were differentiated from 28 samples of wild mussels from a polluted site on the basis of 8 variables: manganese, iron, cobalt, nickel, copper, zinc, cadmium and lead.

Finally in this section, Morris and Greene (1970) measured lead, tin, chromium, cadmium and selenium in wheat blends, the flours milled from them and the subsequent baked products. Selenium was measured by a colorimetric method while the remaining elements were measured by AAS. Although there was a statistically significant variation between geographical regions in the trace element content of some products, the data did not indicate regional variation in trace element nutrition due to consumption of wheat products.

10.4 Atomic emission spectrometry (AES)

Emission spectrometry was the earliest developed multi-element technique and was most widely used in the metal industry where it was particularly useful for determining a few elements repetitively in a well-defined matrix, usually some form of steel (McHard *et al.*, 1980). It was also responsible for much of the early information about trace elements in orange juice (Roberts and Gaddum, 1937; Birdsall *et al.*, 1961; Nikdel *et al.*, 1988). In AES, the excitation source transforms the sample from its initial state as solid, liquid or gas into a plasma of atoms, ions and molecular species. Many aspects are similar to those of FES but electrical arcs or sparks, glow or plasma discharges, or lasers replace the flame as the means of atomization and excitation. These sources provide greater energy than do flames and thus produce more complex spectra that require spectrometers with better resolution. As the emission emerges from the source it is focused on the spectrometer's entrance slit where it is dispersed into its component wavelengths. Multi-element analyses are possible using photomultiplier tubes or diode arrays linked directly to computer-driven data processing systems. The simplest electrical discharge is the d.c. arc between two solid electrodes resulting in a plasma of high velocity ions, electrons and atoms which produces the atomic emission spectra. The complex spectra are further complicated by background emission from the electrodes and by many lines originating from ions of elements with relatively low ionization energies (Willard *et al.*, 1988). In the three-electrode DCP source, the plasma jet is formed between two carbon anodes and a tungsten cathode in an inverted Y configuration. This configuration stabilizes the position of the plasma and the sample excitation area, and provides an effective means of separating the high current plasma from the observation zone. Because the intense lines of excited atoms and ions of the analyte are observed in a region separate from the main plasma, an optimum signal to background noise ratio is obtained producing lower LODs than ICPs for some elements (Willard *et al.*, 1988). In contrast the ICP source (Fig. 10.4), which is generated electromagnetically by the interaction of a fluctuating magnetic field with a stream of ionizable gas (most commonly argon), has no electrode contact and excitation and emission zones are spatially separated. As a result, there is a relatively simple background spectrum that consists of argon lines and some weak band emission from OH, NO, NH and CN molecules. This

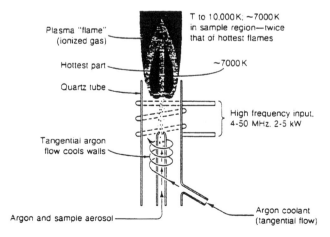

Figure 10.4 Schematic diagram showing inductively coupled plasma, used for atomic emission spectroscopy (adapted from *Quantitative Chemical Analysis*, 1986, Wadsworth, Monterey, California).

low background combined with a high signal to noise ratio of analyte emission, results in reasonable LODs. A good description of ICP-AES and emission spectrometry generally is given by Kimberly et al. (1988) and Vandecasteele and Block (1993). The reader should also consult Boumans (1987) and Montaser and Golightly (1987). ICP-AES has largely superseded DCP-AES, not least because a major drawback of DCPs is their inability to be incorporated into totally automated systems.

10.4.1 Applications using DCP-AES and ICP-AES

McHard et al. (1979, 1980) investigated various approaches for the analysis of orange juice in order to formulate a complete description of elemental composition, and to determine how this composition related to geographical origin. They compared the use of DCP and ICP sources and listed the disadvantages they found with an ICP source as:

1. the nebulizer design was critical in the operation of ICPs (Sharp, 1988a, b, for critical review of nebulizers), since close control of the nebulizer flow was necessary to achieve plasma stability;
2. the observation area to which the optics were exposed was directly in the centre of the plasma and the background emission was high compared to the DCP where very little of the plasma region was observed during analysis;
3. the instrumentation for the multi-element facility of the ICP was expensive and the buyer needed to decide in advance which elements were required.

They therefore chose to use the less expensive, sequential DCP system. However, they acknowledged that it was subject to more or less severe matrix effects, for example the influence of large potassium concentrations on both the atom and ion lines of barium.

The samples investigated by McHard and colleagues were orange juices from Florida (24) and Brazil (74) with a few (6 to 10) samples from Mexico and California for comparison. All samples were prepared for analysis by dry ashing for about 16 h. Frozen orange juice concentrate (20 g) was dried to carbonization under heat lamps and ashed in a muffle furnace at 550 °C until the ashed residue was completely free of visible unoxidized carbon. The final ash was dissolved in 0.1 M nitric acid (50 ml). Standards were formulated to correspond as closely as possible to the element ratios and concentrations usually found in orange juice. The final measured concentrations were adjusted to reflect the dilution to single strength juice. Zinc was used as a reference element since its concentration seemed to be particularly uniform and by ratioing other elements to it, some adjustment could be made for the different solids content of different juices. Pattern recognition was used to discriminate between juices from Florida and Brazil the target elements being barium, boron, gallium, manganese and rubidium. A decision vector for these elements enabled 100% correct classification of the sample population investigated (McHard et al., 1980). Nikdel and co-workers (Nikdel, 1986; Nikdel and Attaway, 1987; Nikdel and Temelli,

1987; Nikdel *et al.*, 1988) used ICP-AES to analyse orange juices from Brazil and Florida to define origin and, possibly, to detect sample manipulation. Authentic Florida (136) and Brazilian (57) juices were prepared for analysis by acid digestion in a microwave system (Nikdel and Temelli, 1987) using either frozen concentrate (5 g) or single strength juice (25 g). Standardization of the ICP-AES system was possible using a low and high standard, containing the multi-element mix in a suitable synthetic matrix, because of the large linear dynamic range of the plasma (5 to 6 orders of magnitude, Fig. 10.1). Seventeen elements were determined and the data subject to pattern recognition using the ARTHUR package. Using variance and Fisher weightings, the following elements had values greater than 1.8 and 0.8 respectively and perhaps could be considered most useful for classification methods: barium, rubidium, calcium, copper, iron, silicon, sodium and boron.

Almost 16 years after McHard's work with DCP-AES, Dettmar *et al.* (1996) used ICP-AES to determine eight trace elements (iron, manganese, phosphorus, zinc, magnesium, potassium, boron and strontium) in orange juices of known origin and combined the data with results for organic components in the same juices. An artificial neural network, Dystal, was used to classify the juices. All the available data were used and varying degrees of success were reported for classification according to (a) adulterated or not by addition of beet sugar and/or pulpwash (89.8% correct), (b) Valencia or non-Valencia varieties (92.5%), and (c) by growth year (88.0%).

In contrast, edible Japanese and Korean seaweeds were analysed for 12 elements by ICP-AES after microwave digestion of the samples (0.250 g) with nitric acid, hydrogen peroxide and hydrofluoric acid (Munilla *et al.*, 1995). The sample digestion time was 20 min. A sea lettuce (*Ulva lactuca*) certified reference material, BCR CRM 279, was also analysed. Despite the seaweed samples being collected in the same season from well-known geographic areas, the area of origin had no influence on the results. There was a correlation between mineral concentrations in similar species.

An inter-laboratory study of the methods used by three laboratories for the analysis of sherry wines was reported (van der Schee *et al.*, 1989). Both AAS and ICP-AES were used for elemental analyses. Organic components were determined by GC and HPLC. The results of this limited trial showed that strongly adulterated samples could be detected by the different laboratories. The elements iron, calcium, barium and magnesium were especially important. Several samples showed strong deviations in only one variable and were usually outliers. These deviations may be caused by processing or agricultural treatment of the vines, e.g. the use of a zinc-containing pesticide.

10.5 Neutron activation analysis (NAA)

Comprehensive overviews of radiochemical methods, particularly NAA, are given by Cornelis (1988), and Willard *et al.* (1988). Neutron activation is a general term for irradiating material with neutrons to create radionuclides. Three steps are involved; neutron bombardment of the sample which causes decay of the analytes with characteristics energies, recording the energy spectrum of the gamma or beta radiation produced, and analysis of the significance of the spectrum features. The energies of the spectral peaks identify the elements present; the areas of the peaks define the quantities of the element (Willard *et al.*, 1988). Two forms of the activation method exist. If the activation of the sample is directly followed by gamma-ray spectrometry it is termed instrumental or non-destructive NAA (INAA). If interferences from other induced activities prevent purely instrumental detection of the activity of interest, a post-irradiation radiochemical separation procedure may be used. After irradiation the sample is dissolved and chemically equilibrated with a relatively large amount (but accurately known and typically about 10 mg) of the element or elements of interest. These are known as carriers. The addition of carriers, as well as adventitious contamination of the irradiated specimen does not affect the accuracy of the method, because added atoms are non-radioactive and do not interfere with the measurements. Cornelis (1988) states that it is very tedious to develop a separation scheme for the simultaneous determination of 10–20 elements. Radioactive tracers offer a method of monitoring and controlling the various steps. However, certain procedures work well only on certain matrices and the use of reference materials should be considered mandatory.

10.5.1 *Applications of NAA*

Instrumental NAA was used by Krivan *et al.* (1993) as well as FAAS and ETAAS, for the multi-element analysis of green coffee. Gamma-ray spectrometry was used to measure the indicator radionuclides for barium, bromine, calcium, cobalt, chromium, caesium, iron, potassium, lanthanum, manganese, sodium, rubidium, scandium and zinc. Potassium, manganese and sodium were determined via medium-lived radionuclides with 70 mg portions of green coffee samples. The remaining elements were determined via long-lived radionuclides, using sample portions of 150–200 mg. Replicate determinations were carried out for some elements using AAS to provide a robust data set. On the basis of the results, manganese was found to be the most interesting element in relation to geographical origin. In 84% of the cases, the difference in concentration of manganese was statistically highly significant. Other elements of interest were cobalt, caesium, sodium, rubidium and the major element carbon.

Teherani (1987) investigated various rice species marketed in Austria measuring selenium, chromium, nickel, rubidium, iron, cobalt, caesium, silver and mercury by INAA. Rice samples (0.2 to 0.3 g) and chemical standards evaporated on to filter paper were sealed in quartz ampoules and irradiated for 24 h. After four weeks the samples and standards were measured. Statistical analyses showed that there were significant differences between samples for selenium, chromium, nickel, rubidium, iron, cobalt and mercury but not for caesium.

The mineral content of Canadian honeys was investigated in relation to geographical origin (Feller-Demalsy et al., 1989). Twenty elements were determined in 40 samples originating from different provinces. The main source of variation was found to be bromine with calcium, sodium, chlorine and magnesium. Discriminant analysis allowed separation of honeys according to their origin.

10.6 Mass spectrometry (MS)

This last section will consider applications using thermal ionization mass spectrometry (TIMS) and inductively coupled plasma-mass spectrometry (ICP-MS). Some consideration will be given to TIMS in terms of the basic principles and the advantages and disadvantages but most emphasis will be placed with ICP-MS. This latter technique has been commercially available for just over a decade and has revolutionized the multi-analyte measurement of isotopes and elements. Good up-to-date reviews are produced annually by Bacon et al. (1995, 1996) covering atomic mass spectrometry, including TIMS and ICP-MS, as well as accelerator MS, glow discharge MS and stable isotope ratio MS.

Thermal ionization MS is generally accepted to be the most accurate and precise method for measuring inorganic stable isotopes. It is based upon the formation of atomic or molecular ions on the hot surface of a metal filament. The ions are subsequently separated in a mass spectrometer and detected. The sample loading on to the filament, when the solvent must be evaporated, and the subsequent filament heating procedure used in the ion source are the crucial steps in TIMS (Kastenmayer, 1996). Thus, although TIMS produces very accurate (particularly when used with isotope dilution analysis) and precise data, the extensive sample preparation (since filament loading procedures are element specific and separation and purification of the element are necessary) makes the technique somewhat impractical for routine use (Vandecasteele and Block, 1993). As a consequence, however, TIMS does not suffer from the interferences that ICP-MS does, particularly for some of the isotopes of elements such as calcium, iron, zinc and selenium.

There are several good publications which describe the principles of ICP-MS in varying detail (Date, 1981; Gray, 1989; Horlick, 1992; Jarvis et al., 1992; Sargent and Webb, 1993). Figure 10.5 shows a schematic diagram of an ICP-mass spectrometer and Fig. 10.6 is a simplified version of the interface region. An induction coil surrounds a series of concentric quartz tubes (the torch) and argon gas is introduced in a tangential flow (Douglas and Houk, 1985). A high-temperature plasma is produced by the interaction of an induced magnetic field and the flow of argon gas. The plasma operates at powers of between 1 and 2 kW, usually at a frequency of 27 MHz, and is at atmospheric pressure. Temperatures in the plasma range from 5000 to 10 000 K. Samples are sprayed into the plasma with a nebulizer. After the liquid sample has been aspirated its

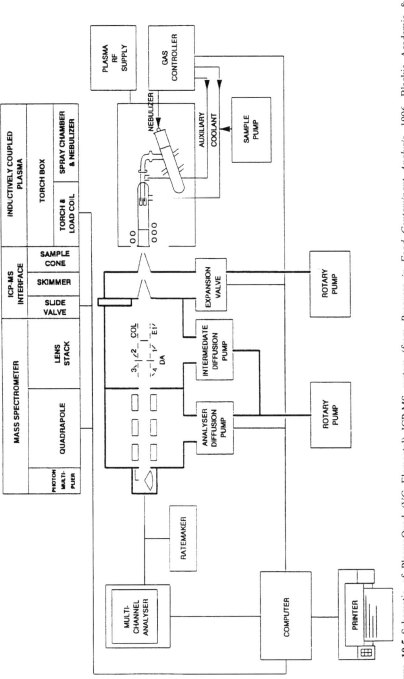

Figure 10.5 Schematic of PlasmaQuad (VG Elemental) ICP-MS system (from *Progress in Food Contaminant Analysis*, 1996, Blackie Academic & Professional).

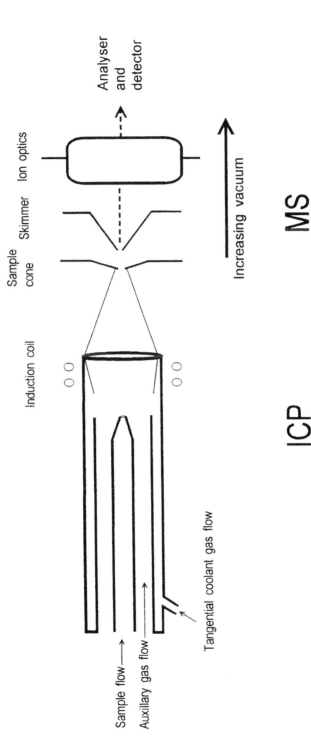

Figure 10.6 Schematic of ICP-MS interface. (From *Progress in Food Contaminant Analysis*, 1996, Blackie Academic & Professional.)

state changes as it travels from the nebulizer to the torch exit (Hieftje and Vickers, 1989). In the nebulizer, aerosol droplets are formed and those greater than 4 μm diameter are lost in the spray chamber (normally water cooled to approximately 10 °C; much lower, -15 °C, if organics are being analysed). The remaining droplets are transported to the plasma via the torch capillary. In the central channel of the plasma a sequence of desolvation, volatilization, and dissociation forms molecular vapour from the droplets, which in turn undergoes excitation and ionization so that a mixture of ions and atoms leaves the torch. The successful development of ICP-MS as a viable analytical system lay in the interface between the argon plasma and the quadrupole mass spectrometer. The former operates at atmospheric pressure while the latter requires an operating pressure of no more than 5×10^{-5} Torr. The ions must be transported from the high-temperature gas source to the low-temperature quadrupole without cooling or perturbing the plasma. This has been achieved by the use of a water-cooled cone with a central extraction aperture of about 1 mm diameter (Fig. 10.6). A reduced pressure (approximately 1 Torr) is maintained behind the sample cone by a mechanical (rotary) pump and this ensures a flow of gas containing ions from the centre of the tail flame of the plasma into the vacuum system (Douglas and French, 1981). Gas is sampled from an area approximately three times the sample cone orifice diameter which represents a large fraction of the total gas in the axial channel (Houk, 1986). A second cone, the skimmer, is located directly beind the sample cone and is positioned to ensure that as much of the sampled beam as possible is transmitted into the second reduced pressure stage (Figs 10.5 and 10.6). Also, to ensure that ions are extracted from the resulting supersonic expansion behind the sampling cone, the skimmer orifice is normally of similar size to that of the sampler orifice (Hieftje and Vickers, 1989). The area between the sampler and skimmer cones is referred to as the expansion chamber and the region behind the skimmer, which contains the ion lenses to collect and focus the beam, is known as the intermediate stage. The vacuum behind the skimmer is maintained when the ICP is not operating by closing a slide valve between the expansion chamber and intermediate stage. The operating pressure of the intermediate stage is about 10^{-4} Torr. The quadrupole is housed in a further stage which operates at 1×10^{-7} to 5×10^{-8} Torr. The vacuum in these two stages is achieved by use of diffusion or turbomolecular pumps. To be observed by the mass spectrometer, an ion must be present in the plasma and survive the extraction process or be formed during the extraction process (Houk, 1986). The interface is the crucial region in ICP-MS and its complexities are not fully understood (Crews, 1996).

The main disadvantages of ICP-MS are the relatively high running costs (mainly due to the large consumption of argon) and the existence of isobaric interferences at some masses. The disadvantages are outweighed by its considerable advantages. It is very sensitive, measurement is rapid (typically ≥ 30 samples per day), sample preparation can be minimal and it has multi-element and isotope capability (Crews et al., 1996).

10.6.1 *Applications using thermal ionization MS (TIMS) and inductively coupled plasma-MS (ICP-MS)*

Horn *et al.* (1993) have investigated the usefulness and practicalities of using strontium isotope ratios for assigning provenance of wines. They proposed that strontium with a specific ^{87}Sr:^{86}Sr isotope signature enters the vine from the soil, and that the same signature is found in grapes, musts and wines.

Unfortunately, details of their method for TIMS measurement of ^{87}Sr:^{86}Sr are not given in their original paper (Horn *et al.*, 1993) but they do state that isotope dilution analysis (IDA) was used to permit nanogram quantities of strontium to be measured with precisions of $\leq 0.1\%$ and that isotope ratios could be determined with precisions of 0.002%. Since the range of strontium concentration in wines was found to be 0.3 to 3.0 mg/kg, it was possible to easily measure millilitre quantities. The authors also state that 'the method allows the determination of isotopic abundance ratios much more precisely than, for example, by ICP-MS where ^{87}Sr:^{86}Sr cannot be analysed with a precision better than approximately 0.01%.' For this approach to work, a thorough knowledge of the geological and pedological conditions of the regions is an essential prerequisite. In addition another limitation may result from the cellarage manipulations such as clarification with bentonite and de-acidification with calcium carbonate which could lead to contamination with extraneous strontium (Horn *et al.*, 1993).

The rare earth elements (REEs) have been suggested as providers of specific patterns in relation to geographical origin. The elements are readily determined in foodstuffs by ICP-MS, usually being present in low concentrations. In addition, wine samples can be measured directly by this technique. Augagneur *et al.* (1996) compared various combinations of instrument and nebulizer systems to determine the REEs between masses 139 and 175 (lanthanum to lutetium). Samples were acidified to give a nitric acid concentration of 0.1% and run undiluted. In addition the effect of ethanol concentration was also investigated. They found that the use of micro-concentric nebulization permitted suppression of matrix interferences thus increasing the sensitivity of the direct analysis of wine. The work was based on a limited number of samples but did find that wines from different vineyards exhibited different REE patterns which might be used as a fingerprint. In contrast, similar patterns were found for different vintages from the same vineyard. The multi-element analysis by ICP-MS of Spanish and English wines (Baxter *et al.*, 1997), including the measurement of REEs, showed that the region of origin of Spanish wines was unequivocally identified following treatment of the data using discriminant analysis. Again, wines were measured directly, using flow injection analysis after dilution of the wine 1:1 with nitric acid (1%). Indium was used as an internal standard and calibration standards were made up in ethanol (5%) and nitric acid (0.5%). Injections (0.5 ml) were introduced into a carrier stream of ethanol and nitric acid which also contained the internal standard. A total of 72 isotopes were determined for 48 elements. Most of the geographic separation for three regions in Spain was achieved by using just seven elements, cadmium, chromium, caesium, erpium, gallium, manganese and ^{86}strontium.

In conclusion, this author feels that the way forward for trace element analysis for food authenticity lies with the multi-element techniques and particularly ICP-MS where isotopic data can also be obtained. If data for a single element are required as part of a suite of other organic analyses, then FASS and ETAAS cannot be surpassed for reliability and availablity. If the main thrust of the approach relies on more than two or three elements then multi-element techniques must be considered. For ease of sample introduction and pre-treatment of samples prior to measurement the ICP techniques are difficult to beat.

References

Alfassi, Z.B. (1994) *Determination of Trace Elements*, VCH Publishers, Weinheim and New York, 39–58.

Augagneur, S., Médina, B., Szpunar, J. and Lobinski, R. (1996) Determination of rare earth elements in wine by inductively coupled plasma mass spectrometry using a microconcentric nebulizer. *Journal of Analytical Atomic Spectrometry*, **11**, 713–21.

Bacon, J.R., Chenery, S.R.N., Ellis, A.T. et al. (1995) Atomic spectrometry update – atomic mass spectrometry. *Journal of Analytical Atomic Spectrometry*, **10**, 253R–310R.

Bacon, J.R., Crain, J.S., McMahon, A.W. and Williams, J.G. (1996) Atomic spectrometry update – atomic mass spectrometry. *Journal of Analytical Atomic Spectrometry*, **11**, 355R–93R.

Bauer, H.H., Christian, G.D. and O'Reilly, J.E. (1978) *Instrumental Analysis*, Allyn and Bacon, Inc.

Baxter, M.J., Crews, H.M., Dennis, M.J., Goodall, I. and Anderson, D. (1997) The determination of the authenticity of wine from its trace element composition. *Food Chemistry*, **60**(3), 443–50.

Birdsall, J.J., Derse, P.H. and Teply, L.J. (1961) Nutrients in Californian lemons and oranges. II Vitamin, mineral and proximate composition. *Journal American Dietetic Association*, **38**, 555–9.

Boumans, P.W.J.M. (1987) *Inductively Coupled Plasma Emission Spectrometry*, Parts 1 and 2, Wiley.

Brause, A. and Raterman, J.M. (1982) Verification of authenticity of apple juice. *Journal of the Association of Official Analytical Chemists*, **65**(4), 846–9.

Cerutti, G., Finoli, C., Vecchio, A. and Benelli, T.G. (1984) Trace elements in wines of different age and origin. *Cattedra di Residui e Additivi Alimentari Vignevini*, **11**(11), 27–30.

Christian, G.D. (1978) Flame spectroscopy, in *Instrumental Analysis* (eds H.H. Bauer, G.D. Christian and J.E. O'Reilly), Allyn and Bacon, Inc, 256–94.

Cornelis, R. (1988) Radiochemical methods, especially neutron activation analysis, in *Quantitative Trace Analysis of Biological Materials, Principles and Methods for Determination of Trace Elements and Trace Amounts of Some Macro Elements* (eds H.A. McKenzie and L.E. Smythe), Elsevier, pp. 263–81.

Crews, H.M. (1996). Inductively coupled plasma mass spectrometry (ICP-MS) for the analysis of trace element contaminants in foods, in *Progress in Food Contaminant Analysis* (ed. J. Gilbert), Blackie Academic & Professional, pp. 147–86.

Crews, H.M., Massey, R.C., McWeeny, D.J. et al. (1988) Some applications of isotope analysis of lead in food by ICP-MS. *Journal of Research of the National Bureau of Standards*, **93**(3), 464–6.

Crews, H.M., Luten, J.B. and McGaw, B.A. (1996) Inductively coupled plasma mass spectrometry, in *Stable Isotopes in Human Nutrition* (eds F. Mellon and B. Sandström), Academic Press, pp. 97–116.

Crighton, J.S., Carroll, J., Fairman, B., Haines, J. and Hinds, M. (1996). Atomic spectrometry update – industrial analysis: metals, chemicals and advanced materials. *Journal of Analytical Atomic Spectrometry*, **11**, 461R–508R.

Date, A.R. (1981) Inductively coupled plasma mass spectrometry. *Spectrochimica Acta Reviews*, **14**, 3–32.
Day, M.P., Zhang, B.L. and Martin, G. J. (1994) The use of trace element data to complement stable isotope methods in the characterisation of grape musts. *Am. J. Enol. Vitic*, **45**(1), 79–85.
Day, M., Zhang, B.L., Martin, G.J., Asselin, C. and Morlat, R. (1995a) Characterisation of the region and year of production of wines by stable isotopes and elemental analyses. *Journal International des Sciences de la Vigne et du Vin*, **29**(2), 75–87.
Day, M.P., Zhang, B. and Martin, G.J. (1995b) Determination of the geographical origin of wine using joint analysis of elemental and isotopic composition. II – differentiation of the principal production zones in France for the 1990 vintage. *Journal of the Science, Food and Agriculture*, **67**, 113–23.
Dettmar, H.P., Barbour, G.S., Blackwell, K.T. *et al.* (1996) Orange juice classification with a biologically based neural network. *Computers Chem.*, **20**(2), 261–6.
Douglas, D.J. and French, J.B. (1981) Elemental analysis with a microwave induced plasma quadrupole mass spectrometer system. *Analytical Chemistry*, **53**, 37–41.
Douglas, D.J. and Houk, R.S. (1985) Inductively-coupled plasma mass spectrometry (ICP-MS). *Prog. Analyst. Atom. Spectrosc.*, **8**, 1–18.
Eschnauer, H., Hölzl, S. and Horn, P. (1994) Isotope signaturs of heavy elements as a parameter of characterising wine: Isotope-Vinogramme. *Vitic. Enol. Sci.* **49**(3), 125–9.
Favretto, L., Favretto, L.G., Marletta, G.P. and Saitta, M. (1989a) Principal component analysis: a chemometric aid for classification of polluted and unpolluted mussels. *Analytica Chimica Acta*, **220**, 135–44.
Favretto, L.G., Marletta, G.P., Bogoni, P. and Favretto, L. (1989b) Chemometric studies of some trace elements in cows' milk. *Z. Lebensm Unters Forsch*, **189**, 123–7.
Feller-Demalsy, M.J., Vincent, B. and Beaulieu, F. (1989) Mineral content and geographical origin of Canadian honeys. *Apidology*, **20**, 77–91.
Fernandez-Pereira, C., Ortega, J. and Martin, A. (1987) Contribution of major and trace metals to the characterisation of Spanish wines. *Alimentaria*, **179**, 39–44.
Fresno, J.M., Prieto, B., Urdiales, R. and Sarmiento, R.M. (1995) Mineral content of some Spanish cheese varieties. Differentiation by source of milk and by variety from their content of main and trace elements. *Journal of Science, Food and Agriculture*, **69**, 339–45.
Gilbert, J. (1996) *Progress in Food Contaminant Analysis*, Blackie Academic & Professional, London, UK.
Gonzales-Larrana, M., Gonzales, A. and Medina, B. (1987) Metallic ions for differentiation of red wines of the three regions of Denomeration of Origin Rioja. *Connaissance de la Vigne et du Vin*, **21**, 127–40.
Gray, A.L. (1989) The origins, realization and performance of ICP-MS systems, in *Applications of ICP-MS* (eds A.R. Date and A.L. Gray), Blackie, Glasgow and London, pp. 1–42.
Hernández, G.G., de la Torre, A.H. and León, J.J.A. (1966) Quantity of K, Ca, Na, Mg, Fe, Cu, Pb, Zn and ashes in doc tacoronte-acentejo (Canary Islands, Spain) musts and wines. *Z. Lebensm. Unters. Forsch.*, **203**, 517–21.
Hieftje, G.M. and Vickers, G.H. (1989) Developments in plasma source/mass spectrometry. *Analytica Chimica Acta*, **216**, 1–24.
Hill, S.J., Dawson, J.B., Price, W.J. *et al.* (1996) Atomic spectrometry update – advances in atomic absorption and fluorescence spectrometry and related techniques. *Journal of Inductively Coupled Plasma Mass Spectrometry*, **11**, 281R–326R.
Horlick, G. (1992) Plasma source mass spectrometry for elemental analyses. *Spectroscopy (Oregon)*, **7**, 22–9.
Horn, P., Schaaf, P., Holbach, B. *et al.* (1993) $^{87}Sr/^{86}Sr$ from rock and soil in vine and wine. *Z. Lebensm. Unters. Forsch.*, **196**, 407–409.
Houk, R.S. (1986) Mass spectrometry of inductively coupled plasmas. *Analytical Chemistry*, **58**, 97–105.
Jarvis, I. (1992) Sample preparation of ICP-MS, in *Handbook for ICP-MS*, Blackie & Son Ltd, 173–224.
Jarvis, K.E., Gray, A.L. and Houk, R.S. (1992) *Handbook for ICP-MS*, Blackie & Son.
Kastenmayer, P. (1996) Thermal ionisation mass spectrometry (TIMS), in *Stable Isotopes in Human Nutrition* (eds F. Mellon and B. Sandström), Academic Press, 81–96.

Kimberly, M.M., DiPietro, E.S. and Paschal, D.C. (1988) Emission spectroscopy, in *Quantitative Trace Analysis of Biological Materials, Principles and Methods for Determination of Trace Elements and Trace Amounts of Some Macro Elements* (eds H.A. McKenzie and L.E. Smythe), Elsevier, 155–76.

Krivan, V., Barth, P. and Morales, A.F. (1993) Multielement analysis of green coffee and its possible use for the determination of origin. *Mikrochim. Acta*, **110**, 217–36.

Latorre, M.L., García-Jares, A., Médina, B. and Herrero, C. (1994) Pattern recognition analysis applied to classification of wines from Galicia (Northwestern Spain) with certified brand of origin. *Journal of Agricultural Food Chemistry*, **42**, 1451–5.

Lopez Martin, J.F. (1979) Determination of trace elements in samples of milk and milk products by neutron activation analysis. *Anales de la Facultad de Ciencias Quimicas y Farmacologicas, Universidad de Chile*, **29/30**, 43–44.

Lundegårdh, H. (1938) The triple analysis method of testing soil fertility and probable crop reaction to fertilisation. *Soil Science*, **45**, 447–53.

Lundegårdh, H. (1943) Leaf analysis as a guide to soil fertility. *Nature*, **151**, 310–11.

Manahan, S.E. (1986) *Quantitative Chemical Analysis*, Brooks/Cole Publishing Company.

Martín-Hernández, C., Amigo, L., Martín-Alvarez, P.J. and Juárez, M. (1992) Differentiation of milks and cheeses according to species based on the mineral content. *Z. Lebensm. Unters. Forsch.*, **194**, 541–4.

McHard, J.A., Foulk, S.J. and Winefordner, J.D. (1979). A comparison of trace elements contents of Florida and Brazil orange juice. *Journal of Agricultural Food Chemistry*, **27**(6), 1376.

McHard, J.A., Foulk, S.J., Jorgensen, J.L. et al. (1980) Analysis of trace metals in orange juice. *American Chemical Society*, Chap. 16 in ACS Syposium Series, **143**, 363–92.

McKenzie, H.A. and Smythe, L.E. (1988). Quantitative trace analysis of biological materials, in *Principles and Methods for Determination of Trace Elements and Trace Amounts of some Macro Elements*, Elsevier.

Montaser, A. and Golightly, D.W. (1987) *Inductively Coupled Plasmas in Analytical Atomic Spectrometry*, VCH Publishers, New York.

Moret, I., Scarponi, G. and Cescon, P. (1994) Chemometric characterisation of classification of five venetian white wines. *Journal of Agricultural Food Chemistry*, **42**, 1143–53.

Morris, E.R. and Greene, F.E. (1970) Distribution of lead, tin, cadmium, chromium and selenium in wheat and wheat products. *Federation Proceedings, Federation of American Societies for Experimental Biology*, **29**(2), 500.

Munilla, M.A., Gomez-Pinilla, I., Rodenas, S. and Larrea, M.T. (1995) Determination of metals in seaweeds used as food by inductively coupled plasma atomic-emission spectrometry. *Analysis*, **23**, 463–6.

Negretti de Brätter, V.E., Brätter, P., Reinicke A. et al. (1995) Determination of mineral trace elements in total diet by inductively coupled plasma atomic emission spectrometry: comparison of microwave-based digestion and pressurised ashing systems using different acid mixtures. *Journal of Analytical Atomic Spectrometry*, **10**, 487–91.

Nikdel, S. (1986) Trace mineral analysis of authentic orange juice by ICP-AES and application of pattern recognition for country of origin classification. *Proceedings of 37th Annual Citrus Processors' Meeting, Lake Alfred, FL*, 36–9.

Nikdel, S. and Attaway, J.A. (1987) Characterization of citrus juices via pattern recognition using trace element contents. *XXV Colloquium Spectroscopium Internationale, Toronto, Canada, 25 June 1987*, Abstract No. B6.2, 78.

Nikdel, S. and Temelli, C. (1987) Comparison of microwave and muffle furnace for citrus sample preparation and analysis using ICP-AES. *Microchimica Journal*, **36**, 240–4.

Nikdel, S., Nagy, S. and Attaway, J.A. (1988) Trace metals: defining geographical origin and detecting adulteration of orange juice. *Food Science Technology*, **30**, 81–105.

Roberts, J.A. and Gaddum, L.W. (1937). Composition of citrus fruit juices. *Ind. Eng. Chem.*, **29**, 574–5.

Sargent, M. and Webb, K. (1993) Instrumental aspects of inductively coupled plasma mass spectrometry. *Spectroscopy Europe*, **5**, 21–8.

Sharp, B.L. (1988a) Pneumatic nebulisers and spray chambers for inductively coupled plasma spectrometry. A review. Part 1 Nebulisers. *J. Analytical Atomic Spectrometry*, **3**, 613–52.

Sharp, B.L. (1988b) Pneumatic nebulisers and spray chambers for inductively coupled plasma spectrometry. A review. Part 2 Spray chambers. *J. Analytical Atomic Spectrometry*, **3**, 939–63.

Taylor, A., Branch, S., Crews, H.M. et al. (1996) Atomic spectrometry update – clinical and biological materials, food and beverages. *Journal of Analytical Atomic Spectrometry*, **11**, 103R–86R.

Taylor, A., Branch, S., Crews, H.M. et al. (1997) Atomic spectrometry update – clinical and biological materials, food and beverages. *Journal of Analytical Atomic Spectrometry*, **12**, 119R–222R.

Teherani, D.K. (1987) Trace elements analysis in rice. *J. Radioanal. Nucl. Chem.*, **117**(3), 133–43.

Thompson, M. and Wood, R. (1995) Harmonised guidelines for internal quality control in analytical chemistry laboratories. *Pure and Applied Chemistry*, **67**, 649–66.

Tölg, G. (1993) Problems and trends in extreme trace analysis for the elements. *Analytica Chimica Acta*, **283**, 3–18.

Tölg, G. and Tschöpel, P. (1994) Systematic errors in trace analysis, in *Determination of Trace Elements* (ed. B.Z. Alfassi), VCH Publishers, Weinheim and New York, 1–38.

Ureña Pozo, M.E., Gimenez Plaza, J. and Cano Pavon, J.M. (1987) Study of trace contents of iron, manganese, copper, lead, zinc, cadmium and cobalt in wines from Malaga. *Alimentaria*, **180**, 83–6.

van der Schee, H.A., Bouwknegt, J.-P., Tas, A.C. et al. (1989) The authentication of sherry wines using pattern recognition: an inter laboratory study. *Z. Lebensm. Unters. Forsch.*, **188**, 324–9.

Vandecasteele, C. and Block, C.B. (1993) *Modern Methods for Trace Element Determination*, John Wiley.

Walsh, A. (1955) The application of atomic absorption spectra to chemical analysis. *Spectrochimica Acta*, **7**, 108–17.

Willard, H.H., Merritt, L.L. Jr, Dean, J.A. and Settle, F.A. Jr (1988) *Instrumental Methods of Analysis*, 7th edn, Wadsworth Publishing Company, California.

Woittiez, J.R.W. and Sloof, J.E. (1994) Sampling and sample preparation, in *Determination of Trace Elements* (ed Z.B. Alfassi), VCH Publishers, Weinheim and New York, pp. 59–107.

11 Pyrolysis mass spectrometry in food analysis and related fields: principles and application
M. LIPP and E. ANKLAM

11.1 Introduction

Conventional analysis of food, other consumer goods and raw materials normally involves preparation steps due to the complex matrix composition of the sample. This is then followed by the application of various chromatographic or spectroscopic techniques, often including mass spectrometry. These procedures are often very time and chemicals consuming.

Pyrolysis-mass spectrometry (Py-MS) is a very fast and sensitive screening characterization technique. Using this method, the sample is placed in high vacuum and heated under controlled conditions. The organic material undergoes rapid decomposition and the low molecular weight products are led into a mass spectrometric device. Applying Py-MS, the pyrolysate is directly analysed, while pyrolysis-gas chromatography-mass spectrometry (Py-GC-MS) allows also a separation of the decomposed samples before detection. The major emphasis in this chapter is the application of Py-MS.

Py-MS has been applied in many scientific disciplines in the past. Besides its use for characterization of various food commodities and of microbial systems, it has been used in fields such as characterization of plant materials including feeding stuff (Mulder *et al.*, 1992; Morisson *et al.*, 1991; Reeves and Galletti, 1933), geochemistry (Meuzelaar *et al.*, 1984), soil (Bracewell and Robertson, 1984), forensic chemistry (Munson, 1995) and polymer sciences (Irwin, 1982; Aries *et al.*, 1985).

The rapid and accurate identification of bacteria, within minutes, rather than hours or even days is of great importance for food quality and other quality control activities. Py-MS has been widely applied to the characterization of microbial systems (Schulten *et al.*, 1973; Irwin, 1982; Meuzelaar *et al.*, 1982; Gutteridge, 1987; Berkeley *et al.*, 1990). Due to its high discriminatory ability, Py-MS has been successfully applied to the inter-strain comparison and classification of a wide range of bacterial species (Meuzelaar *et al.*, 1982; Gutteridge, 1987; Berkeley *et al.*, 1990; Goodacre and Kell, 1993, 1996a) in particular for *Bacillus* (Shute *et al.*, 1984; Sisson *et al.*, 1992), *Corynebacterium* (Meuzelaar *et al.*, 1982), *Escherichia* (Wieten *et al.*, 1984; Goodacre and Berkeley, 1990; Goodacre *et al.*, 1991; Freeman *et al.*, 1995), *Legionella* spp. (Kajioka and Tang, 1984; Sisson *et al.*, 1991), Mycobacteria (Meuzelaar *et al.*, 1976; Wieten *et al.*, 1981a, b), *Salmonellae* (Freeman *et al.*, 1990), *Staphylococcus* spp. (Freeman *et al.*, 1991a; Gould *et al.*, 1991), *Streptococci* (Magee *et al.*, 1989; Freeman *et al.*, 1991b) and *Pseudomonas* (Heyler *et al.*, 1993).

Another application of Py-MS and Py-GC-MS for characterization and proof of authenticity was shown for tobacco blends (Halket and Schulten, 1985a; Simmleit and Schulten, 1986; Schulten, 1986).

11.2 Principles

The analytical technique pyrolysis mass spectrometry can be divided in two steps, usually combined within one instrument. The first step consists of the pyrolysis of the sample being followed by analysis of the pyrolysate using a mass spectrometer.

The important step for pyrolysis is the fast energy transfer to the sample, to heat it up until pyrolysis takes place; the gas so produced has to be transferred to a mass spectrometer. This transfer normally involves a split of the pyrolysate to fulfil the vacuum needs of the mass spectrometer. The whole system has to be maintained at a vacuum level that does not allow intermolecular reactions which could take place within the pyrolysate. For economic reasons, most instruments use a quadrupole mass spectrometer for the analysis of the pyrolysate.

11.2.1 *Pyrolysis techniques*

All pyrolysis techniques are designed for a very fast transfer of energy to heat the sample up to a temperature high enough so that the pyrolysis takes place. Pyrolysis takes place only at the surface of the samples, because the products are gaseous.

For most applications samples will be put on a support. Thus, the crucial points for heat transfer are: from the heat source to the sample support; within the sample support; from the support to the sample; the heat conductivity within the sample. For the overall heat conductivity from the heat source to the sample surface, the bottleneck-principle applies, i.e. the step with the lowest heat conductivity will most influence the system. Using metallic sample supports, the heat conductivity within the sample support is high enough to be neglected. In some techniques the first energy transfer step from the source to the sample by indirect heating is bypassed by directly heating the samples (e.g. by energy transfer with a laser). For indirect heating systems, special care has to be taken to ensure a very high heat conductivity between the heat source and the sample or its support.

Sample preparation. Using an indirect heating system, the thin and uniform spread of the sample on its support is very important. In most applications, the samples are non-metallic material and consequently this heat conductivity will be much lower than for the rest of the system. The bottleneck for such a system is therefore the energy transfer within the sample. As pyrolysis of the sample will take place only in its uppermost layer, a fast heating of the surface can only be achieved by keeping the sample as thin as possible. Any irregularities will result in a non-homogeneous surface temperature, and pyrolysis of the sample will be unsteady. Especially difficult is the preparation of particulate samples. Here the limiting factors for the energy transfer are the interface sample support – and heat transfer within the sample.

For systems with direct heating (i.e. use of laser) the layer thickness of the sample is not as crucial as for indirect heated pyrolysis systems, because the energy transfer takes place from the surface towards the inner part of the sample. However, optical properties of the sample, such as absorption coefficients and/or glancing of the sample has to be taken into account. Different amounts of energy can be transferred in samples that show different absorption coefficients for the irradiation used. In addition, the analysis of particles using direct heating is also difficult. Energy transfer into the sample surface depends not only on the absorption coefficients but also on its reflective properties.

Resistively heated filaments. In systems with filament heating the sample is placed on a support allowing close thermal contact of the heating filament and the sample support. Such filaments exhibit more difficulty in control of their final temperature. In addition, the design of an autosampler is more difficult in order to guarantee the close thermal contact. Recent instrument developments have taken these problems into account and are appropriately designed (Wampler, 1995).

Curie-point. The sample is placed on a ferromagnetic support for Curie-point pyrolysis. Sample exposure to a magnetic field, alternating its direction with high frequency, causes repeated inversion of the magnetization. Due to the hysteresis of the magnetic susceptibility, this alternating magnetization heats the sample support. Sample support will loose its ferromagnetic properties at a material specific temperature, the Curie-point. At the exact temperature of the Curie-point the sample support will become ferromagnetic, thus it is not possible to magnetize the material any longer. As soon as the temperature drops below the Curie-point the sample support will become paramagnetic again and can be magnetized. Thus, Curie-point analysis is used for fast (typical 0.1 s) heating of the sample support to a very well-defined end-temperature, which will be held very precisely. However, the pyrolysis temperature can only be chosen according to the Curie-point of the alloys available (Aries and Gutteridge, 1987b; Aries *et al.*, 1986a).

Laser. Exposing the sample to a laser beam is a very effective way of transferring energy. Using this technique, the sample is mostly heated in the surface layer, thus providing the transfer of energy to the part of the sample where it can be directly used for pyrolysis and avoiding all problems with heat conductivity. The exact amount of energy transferred into the sample is, however, dependent on the optical properties of the sample. The absorption coefficient together with reflective properties limit the amount of energy transferred into the sample surface. Due to the small area covered with a laser beam, the outer area of the sample, not being exposed to the beam, will undergo slower heating (energy transfer by thermal conductivity) and thus cause a delayed pyrolysis (Aries and Gutteridge, 1987a).

Others. The characteristics of a furnace pyrolyser is described in Wampler, 1995. In principle, the sample is introduced into a furnace kept always at a constant temperature. Such instruments are simple to build and use and thus are inexpensive. However, the temperature rise of the sample depends on the nature of the sample, its size and its geometry.

Transfer system. The main task of the transfer system is to guide the molecules in the pyrolysate into the mass spectrometer and to maintain the appropriate vacuum conditions for operating it. Guiding of the molecules can be done by installing a drift tube, that has to be heated to minimize losses by absorption on the wall. The vacuum conditions are normally maintained by a differential pumping system (Meuzelaar et al., 1982).

11.2.2 Ion generation

Electron impact. The molecules have to be transformed into ions for analysis of the pyrolysate by mass spectrometry. This is performed by electron impact ionization for most applications. Using this technique, ions are produced by a heated filament and accelerated to an energy sufficient for ionization of the molecules under investigation to achieve the optimum in ionization efficiency; thus, a maximum in intensity and sensitivity is achieved. However, not all instruments offer these possibilities.

Field ionization. Molecules in the pyrolysate will be ionized by passing an electric field of high strength. This method represents a soft ionization technique and is thus capable of ionizing molecules that would be fragmented by electron impact ionization (Carlsen, 1984).

11.2.3 Mass spectrometer

A quadrupole mass spectrometer is often used for economical reasons. However, a sector field based mass spectrometer will give some distinct advantages, such as better resolution and less discrimination of the heavier molecules. Quadrupole mass spectrometers are easier to maintain and allow a very fast scan of a large mass range. The decision of which mass spectrometer is to be used is guided on the one hand by the specific needs for the planned investigations and on the other hand by budgetary constraints. If there is neither need for a high mass

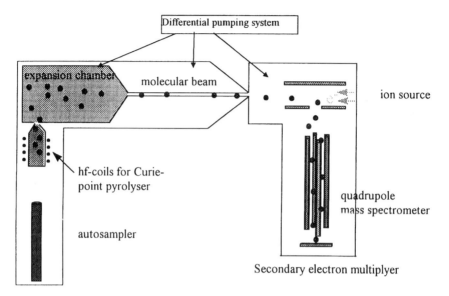

Figure 11.1 Schematic drawing of a typical pyrolysis mass spectrometer with Curie-point pyrolysis and quadrupole mass spectrometer.

resolution nor the determination of molecules at high masses (>400 amu), a quadrupole mass spectrometer will be the instrument of choice (Meuzelaar et al., 1982).

Principles of the instrument. In Fig. 11.1 the principles of the instrument are shown schematically. Normally the instruments are equipped with an autosampler. One analysis takes usually about 20 s for data acquisition, thus the advantage of the short analysis time can be fully used only with an autosampler. An instrument with a Curie-point pyrolyser is shown in Fig. 11.1. The sample is introduced into the pyrolyser which is evacuated and pyrolysis takes place. The pyrolysate expands into an expansion chamber and is subsequently transferred into the ion source. As shown in the scheme below this can be realized via a molecular beam. The molecules will be ionized in the ion source, shown here as electron impact source and analysed with a mass spectrometer, the latter being here a quadrupole system.

11.3 Data evaluation

A general introduction into multivariate data evaluation and the principles of neural networks is given in Chapter 12 of this book. The main difficulty in analysing the Py-MS spectra lies in mapping of complex and multidimensional data into a simple description that can be used for further interpretation. In general, a mass range of about 200 masses is acquired and has to be reduced to two or three dimensions for the benefit of visualization.

Prior to their numerical analysis the data have to be normalized; this is done to exclude the influence of the absolute signal intensities. They depend mainly on the sample preparation and it is in most cases unnecessary information for further data evaluation. Data are usually normalized to the total ion count, i.e. the sum of intensities of all peaks (Harper *et al.*, 1984). However, to avoid pseudo-correlation, signals with a high standard deviation for the calculation of the total ion count can be excluded (Windig *et al.*, 1987).

The normalized data are then subjected to factor analysis with or without factor rotation. Factor analysis is sometimes preceded by selection of significant data by canonical variance analysis, also called discriminant analysis, to save computing time (Harper *et al.*, 1984; Windig *et al.*, 1987; Aries *et al.*, 1986a; Aries and Gutteridge, 1987a). The results of the canonical variance analysis can be visualized by means of cluster analysis, transforming its Mahalanobis distances matrix into a similarity matrix (Gutteridge *et al.*, 1985). Other methods based on factor analysis but suited for supervised learning such as SIMCA (soft independent modelling of classes; Wold *et al.*, 1984) are applied (Harper *et al.*, 1984).

Factor analysis (PCA) (Windig *et al.*, 1987), principal component regression (PCR) (Zarrabi *et al.*, 1991; Goodacre *et al.*, 1994) and neural networks (Goodacre *et al.*, 1993) are applied most often.

11.4 Food analysis by Py-MS

11.4.1 *Application to food without special emphasis on proof of authenticity*

Application of Pyrolysis-mass spectrometry (Py-MS). A comprehensive introduction into the application of Py-MS for characterization of food ingredients such as carbohydrates, proteins, lipids and sterols is given by Meuzelaar *et al.* (1982). Py-MS together with multivariate data handling procedures was investigated as a potential method for the rapid determination of the degree of methylation in pectin. A good discrimination of various degrees of methylation was achieved by this method (Aries *et al.*, 1988). Honeycomb waxes from European, African, and Africanized hybrid honeybees were analysed by Py-MS together with principal component analysis to determine the feasibility of using the technique as a rapid method for bee identification (Beverly *et al.*, 1995). It was found possible to differentiate the European from the African waxes. This technique could probably also be used for characterization of honey containing honeycombs. A number of five different polysaccharide based gums and food thickeners were classified by Py-MS (Aries *et al.*, 1987a). Samples which differ significantly in chemical composition could be easily discriminated. However, samples that differ only within the relative amounts of constituents are difficult to discriminate.

Application of Curie-point pyrolysis-gas chromatography-mass spectrometry (Py-GC-MS). Direct in-source pyrolysis mass spectrometry and pyrolysis gas chromatography mass spectrometry has been used by many researchers for two reasons: first, for fast profiling of complex materials and second, for identification of significant products of the controlled thermal degradation process.

Pyrolysis (Py) field ionization (FI)-mass spectrometry (MS) together with Curie-point pyrolysis gas chromatography-mass spectrometry (Py-GC-MS) was applied for the characterization of roasted coffee, rosehip/hibiscus tea blend, chocolate drink powder, milk chocolate and wheatmeal biscuit (Halket and Schulten, 1988). In this study, emphasis was placed on the rapid profiling of the samples. The same authors have used the Py-GC-MS technique for a rapid analysis of volatiles from orange oil and purple sage perfume (Halket and Schulten, 1985b) and for investigation of caffeine in roasted coffee (Schulten and Halket, 1986).

Application of Py-GC-MS for various foodstuffs was carried out by Baltes, 1985. One reason for using this technique was to study the formation of phenols during coffee roasting and the generation of polycyclic aromatic hydrocarbons in curing smoke. The other reason was to characterize caramel, thickening agents and polymers in food formed by Maillard reaction (Baltes, 1985).

The Py-GC-MS method has been successfully applied (with few exceptions) for classification of unknown caramel colours into four classes (Hardt and Baltes, 1989) and for quality control of rice (Samukawa *et al.*, 1992).

Data analysis of pyrolysis mass spectra has shown that differences in the chemical composition of starches of different origin could be correlated to the differences in degree of degradability (Cone *et al.*, 1992). In this chapter the authors have put more emphasis on the degradability of starch granules than on the use for proof of authenticity. However, the published figures indicate a good classification of starch botanical origin.

11.4.2 *Application to food authentication*

Py-MS is a fast and sensitive technique for classification of food according to its chemical composition. The latter can be correlated mainly to geographical and botanical origin and processing techniques. In the following examples the application of Py-MS is held to prove authenticity of or detect frauds in olive oil and other vegetable fats, whisky, wine, vinegar, orange juice, tea and coffee.

Olive oil. According to Commission Regulations on olive oil and its amendments, olive oils can be classified into nine types. However, only the first six types of olive oils (extra-virgin, virgin, ordinary virgin, virgin lampante, refined, olive oil) are usually supplied to the retail trade. The first three (virgin) oils differ only in the quality of the olives processed. 'Extra-virgin' olive oil is the oil extracted by mechanical means only of sound and ripe fruits of the olive tree.

The free fatty acid content has to be lower than 10 g/kg. 'Extra-virgin' olive oil is a high-priced product due to the high consumer demands and therefore of commercial interest. This leads to a great temptation to adulterate olive oils with other seed oils (Kiritsakis, 1991). Goodacre *et al.* applied Py-MS for the detection of adulteration of 'extra-virgin' olive oil with other seed oils. The major source for variation between the spectra of the samples was shown to be by cultivar differences. However, neural networks could be successfully trained to distinguish between pure and adulterated samples (Goodacre *et al.*, 1992, 1993, 1995). Training of these networks was done on the basis of 24 samples (12 pure and 12 adulterated oils) and a test set of the sample size could then be correctly classified. Despite the low number of samples, generalization of the separation problem of the neural network was indicated by using a completely different training and test set. As an example, the normalized and averaged pyrolysis mass spectra of authentic virgin olive oils and those that have been adulterated are given in Fig. 11.2. The results of the application of neural networks on those mass spectra are shown in Fig. 11.3. Thus, no single oil was

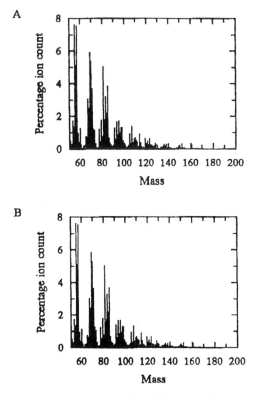

Figure 11.2 Normalized and averaged spectra of (A) all 12 virgin oils from the first set of oils, and (B) all 12 of those olive oils that have been adulterated with lower grade oils, also from the first set of oils.

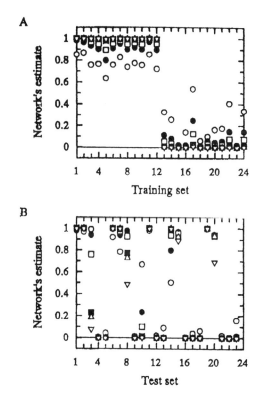

Figure 11.3 Results of the estimates of trained 150-0-1 ANN for both (A) the training set, and (B) the test set (unknowns) of oils. The ANN were interrogated after the RMS error values indicated (0.25–0.001) were reached. ○, 0.25; ●, 0.1; □, 0.05; ■, 0.01; △, 0.005; ▽, 0.002.

included in the training set and the test set. The authenticity of the samples was guaranteed by the authors, producing the olive oils themselves.

Whisky. Seven Scottish malt whisky samples were classified by Py-MS (Aries and Gutteridge, 1987a). The data were analysed by canonical variance analysis as well as by factor analysis. Both methods were able to classify all seven kinds of whisky separately, however, different signals were used for this differentiation by each of these methods.

A more detailed study with emphasis on the assessment of Scotch whisky quality by application of Py-MS was carried out (Reid *et al.*, 1933). The most important factor influencing the quality of Scotch whiskies was found to be the oak casks used for ripening, followed by that of water, barley type, extent of peating and storage conditions. Py-MS was found to be capable of differentiating between 'good quality Scotch whisky' and 'bad quality Scotch whisky'. A set of 14 malt whisky extracts from a specific distillery were prepared and analysed by Py-MS at a Curie-point temperature of 610 °C. Principal component analysis (PCA) performed on the resulting data set revealed excellent sample replication with the first principal component discriminating on the basis of quality (maturation stage). The results are summarized in Fig. 11.4. The authors noted the interesting observation, that the earlywood content of the oak cask correlates with the whisky quality, as derived from the Py-MS analysis.

In another study, Py-MS was applied for the authentication of Scotch whisky in comparison with the proof of authenticity by GC analysis of the higher alcohols (Aylott *et al.*, 1994). The GC analysis showed a very consistent profile of the content of the higher alcohol for different brands, even over several batches. Py-MS, combined with canonical variance analysis was applied on the same samples. The authors could demonstrate that Py-MS offered great ability to separate clusters for the individual producers, thus confirming the results of the GC analysis. They judged the Py-MS as a method that 'has the potential to develop into a useful stand-alone technique for authenticity analysis' (Aylott *et al.*, 1994).

Wine. Py-MS was applied to the classification of European wines, combined with data evaluation by canonical variance analysis, with or without preceding factor analysis and neural networks. The distillation residues of 33 authentic wine samples originating from various EU Member States were investigated. The application of canonical variance analysis did not lead to satisfactory results, however, the combination of canonical variance analysis and factor analysis achieved a good classification by country of origin. By the application

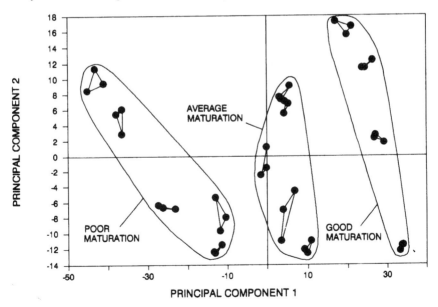

Figure 11.4 Plot of the first two principal components (84% variance) obtained from the analysis of the Distillery 1 whisky extracts.

of neural networks the authors claimed that 90% of the wines were classified correctly, with respect to the country of origin. However, no correlation was found between Py-MS spectra and grape species or area of cultivation by any of the data evaluation techniques. A preliminary study using Py-MS has already been made regarding the classification of red wines (Ottley and Maddock, 1986).

Orange juice. Orange juice is traded as a commodity world-wide on a massive scale, it is therefore not surprising that its adulteration is common (Petrus and Vandercook, 1980). An overview on frauds, adulteration techniques and methods of detection has recently been published (Hammond, 1996). The application of PY-MS in food control including fruit juices was shown (Hussain, 1993).

One of the most common frauds is the incorrect labelling of orange juice by geographical origin. A number of orange juices (58) originating from Brazil and Israel was classified applying Py-MS (Aries *et al.*, 1986b). The majority of the 31 Israeli and the 27 Brazilian orange juice samples could be classified correctly applying canonical variance analysis. A number of masses (20) was selected according to its characteristicity (variance ratio) values, i.e. selection of the 20 most discriminatory masses (Israel vs. Brazil). These results are shown in Fig. 11.5. The clustering of some Brazilian samples within the group of Israeli fruit juices gave cause to doubt their authenticity. Moreover, this investigation succeeded in indicating the potential of Py-MS in differentiating between the producers for each country of origin (Aries *et al.*, 1986b).

Cocoa butter and other vegetable fats. Py-MS was recently applied to the characterization of cocoa butter and other vegetable fats (Anklam *et al.*, 1997). Thirty-six samples of 17 different vegetable fats have been investigated applying factor analysis followed by cluster analysis and neural networks. Using cluster analysis, cocoa butter could be readily separated from all other vegetable fats, excluding kokum butter; applying neural networks for data evaluation, a classification for all samples was achieved. However, this technique failed for one replication (out of three) of two cocoa butters. The authors could show, that the information taken from the Py-MS spectra and used for classification was totally different for the two approaches.

Tea. In a pilot study, 20 tea extracts were classified according to their geographical origin (Assam, Africa, India and Ceylon) (Gutteridge *et al.*, 1984). Despite their heterogeneity, the African samples could be discriminated as a group by application of canonical variance analysis (CVA). The tea extract samples from the other geographical origins clustered distinctively in an homogeneous way. Despite the small number of samples analysed Py-MS, however, seems to be a suitable approach for characterization of various tea samples from different origins.

Vinegar. A number of Italian balsamic vinegar samples (22) from different production processes (13 samples of industrial product and 9 samples traditionally produced) were investigated by Py-MS and compared to the results obtained by applying a sensor technique based on an 'electronic nose' (Anklam *et al.*, 1996). Both techniques were shown to be capable of classifying distinctively the two groups of vinegar. Although the number of samples available for this study did not seem to be sufficient for detailed analysis, both methods indicated possible discrimination of the traditionally prepared samples by the age of samples. In an additional study the application of Py-MS was enlarged to a broad variety of vinegar samples originated from different raw materials (fruit, apple, red and white wine, malt, ethanol, balsamic vinegar) by the same authors (Radovic *et al.*, 1996, unpublished). This method was found to be suitable to classify the vinegar samples according to their origin.

Milk products. Due to the fact that the production of cheeses from ewes' and goats' milk has gained wide consumer acceptance and therefore economic importance, there is a temptation to adulterate such cheeses with cows' milk. Py-MS together with supervised learning was shown to be useful for the assessment of the adulteration of milk from different species (Goodacre and

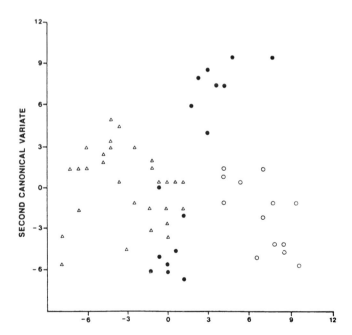

Figure 11.5 Plot of the first two canonical variates (86% variance) for the Israeli and Brazilian juices analysed using the 20 most characteristic masses: ○, Brazil (1983); ●, Brazil (1984); △, Israel (1983).

Kell, 1996b). To determine the percentage of adulteration of either goat or ewe milk with cows' milk to <1% the high dimensional data from Py-MS were analysed either by artificial neural networks (ANNs) or other supervised learning methods exploiting linear regression, i.e. partial least squares (PLS) and principle component regression (PCR). In addition, the same authors have proven the suitability of the Py-MS method for determination of the milk fat content in each of the three milks (cow, ewe, goat). The experiments were carried out with a small number of samples. However, it could be demonstrated that the % fat content could be predicted to $<\pm 0.5\%$ using ANNS with radial basis function (RBFs) (Goodacre and Kell, 1996b).

Py-MS could be successfully applied to distinguish between cheeses of different ripening stages. Preliminary results on 27 samples of the same type of Italian cheese ('Caciotto') have shown that two groups could be distinguished (14 ripened and 13 fresh samples). The results were in excellent correlation with those obtained by detection of chemical marker compounds determined by chromatographic techniques (Corradini et al., 1996, unpublished).

References

Anklam, E., Bassani, M.R., Eiberger, Th. et al. (1997) Characterization of cocoa butter and other vegetable fats by pyrolysis mass spectrometry. *Fresenius J. Anal. Chem.*, **357**, 981–9.

Anklam, E., Lipp, M., Radovic, B. et al. (1996) Characterisation of Italian vinegar by pyrolysis-mass spectrometry and a sensor device ('electronic nose'). *Food Chemistry*, in press.

Aries, R.E., Gutteridge, C.S. and Macrae, R. (1985) Pyrolysis mass spectrometry investigations of reversed phase high-performance liquid-chromatography phases. *J. Chromatogr.*, **319**, 285–97.

Aries, R.E., Gutteridge, C.S. and Evans, R. (1986b) Rapid characterization of orange juice by pyrolysis mass spectrometry. *J. Food Sci.*, **51**, 1183–6.

Aries, R.E., Gutteridge, C.S. and Ottley, T.W. (1986a) Evaluation of a low-cost, automated pyrolysis mass-spectrometer. *J. Anal. Appl. Pyrolysis*, **9**, 81–98.

Aries, R.E. and Gutteridge, C.S. (1987a) Applications of pyrolysis mass spectrometry to food science, in *Application of Mass Spectrometry in Food Science* (ed. J. Gilbert), Elsevier, pp. 377–431.

Aries, R.E. and Gutteridge, C.S. (1987b) Dedicated pyrolysis mass spectrometry: instrumentation and potential applications for industry. *Int. Analyst*, **1**, 10–12.

Aries, R.E., Gutteridge, C.S., Laurie, W. et al. (1988) A pyrolysis-mass spectrometry investigation of pectin methylation. *Anal. Chem.*, **60**, 1498–1502.

Aylott, R.I., Clyne, A.H., Fox, A.P. and Walker, D.A. (1994) Analytical strategies to confirm Scotch whisky authenticity. *Analyst*, **119**, 1741–6.

Baltes, W. (1985) Application of pyrolytic methods in food chemistry. *J. Anal. Appl. Pyrolysis*, **8**, 533–45.

Berkeley, R.C.W., Goodacre, R., Helyer, R.J. and Kelley, T. (1990) Pyrolysis-mass spectrometry in the rapid identification of microorganisms. *Lab. Pract.*, **39**(10), 81–3.

Beverly, M.B., Kay, P.T. and Voorhees, K.J. (1995) Principal component analysis of the pyrolysis-mass spectra from African, Africanized hybrid, and European beeswax. *J. Anal. Appl. Pyrolysis*, **34**, 251–63.

Bracewell, J.M. and Robertson, G.W. (1984) Characteristics of soil organic-matter in temperate soils by Curie-point pyrolysis-mass spectrometry. 1. Organic-matter variations with drainage and mull humification in a-horizons. *J. Soil Sci.*, **35**, 549.

Carlsen, L. (1984) Gas phase Curie-point pyrolysis, in *Analytical Pyrolysis* (ed. K.J. Voorhees), Butterworths, London, pp. 69–94.

Cone, J.W., Tas, A.C. and Wolter, M.G.E. (1992) Pyrolysis mass spectrometry (PyMS) and degradability of starch granules. *Starch/Stärke*, **44**, 55–8.

Freeman, R., Goodfellow, M., Gould, F.K. *et al.* (1990) Pyrolysis-mass spectrometry Py-MS for the rapid epidemiological typing of clinically significant bacterial pathogens. *J. Med. Microbiol.*, **32**, 283–6.

Freeman, R., Gould, F.K., Wilkinson, R. *et al.* (1991a) Rapid inter-strain comparison by pyrolysis mass spectrometry of coagulase-negative staphylococci from persistent CAPD peritonitis. *Epidemiol. Infect.*, **106**, 239–46.

Freeman, R., Gould, F.K., Sisson, P.R. and Lightfoot, N.F. (1991b) Strain differentiation of capsule type 23 penicillin-resistant streptococcus-pneumoniae from nosocomial infections by pyrolysis mass-spectrometry. *Lett. Appl. Microbiol.*, **13**, 28–31.

Freeman, R., Sisson, P.R. and Ward, A.C. (1995) Resolution of batch variations in pyrolysis mass spectrometry of bacteria by the use of artificial neural networks analysis. *A. van Leewenhoek*, **68**, 253–60.

Goodacre, R., Beringer, J.E. and Berkeley, R.C.W. (1991) The use of pyrolysis-mass spectrometry to detect the fimbrial adhesive antigen F41 from Escherichia coli HB101 (pSLM204). *J. Anal. Appl. Pyrolysis*, **22**, 19–28.

Goodacre, R. and Berkeley, R.C.W. (1990) Detection of small genotypic changes in escherichia-coli by pyrolysis mass-spectrometry. *FEM Microbiol. Lett.*, **71**, 133–8.

Goodacre, R. and Kell, D.B. (1993) Rapid and quantitative analysis of bioprocesses using pyrolysis mass spectrometry and neural networks: application to indole production. *Anal. Chim.*, **279**, 17–26.

Goodacre, R. and Kell, D.B. (1996a) Pyrolysis mass spectrometry and its application in biotechnology. *Curr. Opinion Biotechnol.*, **7**, 20–28.

Goodacre, R. and Kell, D.B. (1966b) Use of pyrolysis mass spectrometry with supervised learning for the assessment of the adulteration of milk from different species, and the determination of milk fat content. *Proceedings Food Authenticity 96*, Norwich, UK.

Goodacre, R., Kell, D.B. and Bianchi, G. (1992) Neural networks and olive oil. *Nature*, **359**, 594.

Goodacre, R., Kell, D.B. and Bianchi, G. (1993) Rapid assessment of the adulteration of virgin olive oils by other seed oils using pyrolysis mass spectrometry and artificial neural networks. *J. Sci. Food Agric.*, **63**, 297–307.

Goodacre, R., Neal, M.J., Kell, D.B. *et al.* (1994) Rapid screening for metabolite overproduction in fermentor broths, using pyrolysis mass spectrometry with multivariate calibration and artificial neural networks. *Biotechnol. Bioeng.*, **44**, 1205–16.

Goodacre, R., Kell, D.B. and Bianchi, G. (1995) Food adulteration exposed by neural networks. *Analysis Europa*, **5**, 35–7.

Goodacre, R., Neal, M.J. and Kell, D.B. (1996) Quantitative analysis of multivariate data using artificial networks: a tutorial review and applications to the deconvolution of pyrolysis mass spectrometry. *Zentralbl. Bakteriol.* In press.

Gould, F.K., Freeman, R., Sisson, P.R. *et al.* (1991) Inter-strain comparison by pyrolysis mass spectrometry in the investigation of Staphylococcus aureus nosocomial infection. *J. Hosp. Infect.*, **19**, 41–8.

Gutteridge, C.S. (1987) Characterization of microorganisms by pyrolysis mass spectrometry. *Methods Microbiol.*, **19**, 227–72.

Gutteridge, C.S. and Norris, J.R. (1979) The application of pyrolysis techniques to the identification of micro-organisms. *J. Appl. Bacteriol.*, **47**, 5–43.

Gutteridge, C.S., Swaetman, A.J. and Norris, J.R. (1984) *Analytical Pyrolysis, Techniques and Applications* (ed. K.J. Voorhees), Butterworth, London, pp. 325–49.

Gutteridge, C.S., Vallis, L. and McFie, H.J.H. (1985) Numerical methods in the classification of micro-organisms by pyrolysis mass spectrometry, in *Computer Assisted Bacterial Systematics* (eds M. Goodfellow, D. Jones and F.G. Priest), Academic Press, London.

Halket, J.M. and Schulten, H.-R. (1985a) Rapid characterization of tobacco by combined direct pyrolysis-field ionization mass spectrometry and pyrolysis-gas chromatography-mass spectrometry. *J. Anal. Appl. Pyrolysis*, **8**, 547–60.

Halket, J.M. and Schulten, H.-R. (1985b) Thick-film capillary gas chromatography-field ionization mass spectrometry. A complementary technique for the rapid analysis of volatiles. *J. Chromatogr.*, **322**, 200–205.

Halket, J.M. and Schulten, H.-R. (1988) Fast profiling of food by anlytical pyrolysis. *Z. Lebensm. Unters. Forsch.*, **186**, 201–212.

Hammond, D.A. (1996) Authenticity of fruit juices, jams and preserves, in *Food Authentication* (eds P.R. Ashurst and M.J. Dennis), Chapman & Hall, London, pp. 15–59.

Hardt, R. and Baltes, W. (1989) Analysis of caramel colours. Curie-point pyrolysis-high-resolution gas chromatography/mass spectrometry and simulation of pyrolysis-mass spectrometry. *J. Anal. Appl. Pyrolysis*, **15**, 159–65.

Harper, A.M., Henk, L.C., Meutzelaar, G. et al. (1984) Numerical techniques for processing pyrolysis mass spectral data, in *Analytical Pyrolysis* (ed. K.J. Vorhoes), Butterworths, London, pp. 157–95.

Helyer, R.J., Bale, S.J. and Berkeley, R.C.W. (1993) The application of pyrolysis-mass spectrometry to microbial ecology: rapid characterisation of bacteria isolated from an estuarine environment. *J. Anal. Appl. Pyrolysis*, **25**, 265–72.

Hussain, S.F. (1993) Application of pyrolysis-mass spectrometry as a rapid analytical method for food science. PhD thesis, Reading University, Reading, UK.

Irwin, W.J. (1982) *Analytical Pyrolysis: a Comprehensive Guide*, Marcel Dekker, New York.

Kajioka, R. and Tang, P.W. (1984) Curie-point pyrolysis-mass spectrometry of legionella species. *J. Anal. Appl. Pyrolysis* **6**, 59–68.

Kiritsakis, A. (1991) *Olive Oil*, American Oil Chemists Society, Champaign, IL, USA.

Magee, J.T., Hindmarch, J.M., Burnett, L.A. and Pease, A. (1989) Epidemiological typing of Streptococcus pyrogenes by pyrolysis mass spectrometry. *J. Med. Microbiol.*, **30**, 273–8.

Meuzelaar, H.L.C., Harper, A.M., Pugmire, R.J. and Karas J. (1984) Characterization of coal maceral concentrates by Curie-point pyrolysis mass-spectrometry. *Int. J. Coal Geol.*, **4**, 143–71.

Meuzelaar, H.L.C., Haverkamp, J. and Hileman, F.D. (1982) *Techniques and Instrumentation in Analytical Chemistry*, Vol. 3: Pyrolysis mass spectrometry of recent and fossil biomaterials, Elsevier Scientific Publishing Company, Amsterdam, Oxford, New York.

Meuzelaar, H.L.C., Kistemaker, P.G., Eshuis, W. and Engel, H.W.B. (1976) *Rapid Methods and Automation in Microbiology*, Learned Information, Oxford, pp. 225–30.

Montanarella, L., Bassani, M.R. and Breas, O. (1995) Chemometric classification of some European wines using pyrolysis mass spectrometry. *Rapid Comm. Mass Spec.*, **9**, 1589–93.

Morrison, W.H. III, Scheijen, M.A. and Boon, J.J. (1991) Pyrolysis mass spectrometry of coastal Bermudagrass (Cynodon dactylon (L.) Pers.) and Kentucky-31 tall fescue (Festuca arundinacea Schreb.) cell walls and their residues after ozonolysis and base hydrolysis. *Animal Feed Sci. Technol.*, **32**, 17–26.

Mulder, M.M., Engles, F.M., Schuurmans, J.L.L. and Boon, J.J. (1992) In vitro digested and potassium permanganate delignified maize internode sections studied by histochemistry and analytical pyrolysis mass spectrometry. *Animal Feed Sci. Technol.*, **39**, 335–46.

Munson, T.O. (1995) Forensic applications of pyrolysis-mass spectrometry, in *Forensic Applications of Mass Spectrometry* (ed. J. Yinon), CRC Press, pp. 172–215.

Ottley, T.W. and Maddock, C.J. (1986) Use of pyrolysis mass spectrometry. *Lab. Pract.*, **10**, 53–5.

Petrus, D.R. and Vendercook, C.E. (1980) Methods for the detection of adulteration in process citrus products, in *Citrus Nutrition and Quality* (eds S. Nagy and J.A. Attaway), ACS Symposium Series, Vol. 143, pp. 395–421.

Reeves, J.B. III and Galletti, G.C. (1993) Use of pyrolysis-gas chromatography/mass spectrometry in the study of lignin analysis. *J. Anal. Appl. Pyrolysis*, **24**, 243–55.

Reid, K.J.G., Swan, J.S. and Gutteridge, C.S. (1993) Assessment of Scotch whisky quality by pyrolysis mass spectrometry and the subsequent correlation of quality with the oak wood cask. *J. Anal. Appl. Pyrolysis*, **25**, 49–62.

Samukawa, K., Inoue, A. and Nagano, T. (1992) Characterization of rices by Curie-point pyrolysis gas chromatography mass spectrometry. *Chem. Express*, **7**, 449–52.

Schulten, H.-R. (1986) Pyrolysis-field mass spectrometry – a new method for direct, rapid characterization of tobacco. *Beitr. Tabaksforsch. Intern.*, **13**, 219–21.

Schulten, H.-R., Beckey, H.D., Meuzelaar, H.L.C. and Boerboom, A.J.H. (1973) High resolution field ionization. *Anal. Chem.*, **45**, 191–5.

Schulten, H.-R. and Halket, J.M. (1986) Rapid characterization of biomaterials by field ionization. *Org. Mass Spectrum.*, **21**, 613–22.

Shute, L.A., Gutteridge, C.S., Norris, J.R. and Berkeley, R.C.W. (1984) Curie-point pyrolysis mass-spectrometry applied to characterization and identification of selected bacillus species. *J. Gen. Microbiol.*, **130**, 343–55.

Simmleit, N. and Schulten, H.-R. (1986) Differentiation of commercial tobacco blends by pyrolysis field ionization mass spectrometry and pattern recognition. *Fres. Z. Anal. Chem.*, **324**, 9–12.

Sisson, P.R., Freeman, R., Lightfoot, N.F. and Richardson, I.R. (1991) Incrimination of an environmental source of a case of Legionnaires' disease by pyrolysis mass spectrometry. *Epidemiol. Infect.*, **107**, 127–32.

Sisson, P.R., Kramer, J.M., Brett, M.M. *et al.* (1992) Application of pyrolysis mass spectrometry to the investigation of outbreaks of food poisoning and non-gastrointestinal infection associated with Bacillus species and Clostridium perfringens. *Int. J. Food Microbiol.*, **7**, 57–66.

Wampler, T.P. (1995) Instrumentation and analysis, in *Applied Pyrolysis Handbook* (ed. T.P. Wampler), Marcel Dekker, New York.

Wieten, G., Haverkamp, K., Meuzelaar, H.L.C. *et al.* (1981a) Application of pyrolysis mass spectrometry to the classification and identification of mycobacteria. *Rev. Infect. Diseases*, **3**, 871–7.

Wieten, G., Haverkamp, K., Meuzelaar, H.L.C. *et al.* (1981b) Pyrolysis mass-spectrometry – a new method to differentiate between the mycobacteria of the tuberculosis complex and other mycobacteria. *J. Gen. Microbiol.*, **122**, 109–18.

Wieten, G., Meuzelaar, H.L.C. and Haverkamp, K. (1984) *Gas Chromatography/Mass Spectrometry: Application in Microbiology* (eds G. Odham, L. Larsson and P.-A. Mardh), Plenum Press, New York, pp. 335–80.

Windig, W., McClennen, W.H. and Meuzelaar, H.L.C. (1987) Determination of fractional concentrations and exact component spectra by factor analysis of pyrolysis mass spectra of mixtures. *Chemom. Intell. Lab.*, **1**, 151–65.

Wold, S., Albano, C., Dunn, W.J. III *et al.* (1984) Multivariate data analysis in chemistry. Proceedings of the NATO Advanced Studies Institute on Chemometrics – *Mathematics and Statistics in Chemistry*, Cosenza, Italy. NATO ASI Series C, Vol. 138. D. Reidel, Dordrecht, pp. 17–96.

Zarrabi, K., Durfee, S.L. and Daniel, S.R. (1991) Use of principal component regression to characterise a complex oxidation product mixture. *J. Anal. Appl. Pyrol.*, **21**, 1–14.

12 The principles of multivariate data analysis
M.J. ADAMS

12.1 Introduction

During the past thirty years the analytical and testing laboratory has witnessed spectacular advances in the range and sophistication of available instrumental methods. These developments have meant that the acquisition of characteristic data from a sample is rarely a problem. Instead, the challenge facing many laboratories is how to use and interpret this data most efficiently. To this end in recent years, there has become available a wide variety of statistical and data analysis software packages for use on the now ubiquitous personal computer. In many cases, quite sophisticated data analysis algorithms are integral within the instrument itself, requiring little input or few decisions from the user. Although this trend towards increasing automation in analysis is to be welcomed, a caveat is necessary in that this trend does impose an obligation and responsibility on analysts that they be aware of the limitations as well as the merits of any data analysis and manipulation undertaken (Adams, 1995).

The aim of this chapter is to present an overview and discussion of some of the basic features and techniques of multivariate data analysis, and help to dispel the 'black-box' image often associated with such software. Although space does not permit a rigorous mathematical analysis and description of the many techniques available, it is hoped that the reader will be encouraged to investigate further the many methods available to assist in data interpretation, and become more confident in applying such methods.

12.2 Univariate statistics

One of the most common summary statistics quoted in the literature and in everyday use, is the **mean** value. It provides an efficient way of reducing a list or set of values to a single, most likely value. The mean, or average, value of a set is a **least-squares estimate**, and is calculated by defining an error term, ε, as the sum of squares between each recorded value and the mean, i.e. the sum of residuals squared,

$$\varepsilon = \sum_{i=1}^{n} (x_i - \bar{x})^2 \qquad (12.1)$$

where $\bar{x}$ is the estimate of the mean of n values.

PRINCIPLES OF MULTIVARIATE DATA ANALYSIS

Expanding equation (12.1) and differentiating ε, with respect to $\bar{x}$ gives,

$$\varepsilon = 2 \sum_{i=1}^{n} (x_i - \bar{x}) \tag{12.2}$$

and ε is at a minimum when $d\varepsilon/d\bar{x}$ is zero. Thus at a minimum,

$$2 \sum_{i=1}^{n} (\bar{x} - x_i) = 0$$

that is

$$\bar{x} \sum_{i=1}^{n} 1 - \sum_{i=1}^{n} x_i = 0 \tag{12.3}$$

and rearranging,

$$\bar{x} = \frac{\sum_{i=1}^{n} x_i}{n}, \tag{12.4}$$

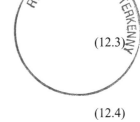

the well known formula for calculating the mean value. Although this derivation may appear somewhat elaborate it does serve to introduce at this early stage the importance of least-squares methods. The fact that least-squares values are readily influenced by **outliers** can be appreciated here by noting that each datum value is equally weighted and rogue values can exert considerable leverage on the result. Thus least-squares estimates are said to be not robust.

The second summary statistic frequently encountered refers to the spread of the data about its mean value, and this is defined as the variance, s^2,

$$s^2 = \frac{\sum_{i=1}^{n} (x_i - \bar{x})^2}{n - 1}, \tag{12.5}$$

a mean-centred sum-of-squares.

More commonly, and because it is expressed in the same units as the original variables, the value quoted is the standard deviation, s,

$$s = \sqrt{\frac{\sum_{i=1}^{n} (x_i - \bar{x})^2}{n - 1}}, \tag{12.6}$$

or the unit-less quantity, relative standard deviation, RSD,

$$\text{RSD} = \frac{s}{\bar{x}} \tag{12.7}$$

The mean, variance, and standard deviation are summary statistics only if we can assume the recorded data is normally distributed.

Table 12.1 presents the results of analysing ten water samples for six analytes (four metals and two classes of organic species). Although the simple univariate

Table 12.1 The concentration (mg kg^{-1}) of four metals and two organic species in ten water

	Ni	Cr	Fe	Ca	PCB	PAH
A	18	32	28	9	52	42
B	22	31	31	10	61	60
C	9	11	13	10	56	58
D	17	30	26	11	58	47
E	19	31	36	12	13	14
F	8	11	14	12	64	63
G	7	9	11	12	9	10
H	25	35	42	13	11	10
I	4	3	7	13	11	13
J	18	32	29	14	12	13
Mean	14.7	22.5	23.7	11.6	34.7	33.0
Standard deviation	7.1	12.3	11.7	1.6	25.0	23.0
% RSD	48.4	54.7	49.5	13.6	72.0	69.6

statistics introduced above efficiently summarize the individual analyte data, these values provide no information about interactions between analytes. To achieve this we need to examine some multivariate statistics.

12.3 Multivariate statistics

Using the data in Table 12.1, consider the relationship between the concentrations of nickel and chromium in the samples. This relationship is illustrated in the scatter plot of Fig. 12.1. If there were no interaction effects between these two measures then the points would be scattered randomly about the mean values. As it is, however, there is an obvious dependence between nickel and chromium concentrations. As the concentration of nickel increases so too does that for chromium. The two variables are said to positively correlate. In order to assign a quantitative value to this relationship between two variables, $x1$ and $x2$, we need to define a multivariate statistic, the covariance, $\text{Cov}_{x1,x2}$,

$$\text{Cov}_{x1,x2} = \frac{\sum_{i=1}^{n}(x1_i - \bar{x}1)(x2_i - \bar{x}2)}{n-1} \quad (12.8)$$

Just as variance describes the spread of data about its mean value, so covariance indicates the bivariate spread about the pair of means. Note that in the univariate case, $\bar{x}1 = \bar{x}2$, and equation (12.8) reduces to equation (12.5). Although covariance is a key parameter in multivariate studies, its interpretation is not immediately obvious, and a more convenient measure of inter-dependence is provided by the correlation coefficient, r,

$$r_{x1,x2} = \frac{\text{Cov}_{x1,x2}}{s_{x1} \cdot s_{x2}} \quad (12.9)$$

The correlation coefficient is a standardized form of covariance. It is a dimensionless quantity and is scaled between -1 and $+1$, so is much easier to interpret than covariance. For the nickel and chromium data,

$$\text{Cov}_{\text{Ni,Cr}} = 84.72 \ (\text{mg kg}^{-1})^2$$
$$r_{\text{Ni,Cr}} = 0.97$$
(12.10)

with a large, positive value for r as expected from our visual observation.

Before continuing with our discussion of multivariate data analysis, a cautionary note is worthwhile. Although correlation values are extensively quoted in the scientific literature and can serve as indicators of inter-dependence between data, this must not be interpreted as implying a causal relationship between the data. In our example, the chromium content is unlikely to increase because the concentration of nickel goes up, more likely they are both linked to some other, unspecified factor. In addition, it should be borne in mind that correlation is a measure of linear dependence and, again, it is a non-robust

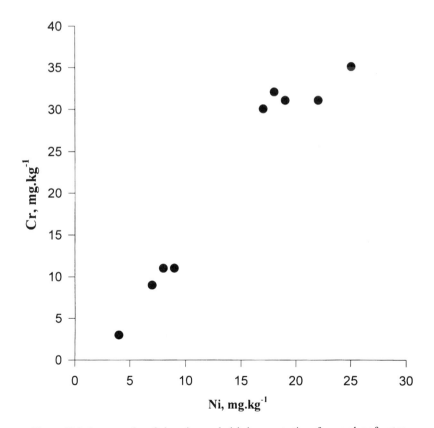

Figure 12.1 A scatter-plot of chromium and nickel concentrations for samples of water.

statistic. This is easily observed in the data illustrated in Fig. 12.2; in all cases the correlation coefficient is identical, highly positive, but the susceptibility of the value to the presence of outliers and non-linear effects is clearly evident.

Covariance and correlation values can be calculated for each pair of variables, and the results displayed in matrix form. For the data from Table 12.1, these values are presented in Table 12.2(a) and (b). Both the covariance matrix and the correlation matrix are square and symmetric about the leading diagonal. In the covariance matrix, the diagonal elements are the values of variance for each variable, and the diagonal for the correlation matrix contains 1s. From Tables 12.1 and 12.2(b) we can conclude that the concentrations of the three transition metals are highly correlated, and that the concentrations of organic species are highly correlated. The concentration of calcium is not highly correlated to any

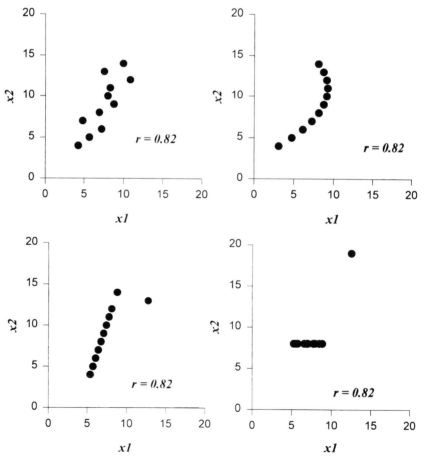

Figure 12.2 Bivariate scatter-plots illustrating that correlation coefficient alone is not a sufficient indicator of a good or suitable model.

Table 12.2(a) Variance-covariance matrix for the data from Table 12.1

	Ni	Cr	Fe	Ca	PCB	PAH
Ni	50.68	84.72	81.57	−1.02	2.68	−8.67
Cr	84.72	151.61	137.28	−2.00	10.39	−16.78
Fe	81.57	137.28	137.79	0.42	−32.32	−47.22
Ca	−1.02	−2.00	0.42	2.49	−29.24	−25.33
PCB	2.68	10.39	−32.32	−29.24	624.01	564.44
PAH	−8.67	−16.78	−47.22	−25.33	564.44	527.78

Table 12.2(b) Variables correlation matrix for the data from Table 12.1

	Ni	Cr	Fe	Ca	PCB	PAH
Ni	1	0.97	0.98	−0.09	0.01	−0.05
Cr	0.97	1	0.95	−0.10	0.03	−0.06
Fe	0.98	0.95	1	0.02	−0.11	−0.17
Ca	−0.09	−0.10	0.02	1	−0.74	−0.70
PCB	0.01	0.03	−0.11	−0.74	1	0.98
PAH	−0.05	−0.06	−0.17	−0.70	0.98	1

other variable measured, and from its relatively low RSD value the calcium content provides little information for this data set.

These observations could lead us directly to the field of pattern recognition and the classification of samples, but before proceeding down this route it is worth examining the subject of data reduction and information extraction.

12.4 Data reduction

Given the relatively small size of the data set in Table 12.1, it may seem unnecessary to apply variable reduction. However, we can use this data for illustrative purposes, and the techniques are valuable for data comprising many hundreds of samples characterized by many, possibly hundreds, of variables. In reducing the amount of data, it is our aim to make its interpretation simpler. To this end we must not only reduce the dimensionality of our problem, but also retain as much of the information content as possible.

Most multivariate data contains considerable redundancy of information. For example, an infrared spectrum may be digitized to several thousand discrete transmission values, but the relevant information in the spectrum can probably be described by tens of values. The degree of this redundancy is indicated by the correlation coefficient; highly correlated data can be described and summarized by fewer variables with minimal loss of information.

In Fig. 12.3(a), the nickel–chromium scatter plot has a pair of new axes added, $m1$ and $m2$, at 45° to our original axes. So,

$$m1_i = a \cdot \text{Ni}_i + b \cdot \text{Cr}_i \qquad (12.11)$$

and describes a linear combination of our original, measured variables. For the 45° line, and normalizing the scale of this axis so that we can compare $m1$ with Ni and Cr, then

$$a = b$$

and

$$a^2 + b^2 = 1 \qquad (12.12)$$

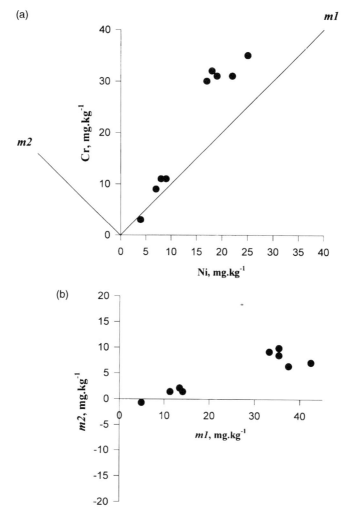

Figure 12.3 (a) New axes, $m1$ and $m2$, can be formed from linear combinations of original variables, and (b) projection of data n to these axes is a simple transformation operation.

Therefore,

$$m1 = 0.707\text{Ni} + 0.707\text{Cr}$$
$$m2 = -0.707\text{Ni} + 0.707\text{Cr}$$
(12.13)

By multiplying our original nickel and chromium concentration values by these coefficients we can project our data on to these new axes, Fig. 12.3(b), and calculate the variables $m1$ and $m2$, Table 12.3. For $m1$ and $m2$ the total variance is the same as for nickel and chromium, but the relative contribution of each variable to this total is now very different. If we elect to ignore $m2$, i.e. reduce the data to one dimension, then $m1$ would by itself describe 92% of the total variance from the nickel and chromium concentration data. This result begs the question 'is any other axis more efficient at summarizing the data?'. The original choice of the 45° line was made for simplicity of calculating the coefficients in producing $m1$ and $m2$. The best single axis would be the line that maximizes the total variance described by the original variables. Any remaining variance would be on the second axis, **orthogonal** to the best line. At the same time, we can expect the correlation between the new variables to reduce to zero as we approach the best line.

The mathematical solution to determining these new, best axes is not trivial, but it is a very common task in data analysis and is simple to evaluate with appropriate mathematical or statistical software. The technique is referred to as principal components analysis, PCA. Starting with the covariance matrix describing the dispersion of the original variables, we extract the eigenvalues and eigenvectors. An eigenvector is simply a list of coefficients (loadings or weightings) by which we multiply the original variables to obtain new variables. They represent the a and b values in our simple nickel–chromium example. An eigenvalue represents the variance of the data when projected on to an axis

Table 12.3 New variables, $m1$ and $m2$, formed by linear combination of Ni and Cr data from Table 12.1, according to $m1 = 0.707\text{Ni} + 0.707\text{Cr}$; $m2 = 0.707\text{Ni} - 0.707\text{Cr}$

	$m1$	$m2$
A	35.35	9.90
B	37.47	6.36
C	14.14	1.41
D	33.23	9.19
E	35.35	8.48
F	13.43	2.12
G	11.31	1.41
H	42.42	7.07
I	4.95	−0.71
J	35.35	9.90
Variance	185.81	16.42
Total variance	202.2	
% of total variance	92%	8%
Correlation coefficient	0.91	

Table 12.4 Principal components analysis of Ni and Cr data of covariance matrix derived from Table 12.1

Covariance matrix	Ni	Cr
Ni	50.68	84.72
Cr	84.72	151.61
PCA		
Factor 1	Factor 2	Factor 3
Eigenvalue	199.76	2.53
Eigenvector	0.494	−0.869
	0.869	0.494

defined by an eigenvector. The larger the eigenvalue, the more important that eigenvector is in describing the data. A principal component is the product of the original data and an eigenvector; the result of projecting the data on to the new axis, it is a new variable. There are as many principal components as there are original variables, but hopefully we need only a few to describe our data with minimal loss of information.

The results of performing PCA on the two-dimensional data for nickel and chromium concentrations are presented in Table 12.4, with the scatter plot of the data in the space defined by the two principal components shown in Fig. 12.4. Note that the sum of the eigenvalues is the same as the sum of variances for the original variables. Now, with our new variables $PC1$ and $PC2$, the percentage

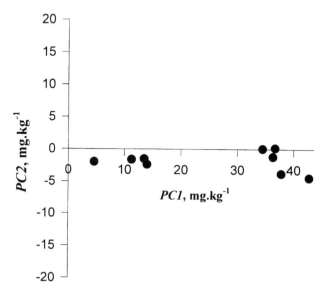

Figure 12.4 Principal components are linear combinations that serve to maximize the variance in a reduced number of dimensions.

Table 12.5 Principal components analysis of the covariance matrix from Table 12.2(a)

	Factor 1	Factor 2	Factor 3	Factor 4	Factor 5	Factor 6
Eigenvalues	1147	331.4	11.2	2.8	1.2	0.8
Cumulative	76.75	98.93	99.7	99.9	99.9	100
% of total eigenvectors	−0.009	0.386	0.181	0.002	−0.618	0.660
	−0.012	0.669	−0.433	0.568	0.104	−0.177
	−0.058	0.629	0.441	−0.528	0.273	−0.232
	−0.034	−0.001	0.091	0.154	0.727	0.662
	0.734	0.086	−0.496	−0.428	0.054	0.147
	0.675	−0.023	0.575	0.438	−0.006	−0.141

contribution of each variable to the total variance is 98.7% and 1.3% respectively, so using only one new variable, $PC1$, we retain nearly 99% of the information from the two original variables. Projecting our original data on to the eigenvectors, Fig. 12.4, indicates two classes of sample, those high in nickel and chromium and those low in these metals.

Having illustrated the principles of performing PCA, we can apply the technique to the full covariance matrix from our original data, Table 12.2(a), and the results are presented in Table 12.5. Examination of the eigenvalues indicates that the first two principal components account for 99% of the variance in the data. Thus our original six-dimensional data can be represented efficiently in the two dimensions defined by the first two principal components, and Fig. 12.5 is the resultant scatter plot. Each of the ten samples appears to belong to one of four groups, but how do we attach a physical interpretation to these principal

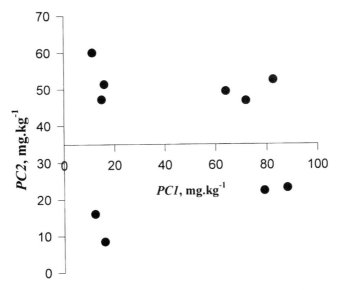

Figure 12.5 A scatter-plot of the ten water samples projected on to the first two principal components.

components? The answer to this lies in the relative magnitude of the individual loadings, or coefficients, in the relevant eigenvectors. The first two eigenvectors, from which the first two principal components are derived, are plotted in Fig. 12.6. This clearly illustrates that the first eigenvector weights most heavily those variables concerned with organic species, and the second eigenvector is weighted towards the transition metal content. In neither eigenvector does the concentration of calcium play a major role in characterizing these samples.

From this analysis and interpretation of eigenvectors we can assign the labels 'organic' and 'metals' to the first principal components, and identify samples A, B and D as comprising a group with high metal and organic concentrations, C and F as high in organic and low in metals, E, H and J as high in metals and low in organic, and samples G and I as low in both species.

Principal components analysis is one of the most powerful and commonly used techniques for reducing the dimensionality of large sets of data, and is particularly valuable when, as here, a chemical or physical interpretation of the results is possible. Many examples of the use and application of PCA are widely available in the literature and near-infrared spectroscopy provides a common process measurement technique for studies of authentication and product adulteration (Downey and Boussion, 1996; Twomey *et al.*, 1995). PCA is an important technique, belonging to that family of data processing schemes referred to as factor analysis (Hopke, 1992). Since principal components are uncorrelated with each other, they find extensive use in developing stable mathematical models for calibration as we shall see later.

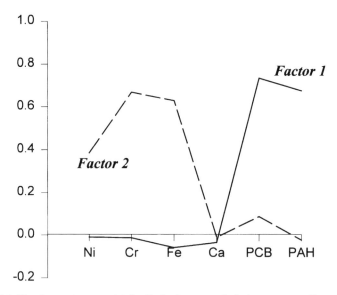

Figure 12.6 The eigenvectors associated with the first two principal components (factors) indicate that the first is weighted heavily towards organic species PCB and PAH, and the second reflects heavy metal content.

12.5 Pattern recognition

The use of PCA to reduce the dimensionality of our data has been demonstrated, and in many cases reduction of complex data to two or three dimensions enables patterns and structure within the data to be observed and identified visually. To this end, PCA can be considered a technique for feature extraction; a means of identifying those variables playing a major part in characterizing our samples. Formal methods of pattern recognition are provided by **cluster analysis, discriminant analysis** and **artificial neural networks**.

12.5.1 *Cluster analysis*

Sometimes referred to as unsupervised pattern recognition, cluster analysis encompasses a wide range of techniques for exploratory data analysis. The principal aim of cluster analysis may be stated as follows: 'Given a number of objects or samples, each described by a set of measured values, we aim to derive a formal mathematical scheme for grouping the objects into classes such that objects within a class are similar but different from those in other classes. The class characteristics are not known a priori, but may be determined from the data analysis'.

For a cluster analysis to succeed then the choice of similarity measure, and the means of grouping objects may be critical. Cluster analysis is not a statistical technique, the results obtained are justified according to their value in interpreting the data and indicating patterns.

There are two major categories of cluster analysis, hierarchical and non-hierarchical. The main difference between the two methods is that while hierarchical methods form clusters sequentially, i.e. starting with the most similar pair of objects and forming higher clusters step-by-step, the non-hierarchical methods evaluate overall distributions of object pairs and then classify them into a given number of groups. Hierarchical methods are the more popular.

The first step in performing a cluster analysis is to derive some quantitative measure of similarity between each of the samples. By far the most common methods of assigning similarity use distance metrics, and the most widely used is the simple **Euclidean distance**. Application of this method is illustrated in Fig. 12.7 for three samples, A, B and C, characterized by two variables $x1$ and $x2$. It is reasonable to infer that A is more similar to B than to C because of the smaller distance A–B than A–C. In two dimensions, the distance between any pair of samples, d_{AB}, is easily calculated by Pythagoras' theorem, which can be extended to multi-dimensional (n-variable) space by,

$$d_{AB} = \sqrt{\sum_{i=1}^{n} (x_{A,i} - x_{B,i})^2} \qquad (12.14)$$

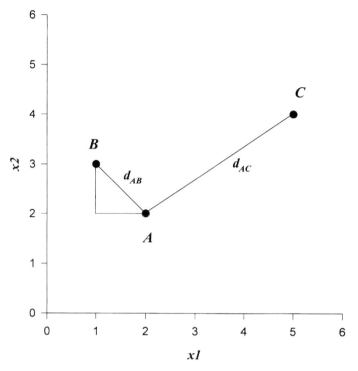

Figure 12.7 The Euclidean distance between objects in bivariate space is simply given by Pythagoras's theorem.

Applying equation (12.13) to the ten sample, six variable data from Table 12.1, results in the 10×10 Euclidean distance matrix shown in Table 12.6.

From such a distance matrix, hierarchical clustering will join and combine objects according to the technique selected. Examination of Table 12.6 indicates that samples E and J are the most similar and are a mutually close pair. These samples are joined, the matrix is reduced by one row and one column, and the

Table 12.6 The Euclidean distance inter-sample matrix derived from the data in Table 12.1

	A	B	C	D	E	F	G	H	I	J
A	0	20.8	31.9	8.6	48.8	36.5	61.8	54.5	63.4	49.7
B	20.8	0	30.4	15.2	66.8	30.1	79.5	71.8	80.0	68.2
C	31.9	30.4	0	26.8	69.4	9.7	67.3	77.5	64.7	69.0
D	8.6	15.2	26.8	0	56.7	29.7	67.4	62.7	68.0	57.4
E	48.8	66.8	69.4	56.7	0	77.5	35.8	10.4	43.1	7.5
F	36.5	30.1	9.7	29.7	77.5	0	76.5	85.2	73.7	77.7
G	61.8	79.5	67.3	67.4	35.8	76.5	0	44.3	8.7	31.6
H	54.5	71.8	77.5	62.7	10.4	85.2	44.3	0	51.9	15.4
I	63.4	80.0	64.7	68.0	43.1	73.7	8.7	51.9	0	39.0
J	49.7	68.2	69.0	57.4	7.5	77.3	31.6	15.4	39.0	0

distances recalculated. A number of options are available for deciding on the distance from our new object, or cluster, EJ to each other object, the most common being single-linkage, complete-linkage, and average-linkage. The basis for calculating each of these distances is illustrated in Fig. 12.8. The single-linkage, or nearest-neighbour, method assesses the similarity between a cluster and an object by the distance from the object to the nearest object in the cluster. The complete-linkage, or furthest-neighbour, method, on the other hand, assesses the similarity between an object and a cluster by the distance from the object to the farthest object in the cluster. Average-linkage assesses similarity by computing the distance between an object and each sample in the cluster and averaging the distances.

The process of forming and joining clusters is repeated until a single cluster containing all samples is obtained, and the result can be displayed as a **dendrogram**. The dendrogram provides a visual summary of the clustering process, presenting a picture of the groups and the proximities inherent in the data. Using the Euclidean-distance matrix and the average-linkage method, our ten-sample cluster analysis results in the dendrogram shown in Fig. 12.9. The vertical axis represents the distance measure and the horizontal axis represents samples and clusters combined. Working our way down the dendrogram, the samples are first contained in two groups, reflecting the presence or absence of high levels of organic species, and then each is split into a further two groups according to

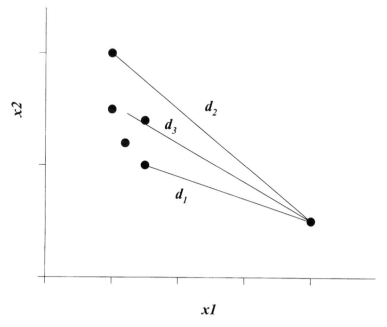

Figure 12.8 Illustration of some common clustering metrics, d_1 single-linkage, d_2 complete-linkage, and d_3 average-linkage.

transition metal content. This result is in agreement with our earlier analysis using PCA.

Generally, the different cluster-object linkage procedures will yield similar results for data having well-separated clusters. If any clusters overlap, however, the results will be different and it is recommended that data be analysed by several clustering algorithms before partitioning the data into distinct subgroups.

The popularity of hierarchical cluster analysis resides in the techniques producing the dendrogram; a visual representation providing a dramatic reduction in dimensionality of the original data. This reduction to a two-dimensional image, however, does distort the data. Working back from the dendrogram to produce the full apparent inter-sample distance matrix gives an indication of the extent of this distortion. A typical application of hierarchical cluster analysis is provided by Guerrero *et al.* in characterizing wine vinegar according to their

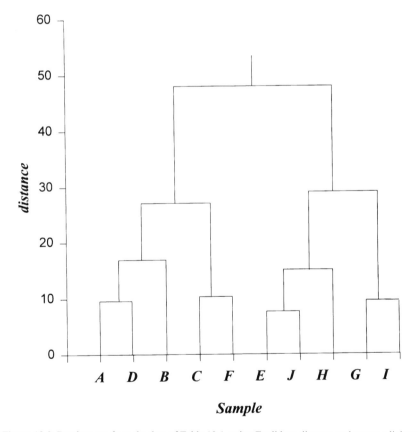

Figure 12.9 Dendrogram from the data of Table 12.1, using Euclidean distance and average-linkage clustering. The number of discrete clusters is subjective and dependent on the distance value selected to cut horizontally across the diagram.

production process (Guerrero et al., 1994). Garcia-Jares et al. provide an interesting example of cluster analysis and other multivariate graphical techniques in a GC-MS study of Spanish wine (Garcia-Jares et al., 1995). A review of display methods for multivariate data is available (Thompson, 1992).

Non-hierarchical clustering methods are generally available with all software packages, and an often quoted method is the *k*-means algorithm which divides m objects characterized by n variables into k clusters. The user selects k objects from the original data matrix to serve as initial cluster centres, then the algorithm proceeds to move objects between the clusters in order to minimize the within-cluster sum of squares of the distances. The results for such an analysis on the data from Table 12.1, using samples A, C, E and G as initial cluster centres, are provided in Table 12.7. These results are in agreement with earlier findings.

12.5.2 Discriminant analysis

In contrast to the subjective, exploratory data analysis features associated with cluster analysis, discriminant analysis (supervised pattern recognition) provides for statistical classification of samples. With supervised pattern recognition, we know in advance how many distinct groups are present in our data and we have representative examples of each type. Armed with this information, the problem facing us is to determine a means by which new, unclassified samples can be assigned to one of the parent groups.

A discriminant function represents a surface dividing our data space into regions. Thus, a binary classifier will partition the data into two regions; samples sharing common properties will be found on one side of the surface and those samples belonging to the second group will be found on the other side.

Linear discriminant functions belong to one of two categories, parametric methods and non-parametric methods. Parametric methods are based on Bayes' Rule and Bayesian statistics, and require a knowledge of the parent statistical

Table 12.7 *k*-Means cluster analysis on data from Table 12.1

Sample	Cluster number
A	1
B	1
C	2
D	1
E	3
F	2
G	4
H	3
I	4
J	3

distribution for the data. From knowledge of the form of this distribution, suitable classification rules can be derived. A simple linear bivariate discriminant function transforms an original set of measurements on a sample into a single discriminant score representing the sample's position on a line defined by the discriminant function. This transformation implies that we minimize the ratio of the difference between the group multivariate means to the multivariate variance within the groups, i.e. we find the axis along which the two groups are separated the most and inflated the least.

Application of linear discriminant analysis is widespread, and recently published examples in the field of foodstuff production include classification of cheeses (Fresno *et al.*, 1995), detecting adulteration in instant coffees (Briandet *et al.*, 1996), and classification of honey according to geographic origin (Sanz *et al.*, 1995).

Non-parametric discriminants make no assumptions regarding the statistical distribution of the data, and the linear learning machine is one of the simplest algorithms. The aim of this method is to devise a linear function of the form,

$$f(x) = \sum_{i=1}^{n} w_i \cdot x_i \qquad (12.15)$$

which separates the samples, where x objects are classified by n measures, each weighted by some coefficient w_i. A positive value for f classifies the sample into one group, and a negative value classifies it into another group. The most common approach to implementing the linear learning machine is to supply initial, guessed, values for w_i and then testing if each sample is classified correctly. When a sample is incorrectly classified then the coefficients are modified according to pre-defined rules and sample testing continues. If a training set is linearly separable, then this algorithm will find a set of coefficients capable of achieving completely correct classification. However, for a training set that is not linearly separable, then the linear learning machine is unlikely to find a discriminant function that minimizes the probability of misclassification.

A simple, non-parametric classification algorithm that is widely available with commercial software is the *k*-nearest-neighbours scheme. Consider the diagram shown in Fig. 12.10 in which there are represented objects belonging to two groups, A and B, and an unknown sample, X, is to be classified. Nearest-neighbour techniques make a classification decision based on the shortest distance between X and each sample in A and B, i.e. a sample is assigned to that class to which it appears closest. For well-separated classes, this technique generally works well, and if rogue or outlier datum points appear to be upsetting the classification then taking the best of three or five distances can be used to implement the classification. In practice, the range of problems to which this method can be applied is rather restricted and it owes its popularity to its conceptual simplicity and ease of implementation.

12.5.3 Neural networks

Neural networks are computer-based simulations of living nervous systems, and they have proved adept at pattern recognition tasks. The basic operation of a single neuron in a neural network can be compared with the algorithm describing the linear learning machine. With reference to Fig. 12.11, a simple neuron accepts a series of input signals, x_i, and produces an output, 'on' or 'off', depending only on the weighted sum of the inputs.

As stated above, the problem with this simple model is that it only works correctly if the training set used to develop the weights is linearly separable, and this failure to solve more difficult problems held up the development of neural networks for many years. Finally, in the mid-1980s it was demonstrated that multi-layer networks of modified neurons could learn and be applied to linearly inseparable problems. The neuron is modified by replacing the 'on/off' step-function output with a gradual, continuous function. By this means, the value of the output from a neuron carries information on the inputs, and decisions as to the degree of change of the input weights can be made, i.e. the system can learn. A common threshold function is the sigmoidal curve illustrated in Fig. 12.12(a).

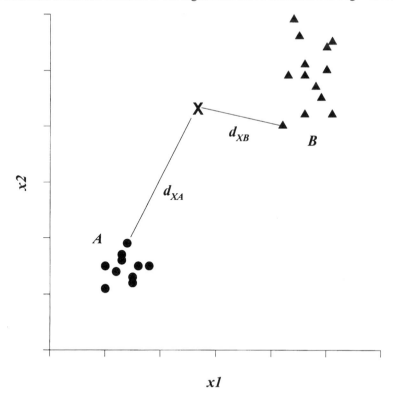

Figure 12.10 The k-nearest-neighbour algorithm would assign sample X to Group B since $d_{XB} < d_{XA}$.

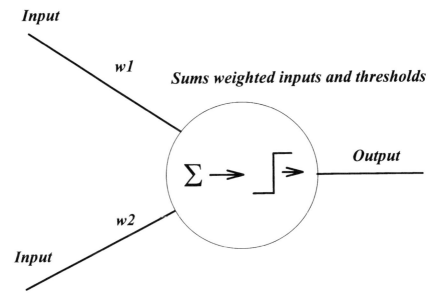

Figure 12.11 A simple, single neuron provides a weighted sum of its inputs and compares this value to some threshold value to provide a '0' or '1' output.

The output from each unit, O_j, is given by the threshold function, f_j, (e.g. the sigmoid function) acting on the weighted sum,

$$f(\text{net}) = \frac{1}{1 + e^{-j \cdot \text{net}}} \qquad (12.16)$$

In addition, these neurons can be organized as a multi-layer network, Fig. 12.12(b). The use and operation of neural networks is simple with commercial software. A pattern from a training set is input to the network and the net's response calculated. Comparison of this output with the required, correct response enables the weights throughout the network to be modified in order to move the output value towards the correct response. The weight-modification procedure implemented within the neural network provides it with its learning ability, and is based on some error function that represents the difference between the observed output and the correct output. To learn successfully the output of the net should approach the desired output by reducing the value of this error function and this is achieved by back-propagation of the error from one layer to the previous one. The mathematical rules for this operation are detailed and discussed by Beale and Jackson (1990).

12.6 Calibration

Modern instrumental methods of analysis do not provide absolute results, instead some recorded response is proportional to, and a function of, analyte concentration. It is the role of calibration to determine this functional relationship, with the aid of known concentration standards, and to predict the concentration of analytes in test samples. Thus efficient and effective calibration models are fundamental to achieving satisfactory quantitative analysis.

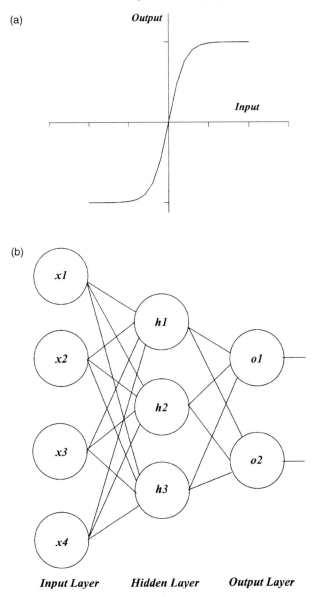

Figure 12.12 Major advances in neural networks have been brought about by using a continuous transfer function such as the sigmoid curve (a) and (b) by including hidden layers, $h_1 \ldots h_3$. in a network design.

By far the most common method of deriving practical calibration models is by least-squares regression analysis, despite the known problems with these methods such as their susceptibility to outliers. It is also frequently assumed, or arranged, that the response varies as a linear function of analyte concentration. The least-squares method determines the line of best fit between dependent and independent variables by minimizing the sum of the squares of the residuals, i.e. the differences between actual and predicted values. Thus, for the simple case of a single independent variable, x, and a dependent variable, y, described by the linear model

$$\hat{y}_i = a_0 + a_1 \cdot x_i \tag{12.17}$$

then the error term to be minimized is given by,

$$E = \sum_{i=1}^{n} (\hat{y}_i - y)^2 \tag{12.18}$$

where $\hat{y}$ and y are predicted (model) and actual (known) values for n samples.

Substituting equation (12.17) into equation (12.18), differentiating with respect to a_0 and a_1, and equating the results to zero, the minimal condition, leads to the so-called normal equations

$$a_0 \cdot n + a_1 \sum_{i=1}^{n} x_i = \sum_{i=1}^{n} y_i$$

$$a_0 \sum_{i=1}^{n} x_i + a_1 \sum_{i=1}^{n} x_i^2 = \sum_{i=1}^{n} (x_i \cdot y_i) \tag{12.19}$$

from which the values of a_0 and a_1 can be derived.

It is evident from the first of these two simultaneous equations that the best-fit line passes through the centre of the data, $\bar{x}$ and $\bar{y}$,

$$a_0 = \bar{y} - a_1 \cdot \bar{x} \tag{12.20}$$

Therefore, if the original values are mean-centred (each variable value has its variable's mean value subtracted) then the least-squares, best line will pass through the origin and calculation of the intercept value is unnecessary.

So, if

$$\tilde{y}_i = y_i - \bar{y} \quad \text{and} \quad \tilde{x}_i = x_i - \bar{x} \tag{12.21}$$

then the linear model is,

$$\tilde{y}_i = a_1 \cdot \tilde{x}_i \tag{12.22}$$

and the error is,

$$E = \sum_{i=1}^{n} (\tilde{y}_i - y_i)^2 \tag{12.23}$$

and minimizing E gives,

$$a_1 = \frac{\sum_{i=1}^{n}(\tilde{x}_i \cdot \tilde{y}_i)}{\sum_{i=1}^{n}\tilde{x}_i} \quad (12.24)$$

The least-squares principles can be easily extended to higher-order polynomials, e.g. quadratic, cubic, etc., and multivariate measures, and the most appropriate model can be selected or confirmed by a variety of criteria. A most useful and simple means is accomplished by visual inspection of the residuals. For a suitable model, the residuals will be randomly distributed about zero, Fig. 12.13. Visual examination of the calibration graph and residuals plot is always recommended no matter what other mathematical methods may be used, as this can immediately indicate outliers. Statistical testing of goodness-of-fit is also common, by testing successively higher order models and applying appropriate significance tests.

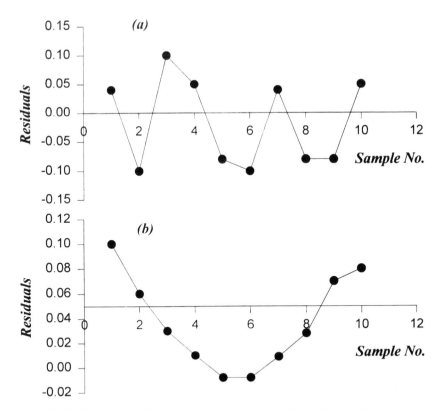

Figure 12.13 (a) Applying a linear model to linear data should provide residual errors that are randomly and evenly distributed about zero; (b) while a linear model with non-linear (quadratic) data will exhibit residuals with a discernible pattern.

For prediction purposes, and after all this is the aim of performing a calibration, then the efficacy of a model is best judged by its performance measured as the error of calibration and, more importantly, the error of validation. These errors are given by RMSEC (root mean square error of calibration) and RMSEV (root mean square error of validation) respectively, defined as,

$$\text{RMSEC} = \sqrt{\frac{\sum_{i=1}^{n}(\hat{y}_i - y_i)^2}{n}} \qquad (12.25)$$

where n is the number of samples in the calibration or training set, and

$$\text{RMSEV} = \sqrt{\frac{\sum_{i=1}^{m}(\hat{y}_i - y_i)^2}{m}} \qquad (12.26)$$

where m is the number of samples in the testing or validation set.

The process of validation assumes we have a set of samples, of known concentration of analyte, that has not been used in developing our model, and which we can use to test the predictive ability of our model. RMSEC and RMSEV are valuable indicators in developing calibration models, they indicate how good a model is in terms of fitting the calibration data and indicate the model's effectiveness for prediction, particularly if overfitting is a problem.

Table 12.8 provides UV absorbance data recorded at seven wavelengths (A1 to A7) from 14 solutions containing known amounts of the amino acid tryptophan. Each solution contains unknown amounts of other absorbing species, and our aim is to develop a suitable calibration model to allow prediction of tryptophan concentration from multi-wavelength absorbance data. Although we could elect to develop our model with all 14 samples, the result would not provide any valid indication of the model's predictive ability. Instead, it is better

Table 12.8 Seven-wavelength absorbance data for solutions of known tryptophan concentration. Samples marked (*) are used for validation

Tryptophan (mg kg^{-1})	A1	A2	A3	A4	A5	A6	A7
2.0	0.63	0.29	0.32	0.44	0.30	0.07	0.08
4.0	0.56	0.27	0.42	0.47	0.60	0.12	0.07
*6.0	0.56	0.30	0.39	0.50	0.28	0.04	0.05
8.0	0.55	0.33	0.50	0.51	0.22	0.05	0.02
*10.0	0.57	0.35	0.45	0.48	0.22	0.06	0.02
12.0	0.27	0.31	0.43	0.32	0.16	0.06	0.08
*14.0	0.28	0.38	0.42	0.26	0.06	0.02	0.01
16.0	0.47	0.44	0.55	0.46	0.18	0.06	0.05
*18.0	0.50	0.55	0.58	0.52	0.17	0.11	0.08
20.0	0.55	0.57	0.65	0.51	0.17	0.07	0.08
22.0	0.50	0.55	0.67	0.52	0.14	0.10	0.03
*24.0	0.46	0.64	0.90	0.52	0.12	0.08	0.10
26.0	0.74	0.74	0.90	0.78	0.31	0.09	0.07
28.0	0.75	0.79	0.94	0.77	0.26	0.02	0.09

to use some of the data to develop the model and the rest for testing and validation purposes. There are many methods for splitting the data into calibration and validation sets. We could develop a model with 13 samples and test the fourteenth, and repeat this a further 13 times – the 'leave-one-out' method. Leave-two-out and leave-10% methods have also been proposed. For illustration purposes only, the simple technique of splitting the data will be applied here. In this particular case we could decide that nine samples will be used for calibration and the remaining five for validation are adequate. Bearing in mind that calibration models should be used for interpolation and not extrapolation we would wish to include extreme concentration values in the calibration set.

Multiple linear regression, MLR, is a commonly used statistical tool for curve-fitting and calibration model development. Of the many variants of MLR, stepwise forward regression is typical. Each independent variable is examined in turn and added to the calibration model according to its partial correlation to the dependent variable. The variable addition process stops when some pre-specified F-ratio value (the F-to-enter value) is reached. For our seven variable, nine sample data then MLR gives the following equation for tryptophan concentration Tr,

$$Tr = 3.04 - 26.91A1 + 59.34A2 + 3.57A5 - 34.73A7 \qquad (12.27)$$

It is interesting to speculate on what if we believe all the wavelengths of measurement to carry information about the samples and, hence, include all independent variables in the model. The result in this case is given by,

$$Tr = 6.40 - 3.70A1 + 80.61A2 - 21.41A3 + 11.18A4 \\ + 9.79A5 - 13.63A6 - 56.44A7 \qquad (12.28)$$

The performance of these two models is illustrated in Fig. 12.14(a) and (b) along with the computed RMSEC and RMSEV values; the latter calculated on the five samples not included in formulating the regression model.

It is of particular interest to note that in each case, in this example, that RMSEC values decrease with increasing squared correlation (r^2) values, but RMSEV values increase. The best fitting model has the poorest predictive performance. Adding more terms to the model produces a better fit but not only fails to improve its predictive ability but actually worsens it. A recently published application of MLR is provided by a study of the phenolic pigment composition of teas (McDowell et al., 1995).

It is a feature of MLR models that as each new term is added, so the coefficients of terms already in the model change. This is a consequence of the variables not being truly independent, they are correlated, and it leads to unstable models. Stable, whole spectra, multi-variate models can be developed if the variables are uncorrelated. Such models are referred to as orthogonal.

We have seen earlier that original, measured variables can be transformed to orthogonal, uncorrelated new variables by PCA, and it is not surprising, therefore, that regression analysis using principal components as the independent variables has gained popularity, i.e. principal components regression (PCR). A principal components analysis of our mean-centred, calibration data is provided in Table 12.9. The greatest proportion of the total variance, as expressed by the

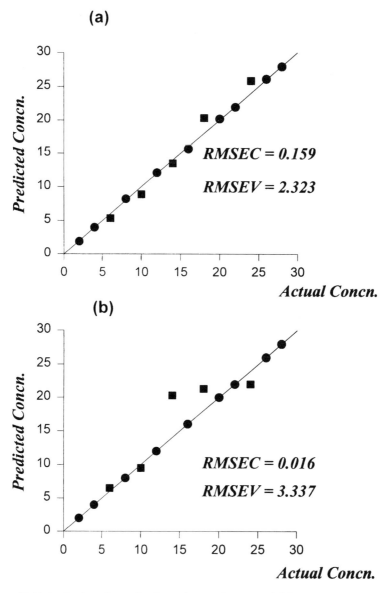

Figure 12.14 Application of stepwise forward regression (a), and full variable, multiple linear regression, MLR (b), to the calibration data from Table 12.8. Although MLR provides a significantly better fit for the calibration model (●) it is worse as a prediction model for validation data (■).

Table 12.9 Principal components analysis of the calibration data from Table 12.8, and the regression coefficients for least-squares fit of the principal components to the concentration data

Eigenvalues	0.114	0.028	0.006	0.001	6.8×10^{-4}	1.3×10^{-4}	1.3×10^{-5}
Eigenvectors	Factor 1	Factor 2	Factor 3	Factor 4	Factor 5	Factor 6	Factor 7
	0.32	0.49	0.69	0.14	−0.10	−0.26	0.28
	0.57	0.20	0.10	0.60	−0.17	−0.06	0.48
	0.62	0.16	0.40	−0.30	0.09	−0.25	0.52
	0.43	0.23	0.13	−0.48	0.12	0.63	0.33
	−0.05	0.80	0.55	0.11	0.06	−0.14	0.13
	−0.02	0.07	0.15	0.04	−0.87	0.37	0.29
	0.01	0.03	0.05	0.53	0.43	0.56	0.46
Regression coefficients	24.81	22.89	24.61	14.90	22.21	20.65	95.03

eigenvalues, is contained in the first two or three eigenvectors, and, hence, a two- or three-factor model may be sufficient to model the tryptophan concentration. If we form the seven principal components and perform a least-squares regression of these on to the dependent variable, i.e. tryptophan concentration, then the regression coefficients are as given in Table 12.9. Because the principal components are orthogonal, then the regression coefficients do not change as more factors are entered into the model, and we can see from Fig. 12.15(a) the effects on RMSEC and RMSEV as more factors are included. Although the value of RMSEC continuously declines as more factors are added, RMSEV values dip to a minimum at a two-factor model given by,

$$Tr = 13.33 + 24.81 PC1 + 22.89 PC2 \qquad (12.29)$$

where $PC1$ and $PC2$ are the first and second principal components. A plot of predicted concentration (using equation 12.29) vs actual concentration is presented in Fig. 12.15(b).

It may not be the case, of course, that the principal components corresponding to the largest eigenvalues are the best for a calibration model. After all, each principal component reflects successively decreasing variance in the data, and features correlating with the dependent variable may account for only a small proportion of the recorded variance. For this reason, having formed the principal components it is usual to calculate the correlation between each and the dependent variable, and then add them to the calibration model in order of decreasing correlation. One recent paper describes the use of PCR for the non-destructive determination of moisture, fat, and protein in fish fillet (Isaksson *et al.*, 1995).

This technique of eigenvector analysis and adding orthogonal terms to the model in order of their predictive ability is addressed by partial least-squares regression (PLSR). The major operational difference between PLSR and PCR is that in PLSR the data matrix includes the dependent variable. A complete mathematical description of the technique, as well as suitable algorithms, is available (Martens *et al.*, 1996). The results for our tryptophan data using a PLSR model are illustrated in Fig. 12.16(a) and (b), and are similar to those obtained previously by PCR.

Like PCR models, PLSR calibration models have been extensively reported in the literature, for determining the moisture content of mushrooms (Roy et al., 1993), the blending of red and white wines (Garcia-Jares and Maderia, 1993), and for determining the fruit type in jams (Defernez and Wilson, 1995).

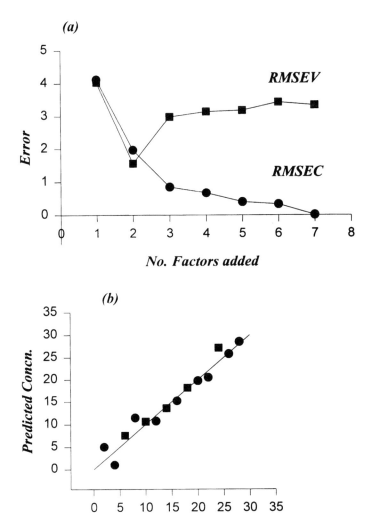

Figure 12.15 (a) RMSEC and RMSEV values for a principal components regression model as a function of number of factors in the model; (b) predicted vs actual tryptophan concentrations using a linear model with two principal components.

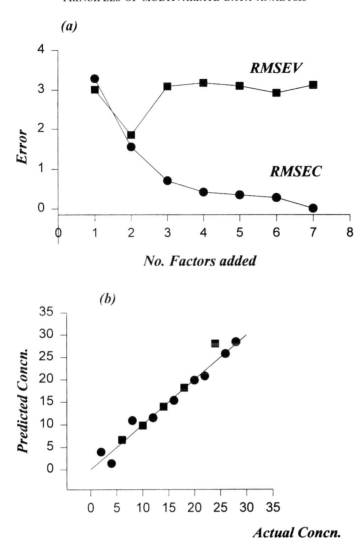

Figure 12.16 (a) RMSEC and RMSEV values for a partial least squares regression model as a function of number of factors in the model; (b) Predicted vs actual tryptophan concentrations using a linear model with two factors.

12.7 Conclusions

The developments and applications of chemometrics ('the mathematical and statistical manipulation and analysis of chemical data') have been considerable. In this short account we have only been able to examine an overview of some of the more common and widely used techniques. Current research in this area extends from novel designs and applications of neural networks, through the development of genetic algorithms for automated optimization, to the application of expert systems for computerized data interpretation. There is increasing interest in chemometric procedures for process monitoring and quality control, and the use of multivariate metrics for on-line measurement is becoming widespread in many manufacturing and process industries.

Whatever the developments, however, the onus for their correct use lies with scientific staff and this, in turn, requires staff to be familiar with the choice, use and interpretation of appropriate methods.

References

Adams, M.J. (1995) *Chemometrics in Analytical Spectroscopy*, RSC, Cambridge.
Beale, R. and Jackson, T. (1990) *Neural Computing: an Introduction*, Adam Hilger, Bristol.
Briandet, R., Kemsley, E.K. and Wilson, R.H. (1996) Approaches to adulteration detection in instant coffees using infrared spectroscopy and chemometrics. *J. Sci. Food Agric.*, **71**(3), 359–66.
Defernez, M. and Wilson, R.H. (1995) Mid-infrared spectroscopy and chemometrics for determining the type of fruit used in jam. *J. Sci. Food Agric.*, **67**(4) 461–7.
Downey, G. and Boussion, J. (1996) Authentication of coffee beans variety by near-infrared reflectance spectroscopy of dried extract. *J. Sci. Food Agric.*, **71**(1) 41–9.
Fresno, J.M., Prieto, B., Urdiales, R. *et al.* (1995) Mineral content of some Spanish cheese varieties. Differentiation by source of milk and by variety from their content of main and trace elements. *J. Sci. Food Agric.*, **69**(3) 339–45.
Garcia-Jares, C.M., Garcia-Martin, M.S., Carro-Merino, N. and Cela-Torrijas, R. (1995) GC-MS identification of volatile components of Galacian white wines. *J. Sci. Food Agric.*, **69**, 175–84.
Garcia-Jares, C. and Maderia, B. (1993) Research on white and red wine blending in the production of rosé wines by means of the partial least squares method. *J. Sci. Food Agric.*, **63**(3) 349–54.
Guerrero, M.I., Heredia, F.J. and Troncoso, A.M. (1994) Characterisation and differentiation of wine vinegars by multivariate analysis. *J. Sci. Food Agric.*, **66**(2) 209–12.
Hopke, P.K. (1992) Factor and correlation analysis of multivariate environmental data, in *Methods of Environmental Data Analysis* (ed. C.N. Hewitt), Elsevier Applied Science, pp. 159–80.
Isaksson, T., Togersen, G., Iversen, A. and Hildrum, K.I. (1995) Non-destructive determination of fat, moisture and protein in salmon fillets by use of near-infrared diffuse spectroscopy. *J. Sci. Food Agric.*, **69**(1) 95–100.
Martens, H. and Naes, T. (1996) *Multivariate Calibration*, John Wiley and Sons, Chichester.
McDowall, I., Taylor, S. and Gay, C. (1995) The phenolic pigment composition of black tea liquors – Part I: Predicting quality. *J. Sci. Food Agric.*, **69**(4), 467–74.
Roy, S., Anantheswaran, R.C., Shank, J.S. *et al.* (1993) Detection of moisture content of mushrooms by vis-NIR spectroscopy. *J. Sci. Food Agric.*, **63**(3) 355–60.
Sanz, S., Perez, C., Herrera, A. *et al.* (1995) Application of a statistical approach to the classification of honey by geographic origin. *J. Sci. Food Agric.*, **69**(2), 135–40.
Thompson, J.M. (1992) Visual representation of data including graphical exploratory data analysis, in *Methods of Environmental Data Analysis* (ed. C.N. Hewitt), Elsevier Applied Science, pp. 213–58.
Twomey, M., Downey, A. and McNalty, P.B. (1995) The potential of NIR spectroscopy for the detection of the adulteration of orange juice. *J. Sci. Food Agric.*, **67**(1) 77–84.

Index

Page numbers appearing in **bold** refer to figures and page numbers appearing in *italic* refer to tables.

Accuracy, 264, *265*
Acetaldehydes, 29, *143*, 146
Acetate, 46
Acetic acid, 28, **43**, 52, 63, 70, *142*, 147
Acetyl-CoA synthetase, *142*
Acid-base catalysis, 146
Acids, 89
Acquisition time, 38, 39
Activators, 138
Added water in wines, 25
AFS, *271*, 273
Agarose, 182, 205
Agarose gel electrophoresis, 193, 212
Agarose gels, 205
Agave, 20
Age
  at death, 194
  of vinegar, 303
AIJN Code of Practice, 67
Air-acetylene, 273
Alanine, **43**, 52
Alavesa, 276
Albacore, 188
Alcohol dehydrogenase, *143*, 143, 234
Alcohols (ethanol and propanol), 44
Alcohols, 57
Aldehyde dehydrogenase, 143, *143*
Aldehydes, 57
Aldose, 99
Aliphatic esters, 71
Alkaline phosphatase, 186, *247*
Alkaline phosphate, 228
Alta, 276
Aluminium, 273, 276
Amido Black 10B, *224*
Amines, 237
Amino acids, 37, 44, 48, 49, 51, 52, 229, 237
γ-Amino butyric acid, **43**
2-Amino-4-fluorobenzoic acid, 132
3-Amino-4-hydroxybenzenesulphonic acid (AHBA), 132
Ampholytes, 214
Amyl acetates, 29
Amylase, 222
α-Amylase, 101, 104
Amylases, 159
Amyloglucosidase (AGS), *143*, 153, 159

Amylopectin, 100
Amylose, 100
Anabolic reactions, 172
Anetholes, 22, 29, 59, 71
Anharmonic oscillators, 78
Animal fat, 91
Animal organs, 138
Anion exchange resin, 105
Antelope, 186
Antibodies (immunoglobulins), 183, 218, 241
Antigen, 218
Apple authenticity, 66
Apple juice, 18, *19,* 43, 46, **47**, 97, 110, 121, 133, 135, 277
  concentrates, 127
Apples, 25, 27, 42, *67*, 83, 158
Apricot, *20*
  kernel oil, 28
Arabinose, 52
Arbitrary primers, 196
Arginine, 52
Argon plasma, 286
Aromatic amino acids, 46
Aromatic protons, 46
Artificial neural net, 91
Artificial neural networks (ANNs), 87, 281, 304
Ascorbate oxidase, *142*
L-Ascorbic acid, *142*, 148
Asparagine, **44**
Asparagine, 46
L-Aspartic acid, 148
Atomic absorption, 65
Atomic absorption spectrometry (AAS), 273
Atomic emission spectrometry (AES), 278
Attenuated total reflectance (ATR), 80, 86, 87
Authentice range, 121
Authenticity research, 2
Automated systems, 280
Automatic sampler, 61
Autoradiography, 226, 228
Average-linkage, 321
Avocado, 10, *20*
Azodye, 222

Bacteria, 292
Baja, 276

Bakery Goods, 149
Baking Industry, 234
Balsamic vinegar, 303
Banana, *20*
Barium, 277, 280, 281, 282
Barley, 18, 234
Barley malt, 18
Barley type, 300
Baseline correct, 41
Basmati US long grain, 89
Battered cod, 190
Bayesian technique, 91
Beef, 21, 58, 182, 183, 186, 193, 196, 263
Beer, 20, 146, 147
Bees, 104
　identification, 297
Beet, 70
Beet/cane invert sugar, 121
Beet/cane sugar, 98
Beet medium invert sugar, 49
Beet medium invert syrup, 51
Beet sucrose, 65
Beet sugar, 64, 66
Beluga sturgeon, 194
Benzal chloride hydrolysis, 28
Benzaldehyde, 22, 28, 59
Beta radiation, 282
Beverages, 69, 237
Bifidus-promoting capacities, 104
Big game animals, 186, 199
　meat, 195
Bigeye, 188
Biofluids, 40
Biological activity, 184
Bitter almond oil, 28
Black leaf tea, 92
Blackberry, *20*
Blood plasma, 37
Blotting of gel patterns, 227
Blueberry, *20*
Bluefin tuna, 183, 188, 192
Bonitos, 196
Borate complex, 132
Boron, 280, 281
Boscop, **43, 44, 45**, 46
Botanical origin, 59, 298
Bourbon whiskies, 69
Bovine Y chromosome, 194
Bramley, 46, **47, 49**
Brandy, 146
Brassica, 235
　hybrid purity, **223**
　proteins, 234
Brazil, *67*, 280, 302
Bread, 147
Broad-band decoupling, 61
Bromine, 282, 283
'Buckets', 41

Buffalo, 193, 195, *261*
Bulked segregant analysis, 197
n-Butanol, 143
Butter, 146
Butterfat, 91

^{13}C, 60
^{13}C/^{12}C, 16
^{13}C content, 66, 70
^{13}C signals, 52
^{13}C-NMR spectra, 51, 53
^{14}C, 18
Cabernet Franc, 50
Cabernet Sauvignon, 50
Cadmium, 278
Caesium, 282
Caffeine, 22, 86
Calcium, 276, 277, 281, 282
Calcium fluoride, 85
Calibration, 326
California, 280
Calvin cycle (C-3), 16
Camel, *262*
Canadian honeys, 283
Candied fruit, 18
Cane, 69
Cane sugar, *19*, 66
Canned food, 147
　meat products, 186, 187
　salmon, 190
　tuna, 192, 196
Canola oil, 90
Canonical variance (variates) analysis (CVA), 47, 48, 58, 277, 297, 300, 301, 302
Canonical variates discriminant analysis, 89, 90
Cantaloupe, *20*
Canyon Diablo Troilite, *15*
Capillary blotting, 227
Capillary electrophoresis (CE), 131, 228
　in polymer solutions, 23
Capillary isoelectric focusing, 232
Capillary zone electrophonesis (CZE), 104, 132, 230
Carambola, *20*
Caramel, 298
Carbohydrate/starch, 138
Carbohydrate(s), 97, 223, 297
Carbon, 282
Carbonation, 14
Carbonyl chemical shift, 56
Carbonyl signals, 54
CARREZ reagents, 138
Carriers, 282
Carvacrol, 29
Caseins, 257
Cassava, 98
Cassia oil cinnamaldehyde, 29

Catabolic reactions, 171
Catalase, 137
Cation exchange resin, 105
Cattle, 195, *261*, 263
Causal relationship, 311
Caviar, 194
Cellobiose, 129
Cellulase, 129, **131**
Cellulose acetate, 182
Cereal grains, 222
Cereals, 69, 88, 235
Ceric ammonium nitrate, 28
Certified reference materials, 61
CGC-FID, 105, 121, 126, **130**
CH group, 40
CH_2 group, 40, 46
CH_3 group, 40
Champagne, 146
Chaptalization, 64
Cheese(s), 137, 146, 237, 256
    from ewes' milk, 303
    from goats' milk, 303
Chemical isotopic fractionations, 26
Chemical modifiers, 272
Chemical shift(s), 38, 52, 61
Chemometrics, 37
Cherry, *20*
Chicken, 182, 183, 186, 193, 198
    bones, 9
    nuggets, 190
    and turkey, *261*
Chicory, 86, 98, 152
Chiral additives, 231
Chiral capillary gc, 23
Chiral compounds, 230
Chiral substances, 23
Chlorogenic acid, **43**, 44, **45**, 46, 86
Chocolate drink powder, 298
Cholesterol, 140, *143*, 149
Cholesterol oxidase, *143*
Chromatographic methods, 184
Chromium, 276, 278, 282
Chromium trioxide, 28
Cider, 70
Cider vinegars, 21
Cinnamaldehyde, 22
cis-3-Hexenol, 29
Citramalic acid, **43**
Citrate, 46
Citrate lyase, *142*
Citrate synthase, *142*
Citric acid, 20, 46, 52, 133, 149
Citrus honeys, 17
Citrus juice(s), 46, 97, 111, 129, 133
Classification
    by growth year, 281
    of cheeses, 324
    of food samples, 37

Climatic conditions, 65
Climatological factors, 7
Clove eugenol, 22
Cluster analysis, 58, 277, 297, 302, 319
    hierarchical, 319
    non-hierarchical, 319
Cobalt, 276, 282
Cocoa butter, 302
Coconut, *20*
Cod, 59
Coenzymes, 168
Coffea arabica, 84, 86, 92
Coffea canephora variant *Robusta*, 84, 86, 92
Coffee, 84, 92, 298
    instant, 85
    roasted, 298
Combination bands, 78
Combinations of elements, 272
Competitive enzyme immunoassay, 248, **248**
Competitive methods, 253
Complete-linkage, 321
Compositional analysis, 233
Computer-aided medical diagnosis, 37
Concentrates, 40
Conductivity, 229, 238
Consumers, 11
Control, 193
Cooked meat(s), 187, 193, 263
Coomassie Blue, 223
Coomassie Blue G250, *224*
Coomassie Blue R250, *224*
Co-ordination of surveillance activities, 4
Copper, 276, 277, 281
Coriandrum sativum, 23
Corn, 69, 98
Corn oil, **55**, 56, 90
Corn syrup, 17,*19*
Corned beef, 186
Correlation, **48**
    coefficient, r, 310
    matrix, 42
    method, 42
Cottonseed oil, 90
p-Coumaric acid, **45**
Counter immunoelectrophoresis, 219, 234
Countercurrent electrophoresis, *257*
Coupling constant, 38
Covalent catalysis, 146
Covariance, **48**, 310
    matrix, 42
    method, 42, 46, **47**
Cow(s), 186, 193, 194, 198, *243*
Cows' milk, 237, 256
Cranberry juice, *19*, 121
Crassulacian acid metabolism (CAM), 16
Creatine kinase, *142*, 183
Creatine/creatinine, *142*, 150
Creep reaction, 143, **174**

Cross reactivity, 243
Crossed immunoelectrophoresis, 219, 246, *257*
Cubic spline, 264
Cultivar, 47, 57
  differences, 299
  identification, 235
Curie-point, 294
Curie-point pyrolysis-gas chromatography-mass spectrometry (Py-GC-MS), 298
Customs officers, 182
Cut-off limit, 4
Cut-off points, 67
Cyclodextrins, 230
Cytochrome b, 190
Cytochrome c, 216

2-D (two-dimensional) NMR, 41, 51
2D-PAGE, 220
D_2O, 38
D-3-hydroxybutyric acid, *142*, 154
Dahlia tubers, 98
Dairy products, 91, 146
Data
  acquisition time, 38
  evaluation, 296
  reduction, 313
DCP-AES, *271*
De-aeration, 209
Deer, 193
  species, 196
Delay time, 38
Dendrogram, 321
3-Deoxyglucosulose, 112
Deproteinization, 138
Derivatization, 61
Desserts, 71
Detecting adulteration in instant coffees, 324
Detection, 238
Detection limits, 40, 135
Deterioration of food, 236
Deuterated solvents, 37
Deuterium NMR, 37
Dextrose polymers, 114
Diastase, 137
Dietary fibre, 104
Diethylene glycol, 52
Diffuse reflectance (DRIFT), 80, 83, 85
Diffusion, 15
Digital resolution, 38, 39
Diglycerides, 37, 52, 56, 57
Dimensionality of data, 318
Dipole moment, 76
Dipstick and immunodot tests, 251
Dipstick enzyme immunoassay, 251
Disaccharides, 116, 123, 128
Disadvantages of ICP-MS, 286
Disc gel electrophoresis, 210

Discriminant analysis, **48**, 51, 83, 85, 89, 277, 283, 297, 323
Discriminant procedures, 93
Dishonest competitors, 2
Distilled spirits, 18
Dithiothreitol, 212
DNA, 184, 226, 229
DNA polymerase, 187
2-Dodecylcyclobutanone, 9
Dolphin products, 191
Domestic management of whale products, 192
Domestic species, 183
Donkey, *262*
Double diffusion, 245
Dough, 147
DSS (sodium 2,2-dimethyl-2-silapentane-5-sulphonate), 38
Ducks, 59
Due diligence legislation, 1
Durum wheat lines, 235
Durum wheat-pasta, 88
Dynamic range, 39

Ecological studies, 195
Edible oils, 87
EDTA, 138
Efficiency of washout systems, 272
Egg(s), 158, 162, 196
  liqueur, 149
Eigenvalues, 315
Eigenvectors, 42, 315
Eland, *261*
Electrical arcs, 279
Electroblotting, 220
Electrochemical refractive index, 229
Electroendosmosis, 212, 215, 232
Electroendosmotic flow, 230
Electron impact, 295
Electron spin resonance (ESR), 9
Electronic sensor technology (electronic nose), 12
Electronic transitions, 77
Electro-osmosis, 131
Electrophoresis, 184
Electrophoretic induction, 229
Electrophoretic methods, 184, 204
Electrophoretic migration, 230
Electrothermal atomization, **275**
ELISA, 184, 218
Elk, 187
Emu shish kebab, 191
Enantiomers, 23, 231
End-point kinetics, 138
Endangered species, 182
Enforcement authorities, 1
English locations, 46
Enzymatic analysis, 137
Enzymatic isotopic fractionations, 26

Enzyme immunoassay (EIA), 139, 184, 247
  procedures, 260
Epicatechin, **45**, 46
Epitope, 242, 250, 255
Equatorial latitudes, 24
Equilibrium isotope effects, 15
Error of calibration, 330
Error of validation, 330
Erythritol, 52
Esterase, *143*, 222
Esters, 138
Estragole, 29
Estragols, 71
ETAAs, *271*, 274
Ethanol, **43**, 46, 52, 63, 64, *143*, 150
Ethidium bromide, 226
Ethyl acetate(s), 29, 71
Ethyl butyrate(s), 22, 29, 71
Ethyl caproates, 29
Euclidean distance, 319
Eugenol, 29
European wines, 301
Evaporation, 15
Ewes' milk, 256
External referencing method, 61
Extra virgin olive oil, 87, 90, 298
Extraction, 61
Extracts, 40

FAAS, *271*, 274
Factor analysis (PCA), 297, 301, 302, 318
Factorial discriminant analysis (FDA), 50, 89, 92
False negatives, 256
False positives, 256
'Far infrared', 76
Fat-containing foods, 10
Fatty acids, 55
  composition, 52
Feeding stuff, 292
Feijoa, *20*
Feta, 257
Field-frequency lock, 37
Field-frequency locking device, 61
Field homogeneity, 40
Field ionization, 295
Fingerprint oligosaccharide(s), 101, *108*, 114, 115, 124, 129, 135
Fingerprinting, 121, 199
Fingerprints, 83, 104, 112, 185
Fish, 58, 59, 196
  cakes, 237
  fillet fat, 333
  fillet moisture, 333
  fillet protein, 333
  fingers, 237
  species, 196
Flame emission spectrometry (FES), *271*, 273

Flavonoids, 237
Flavour changes, 236
Flavourings, 59
Florida, *67*
  orange juice, 277, 280
Flow-through 'immunofiltration', 251
Fluorescein isothiocyanate, 226
Fluorimetry, 138, 177,
Fluorine lock, 61
Food
  irradiation, 8
  producing companies, 1, 4
  quality, 292
  thickeners, 297
Forensic chemistry, 292
Forensic work, 198
Forensically informative nucleotide sequencing (FINS), 187, 188, 191, 193
Formate dehydrogenase, *142*
Formic acid, **45**, 46, *142*, 151
  derivative, 28
Formol value, 148
Fourier transform, 41
Fourier transform (ft) nmr spectra, 38
Four-parameter-logistic model, 264
Free fatty acids, 52, 59
Free induction decay (fid), 38, 41
Freeze-dried samples, 38
Freeze-thawed fruit, 84
Frequency domain, 38
Fresh cod, 58
Fresh fruit, 84
Fresh meat, 137, 193
Frozen meat, 137
Frozen–thawed cod, 58
Frozen-thawed meat, 92
Fructose, 44, **44**, 51, 52, 86, 90, 98, 111, 121
D-Fructose, 101, 141, *142*, 152
Fructofuranose, 103
β-Fructosidase, *143*
Fruit acids, 158
Fruit alcohols, 59
Fruit juices, 37, 40, 42, 51, 59, 65, 89, 105, 106, 119, 121, 132, 146, 147, 158
Fruit purees, 83, 89
Fruit sugars, 66
Fruit type, 67
  in jams, 334
Fruit and vegetable mixtures, 12
Fruits, 51, 83, 147, 222, 235
FTIR, 83
2-Furaldehyde, 112
Furnace pyrolyzer, 295
Furthest-neighbour, 321

Galactose, 45, 52
α-Galactose, **44**
β-Galactose, **44**

D-Galactose, *143*
Galactose dehydrogenase, *143*
α-Galactosidase, *143*
β-Galactosidase, *143*
β-D-Galactosidase, *247*
Galacturonic, 52
Galicia wine, 277
Gallium, 280
Game species, 183
Gamma radiation, 282
Gander liver, 199
GC-IRMS, 23
Geese, 59
Gel electrophoresis, 182
Gel-filled capillaries, 232
Genetic diseases, 198
Genetic engineering, 182
Genetically modified organisms (GMOS), 11
Genomes, 185
Gentiobiose, 52
Geographical conditions, 65
Geographical origin, 59, 65, 67, 280, 282, 283, 287, 298
Geological nature of the soil, 276
Geological point of view, 276
Geraniol, 71
Gin, 69
γ-Globulin, 218
Glow discharges, 279
Glucoamylase, 101, 104
Gluconate kinase, *142*
Gluconic acids, 52
D-Gluconic acid, *142*
D-Gluconic/D-glucono-δ-lactone, 152
Glucose, 44, 51, 52, 86, *98*, 111, 121
α-Glucose, **44**
β-Glucose, **44**
D-Glucose, 101, 141, *142*, 152, 153
Glucose oxidase, *247*
Glucose/fructose ratio, 153
Glucose-6-phosphate dehydrogenase, 141, *142*
α-Glucosidase, *143*
α,β-Glucosidase, 104
Glucuronic, 52
L-Glutamate, 141
Glutamate dehydrogenase, 141, *142*
Glutamic acid, **43**, 52
L-Glutamic acid, *142*, 153
Glutathione, 138
Glycerokinase, *143*
Glycerol, 52, *143*, 154
Glycoproteins, 225, 241
Goat, 186, 193, 195, *243*
   meat, 193
Goats' milk, 237, 256
Goechemistry, 292
Goodness-of-fit, 329
Goose liver, 199

Grain, 59
Granny smith, **43**, **44**, **45**, 46
Grant's gazelle, *261*
Grape juice, 18, 19, 43, 67, 121, 161
Grape musts, 59, 65, 275
Grape seeds, 50
Grapes, 25, 42, 287
Grapefruit, 42, 49, *67*, 135
Grapefruit authenticity, 66
Grapefruit juice, 18, *19*, 67, 89, *108*, 110, 111, 114, 121
Grapeseed oil, 90
Grapevine, 50
Graphite furnace, **274**
Green coffee, 282
Groundnut oil, **55**
Groundwater, 24, 26
Guaiacol, 21, *70*
Guanidine hydrochloride, 212
Guar gums, 91
Guava, *20*
Gums, 297

$^1H/^{13}C$ chemical shift correlation, 41
$^1H/^{13}C$/2-D NMR, 50
^1H COSY, 41
^1H NMR, 49
^1H-NMR spectra, 51
   500 MHz, 40, 46
   600 MHz, 42
^2H-NMR, 61
^2H-NMR spectrum, 70, **71**
Haemoglobin, 216
Haploid locus, 190
Haploid micro-organisms, 197
Haptens, 241
Hard wheat flours, 234
Hatch-slack (C-4), 16
HCA (hierarchical cluster analysis), 50
Heat abused, 129, 130
Heat transfer, 293
Heat-stable proteins, 183
Heavy chain, 241
Heavy metal ions, 138
Hertz, 38
Heterozygosity, 190
Hexokinase, 141
HFCs, 121
High altitudes, 24
High-field region, 44
High fructose corn syrup, 121, 125
High fructose inulin syrup, 121
High fructose starch syrup, 98
High fructose syrups (HFS), 101, 107, 114, 119, 121, 126, 133
   in citrus juices, *109*
   from inulin, 127

High performance anion exchange chromatography with pulsed amperometric detection (HPAE-PAD), 101, 105, 106, 109, 115, 118, 126
High specific enzymes, 141
Historical changes, 7
HMF, 129
Honey, 17, 69, 97, 105, 106, 109, 118, 119, *128*, 132, 133, 135, 137
  geographic origin, 324
Honeycomb waxes, 297
Honeydew melon, *20*
Horse, 182, 193, *261*, 263
Horseradish peroxidase, 228
Horse-radish sauce, 12
Human Genome Project, 220
Human tea tasters, 92
Hunting regulations, 195
Hybrid variants, 234
Hybridization, 185, 187, 193
Hybridomas, 243
Hydrochloric acid, 101
Hydrodynamic radius, 211
Hydrogen sira, 26
Hydrolase enzyme, 128
3-Hydroxybutyrate dehydrogenase, *142*
Hydroxyethyl cellulose, 231
5-Hydroxymethyl-2-furaldehyde (HMF), 112
Hydroxy methyl proline, **43**
Hydroxyl site, 63

^{125}I-labelled lectins, 226
Ice-cream, 71
ICP-AES, *271*
Immobilines, 217
Immunoblot techniques, 252
Immunodiffision, 245, 260
Immunoelectrophoresis, 218, 246
Immuno electrophoretic methods, 237
Immunofiltration assay, 252
Immunogen, 241
Immunological methods, 184
Immunology, 183
Impala, *261*
Incubated eggs, 155
Indirect UV, 237
Inductively coupled plasma for atomic emission, **279**
Inductively coupled plasma-mass spectrometry (ICP-MS), *271*, 283
Inexpensive sweetners, 97
Influences on data evaluation, 6
Infrared spectroscopy
Infrared spectrum, 313
Inhibitors, 138
Instant coffee, 851
Intensity reference, 38
Interferences, 176

Interferogram, 79
Internal pattern of carbon SIRA, 23
Internal referencing procedure, 61
Internal standard, 17
  approach, 233
International management of whale products, 192
International Whaling Commission, 192
Interpreting databases, 4
Inulin, 98, 104, 152
Inverse probes, 40
Inversion, 152
Iodine value, 53
Ion generation, 295
Ionones, 23
Iota-carrageenan, 91
IPG (immobilized pH gradient) gels, 217
Iron, 276, 277, 281, 282
Irradiated Brown Shrimp, 8
IS, 107, 118, 121, 133
  in citrus juices, *109*
Isoamylacetate, 71
Isobaric interferences, 286
D-Isocitrate, 149
D-Isocitrate dehydrogenase, *144*
D-Isocitric acid, *144*, 155
Isoelectric focusing (IEF), 214
Isoelectric points, 214
Isoleucine, **43**
Isomaltose, 101, 114, 126, *134*
Isomaltotriose, *134*
Isotope dilution analysis, 283
Isotopic content, 59
Isotopomer, 60
Israel, 17, *67*, 302
Italian cheese, fresh, 304
Italian cheese, ripened, 304
Italian regions, 57, 278
Italian varieties, 56

Jam, 83, 90
Japanese seaweeds, 281
Jerusalem artichoke, 98

K (k) nearest neighbours, 51, 277, 324
k-means algorithm, 323
Kangaroo, *262*, 263
Kelose, 114, 115
Ketchup, 147
Ketose, 99
6-Ketose, 114
Kinetic isotope effects, 15
Kiwi fruit, *20*
Korean seaweeds, 281
Kumquat, *20*

Lactate, 46
D-Lactate dehydrogenase, *144*

L-Lactate dehydrogenase, *144*
Lactic acid, **43**, 52
D-Lactic acid, *144*, 155
L-Lactic acid, *144*, 158
Lactones, 23
Lactose, *143*, 156
Lactulose, *143*, 152, 157
Lamb, 58, 196
Lanthanum, 282
Larvae, 196
Lasers, 279, 279
Lead, 278
Least-squares estimate, 308
Least-squares methods, 309
Least-squares regression analysis, 328
'Leave-one-out' method, 331
L-α-Lecithin, *143*, 158
Lectins, 225
Legislation, 6
Legumes, 235
Lemon juice, *19*
Leucine, **43**
Light chain, 241
Lignin, 70
Lime, *20*
Limit of detection (LOD), 272
Limonene, 71
Linalool, 22, 23, 29, 71
Linalyl acetates, 22, 29
Line broadening factor, 41
Linear dependence, 311
Linear discriminant analysis (LDA), 47, 48, 84, 87
Linear discriminant functions, 323
Linear regression, 304
Line-widths, 38
Linoleic acid, 90
Linoleic oil, 53, **54**
Linolenic acid, 56, 57
Linolenic oil, 53, **54**
Lipases, *143*, 235
Lipid content, 85
Lipids, 297
Liquid chromatography, 184
Lithium, 276
Livers, 59
Long chain alcohols, 58
Low background, 280
Low-field region, 46
Luciferase, *247*
Luciferin, 228
Luminescence techniques, 177
Luminol, 228

Mackerel, 190
Magnesium, 276, 277
Magnetic field homogeneity, 37
Magnetic field stabilization, 37
Magnetization, 294
  transfer, 58
Magnetization, 294
Mahalanobis distance discriminant analysis, 90
Mahalanobis distance discrimination function, 92
Mahalanobis distances matrix, 297
Maize hybrids, 197
Major components, 40
Maker-free methods, 244
Malaga wines, 277
Malate, 46
L-Malate, *143*
D- Malate dehydrogenase, *144*,
L-Malate dehydrogenase, *143*, *144*
Malic acid, 17, 20, **44**, 46, 48, 52
D-Malic acid, 49, 50, *144*, 158
L-Malic acid, *144*, 158
Malt, 70
Malting quality, 234
Maltoheptaose, 101, *134*
Maltohexaose, *134*
Maltopentaose, 116, *134*
Maltose, 52, 114, 115, 133, *134*, *142*, *143*, 159
Maltotetraose, 114, 133, *134*
Maltotriose, 52, 114, 115, *134*
Manganese, 276, 277, 280, 282
Mango, *20*
Mannitol, 52
D-Mannose, 141
Manufacturing industries, 336
Manufacturing technology, 278
Maple syrup, 14, 17, 69, 97, 105, 106, 109, 116, 119, 132, 133, 135
Margerine, 91
Mass spectrometer, 229
Matrix modification, 274
Mayonnaise, 147
Mean, 308
Measurement precision, 5
Meat, 21, 251, 259
  lipids, 59
  pies, 237
  products, 164, 251, 259
  species, 91, 182
Medium invert sugar, 98, 103
Membrane filtration, 115
2-Mercaptoethanol, 212
Mercury, 282
Merlot Noir, 50
Mescal, 20
Metal ions, 138
Meteoric water, 25, 27
Methanol, 52
Method validity, 1
Methyl groups, 21
Methyl salicylates, 22, 29
Methyl site, 63

Methylene, 21
Methylene site, 63
Mexico, 280
Micellar electrokinetic capillary
    chromatography (MECC), 230
Micelles, 231
Michelson interferometer, 80
Microbial contamination, 137, 155
Microbial decomposition, 162
Microbiological activity, 46
Micro-organisms, 138
Microsatellite loci, 199
Microsatellites, 198
Microscopy, 12
Microtitre plate assays, 250
'Mid infrared', 76, 79
Milk, 137, 251, 256, 263
Milk chocolate, 298
Milk products, 256, 303
Milk proteinase, 236
Minor components, 40
Mint terpenoids, 29
MIS, 107, 133
    in citrus juices, *109*
    in orange juice, *108*
Mitochondrial genes, 190
Mixed pickles, 147
Model's effectiveness for prediction, 330
Modified sugar syrups, 65
Molecular weight calibration, 221
Monoclonal antibodies, 243
Monoglycerides, 37
Monosaccharide(s), 37, 99, 116, 118
Monospecific antiserum, 255
Montilla wine, 277
Mosel-Saar-Ruwer, 51
Mouse, *243*
Mule deer, 187
Multi-element analyses, 279, 282
Multiple linear regression (MLR), 331
Multiplet, 40
Multivariate data analysis, 41, 308
Muscle protein, 164
Mushrooms moisture content, 334
Mussels, 278
Musts, 287
Mutton, 183
Myofibrillar proteins, 58
Myoglobin, 216
Myo-inositol, 52, 100

^{15}N, 60
Naphthol, 222
Naphthylamine, 222
National governments, 2
National surveillance, 3
Natural abundance, 61

Natural flavourings, 70
Natural status of flavouring, 59
'Near-infrared', 76, 79
Nearest-neighbour, 321
Nebulizer, 280
Nectarine, *20*
Nephelometry, 245
Networks, 2
Neural networks, 297, 299, 301, 302, 325
Neutron activation analysis (NAA), *271*, 281
Neutron bombardment, 282
New technology, 7
Nickel, 282
Nickel vessel, 25
NIR reflectance, 88, 92
    technique, 91
NIR transflectance spectra, 91
NIR transmission and reflectance spectroscopy,
    90
Nitrate, *142*, 159
Nitrate reductase, *142*
p-Nitrophenyl oligoglucosides, 138
Nitrous oxide-acetylene, 273
NMR-multivariate analysis, 50
NOESY-presat pulse sequence, 40
Non-competitive assay, 263
Non-competitive EIA, 248, **249**
Non-competitive methods, 253
Non-parametric methods, 323
Noodles, 149
Normalized data, 297
Normalized spectra, 41
Northern latitudes, 24
Novel genetic markers, 197
Nuclear magnetization, 39
Nuclear Overhauser effects, 61
Nucleic acid(s), 212, 223
Nucleotide sequence, 188
Nutritional value, 104

Oak acorn fed pigs, 91
Oak casks, 300
Oats, 18
Offals in comminuted meat, 12
Oil of bitter almond, 22
Oil of cassia, 22
Oils, 40, 90
Olefinic protons, 52
Oleic acid, 53, **54**, 56, 90
Oleic/linoleic acid, 56
Oligo-β-fructosans, 152
Oligonucleotide primers, 193
Oligonucleotides, 186, 187
Oligopolysaccharide, 143
Oligosaccharide(s), 37, 51, 52, 84, 99, 104,
    106, 112
Olive oil(s), 53, **54**, 87, 298

Orange authenticity, 66
Orange juice, 18, *19*, 49, 67, 89, 97, 110, 121, **130**, 152, 155, 237, 298, 302
  concentrates, 51
  geographical origin, 302
Orange oil, 298
Oranges, 25, 27, 42, *67*,135
Organic acids, 20, 37, 52, 84
  in wine and coffee, 237
Origin of ethanols, 69
Orthogonal models, 315, 331, 333
Oryx, *261*
Outliers, 309
Overfitting, 330
Overlap, 40
Oversampling technique, 39
Overtones, 79
Oxalate, *142*
Oxalate decarboxylase, *142*
Oxalic acid, 160
Oxidase, 137
Oxidation, 28
Oxonium ion, 104

Paddlefishes, 194
PAGIF, 214
Palm, 98
Papaya, *20*
Parametric methods, 323
Parametric statistics, 93
Paratope, 242
Parchment, 86
Parsnip, **13**
Partial least squares (PLS), 87, 304
  regression (PLSR), 84, **85**, 333
Partial linear fit, 42
Parvalbumins, 183
Passion fruit, *20*
Pasteurization, 137
Pasteurized milk, 157
Pate de foie gras, 59, 199
Pattern analysis, 276
Pattern recognition, 11, 49, 319
  analysis, 277
PC scores, 47
PCA, 51, 277
  method, 46
PCR/DNA, 182
Peach, *20*
'Peak-picking routines, 41
Peanut oil, 90
Pear, *20*
Pecorino, 257
Pectin, 84, 297
Pedological, 276
Pee Dee Belemnite (PDB), *15*
Peppermint oil, 23
Peptides, 138, 229

Perchloric acid, 28
Periodic acid-Schiff staining, 225
Peroxidase, 137, 222, *247*
Persimmon, *20*
pH, 38
Phase, 41
Phenethyl acetates, 29
Phenolic compounds, 138
Phenolics, 37, 46, 57
Phenylethanols, 29
Phenyl propanoid, 71
Phloretin, **45**
Phloridzin, **45**
Phosphatase, 137
Phosphoglucomutase, 234
Phosphoglucose isomerase, *142*
Phospholipase, *143*
Photometry, 138, 177
Photostimulated luminescence, 8
Phycocyanine, 216
Phylogenetic analysis, 191
Pickled herring, 190
Pig, 186, 193, 198, *261*
  carcass quality, 91
Pineapple, 42, 135
Pineapple juice, 18, *19*, 67, 110,
Plants, 138
  breeding programmes, 234
  species, 7
Plasma discharges, 279
Plasma stability, 280
Plasmid vectors, 186
PLF, 51
PLSI, 92
Plum, *20*
Poaching, 194, 195
Pollen analysis, 12
Polyacrylamide gel electrophoresis (PAGE), 205
  apparatus, 207
Polyacrylamide, 182, 205
Polyclonal antibodies, 243
Polymerase chain reaction (PCR), 182
Polymerization, 209
Polymers, 229
Polyphenol species, 46
Polyphenols, **45**, 50, 237
Polysaccharide(s), 85, 99, 297
Pomace oils, 58
Pomegranate, *20*
Pork, 21, 58, 182, 183, 196, 263
  in beef, 193
  liver, 199
Positive release, 4
Potassium, 49, 50, 276, 277, 282
Potassium bromide, 81, 85
Potatoes, **13**, 18, 69, 98, 222, 235
Potentiometry, 138

Poultry, 196, 263
Precipitin arc, 219
Precision, *265*
Preconcentration methods, 274
Presaturation method, 39
Preservatives, 237
Preserved skins, 190
Prevention of contamination, 272
Prickly pear, *20*
Principal components, 86, 89, 316
Principal components analysis (PCA), 11, 42, 48, **48**, 49, 50, 276, 315, 318
Principal components regression (PCR), 297, 304, 332
Probe, 40, 186
Process industries, 336
Process monitoring, 336
Processed meat, 182
Processing or agricultural treatment of the vines, 281
Processing techniques, 298
Proline, **43**, 49, 51, 52
1,3-Propanediol, 52
Propanol, **43**
n-Propanol, 143
Protein(s), 18, 138, 182, 223, 229, 297
  molecular weight, 211
  rice plants, 235
Proteinases, 235
Proteolytic activity, 137
Protocol for PAGE, *208*
Proton decoupling, 40
Proton dedicated probe, 40
Prune, *20*
Pulp wash, 49, 89, 281
Pulse length, 38
Pulse power, 39
Pure apple juice, **131**
Purification, 61
Purple sage, 298
Pyrolysis, 25
  techniques, 293
Pyrolysis-gas chromatography-mass spectrometry (Py-GC-MS), 292
Pyrolysis-mass spectrometry (Py-MS), 292
Pyruvate, 160

Quadrupole mass spectrometer, 286, 293, 295
Qualitative immunoelectrophoresis, *257*
Quality control, 336
Quality of spectrum, 40
Quantification, 37, 38
Quantitative deuterium determinations, 61
Quantitative results, 224
Quantitative trait loci, 198
Quinic acid, **43**, 44

Rabbit, 193, *243*, *261*, 263

Radioimmunoassay, 218
Radiometric detectors, 229
Raffinose, 100, *143*, 152, 161
RAPD (randomly amplified polymorphic DNA), 196
Rapeseed oil, 53, **54**, **55**, 90
Rapid immunochemical test, 251
Rapid test methods, 141
Rare earth elements, 273, 287
Raspberry, 46, **85**
Raspberry juice, *19*
Raw meat, 182, 263
Reaction rate, 138
'Real' samples, 276
Record purposes, 227
Red wines, 334
Reference materials, 282
Refined oil, 58
Refined olive oils, 88
Region of origin, 56
Regions, 287
Regression, 93
Relative standard deviation, RSD, 309
Relaxation time, 38
Repeatability, 5, 264
Reproducibility, 5, 38
Resistance to fungal diseases, 234
Resistively heated filaments, 294
Resolution, 216
Restriction endonucleases, 187
Restriction enzyme analysis, 187
Restriction fragments, 229
Restriction site polymorphisms, 195
Reversion, 103, 104
rf (radio-frequency) pulse, 38
rf excitation pulse, 40
  90° pulse, 39
rf irradiation, 40
Rhamnose, 52
Rheingau, 51
Rheinhessen, 51
Ribose, 52
18S Ribosomal RNA gene, 193
Rice, 89, 298
  species, 282
Rioja, 276
Ripeness of fruit, 83
RMSEC (root mean square error of calibration), 330
RMSEV (root mean square error of validation), 330
RNA, 226, 229
Roasted coffee, 298
Rocket immunoelectrophoresis, 219, 246, *257*
Roquefort, 257
Rosehip/hibiscus tea blend, 298
Rubidium, 276, 277, 280, 281, 282
Rum, 69

Russet, 46, **47**, **49**
Russian sturgeon, 194

Salted cod, 190
Sample density, 210
Sample preparation, 61, 293
Sampling, 3
Sandwich EIA, 249, 263
Sapodilla plum, *20*
Saturated oil, **54**
Saturation, 40
Sausages, 91, 237
Scandium, 273, 282
Scotch malt whiskies, 69
Screening, 233
  for genetic variation, 197
SDS-PAGE, 211, *224*
Selenium, 278, *282*
Semolinas, 89
Sensitive test methods, 141
Sensitivity, 38, 177, 252, 253, *265*
Sensory data, 91
Serum albumin, 138
Sevruga, 194
Sex specific primers, 194
Sex specific probes for non-mammalian species, 197
Shared reactivity, 255
Sheep, 186, 193, 195, 198, *243*, *261*, 263
  meat, 193
Sheeps' milk, 237
Sherry wines, 281
Shikimic acid, 46
Shim settings, 40
Silanol groups, 131
Silicon, 273, 281
Silver staining, 223
  protocols, *225*
SIMCA (soft independent modelling of class analogies), 89, 277, 297
Single analyte analyses, 275
Single radial diffusion, 245
Single strand conformational polymorphisms, 195
Single-linkage, 321
Singlet, 40
Size fractionation, 231
Slot blot, 185
Smoked chicken, 186
Smoked salmon, 190
SNIF-NMR, 36
Sodium, 276, 277, 281, 282
Sodium benzoate, 49
Sodium deoxycholate, 231
Sodium tungstate, 18
Soft wheat flours, 234
Sophisticated adulteration, 4
Sorbic acid, 52

Sorbitol, 45, 52, 121, 210
D-Sorbitol, *142*
D-Sorbitol/xylitol, 161
Sorbitol dehydrogenase, *142*
Southern blot analysis, 187, 227
Southern latitudes, 24
Soya oils, **55**, 56, 90
Soybean, 21, 161
Soybean genomic map, 197
Soybean oil, 90
Spain, 17
Spanish and English wines, 287
Spanish cheese, 277
Spanish wine, 276, 323
Sparkling wine, 14
Spartan, 46, **47**, **49**
Species, 59, 90, 195
  identification, 233, 236, 250
  origin, 186
Species specific primers, 193
Specific haplotypes, 195
Specificity, 254, *265*
Spectral transformation, 93
Spectral width, 39
Spin quantum number, 60
Spirits, 69
Splitting pattern, 36
Squalene, 58
Squared Mahalonobis distance, 47
Stability, 138
  changes, 236
Stable isotope ratio analysis (SIRA), 14
Stacking, 210
Staining procedures, 222
Standard deviation, s, 309
Standard Mean Ocean Water (SMOW), *15*
Standard proteins, 217
Starch, 86, 101, *142*, *143*, 153, 159, 162, 182
  botanical origin, 298
Statistical parameters for EIA, 263
Steers, 194
Stepwise forward regression, 331
Stereo-specific enzymes, 144
Sterilization, 137, 193
Sterilized milk, 157
Sterols, 37, 57, 58, 297
Storage conditions, 300
Strawberry, 90
Strawberry juice, *19*, 121
Strawberry pulp, 89
Strawberry purees, 46
Streptavidin, 186
Strontium, 276
  isotope ratios, 287
Structure determination, 36
Sturgeon, 194
Succinic acid, **43**, 52, *143*, 162
Succinyl-CoA synthase, *143*

Sucrose, 44, **44**, 46, 49, *98*, 100, 111, 121, 138, *142*, *143*, 152, 163, 210
Sugar beet, 17
Sugar cane, 17
Sugar nitrate ester, 27
Sugars, 14, 48, 59, 84, 86, 89
Sulphite, *143*, 163
Sulphite oxidase, *143*
Sulphur dioxide, 83
Sunflower oil, 53, **54**, **55**, 56, 90
Superconducting magnets, 36
Supervised pattern recognition, 323
Surveillance considerations, 3
Synthetic ascorbic acid, 20
Synthetic essential oil, 29
Synthetic status of flavouring, 59

T_1, 39, 58
T_2, 58
Tamorillo, *20*
Tangerine, *20*
Tartaric acid, 20, **44**, 52
Taste panel, 11
Tea, 92, 298
    geographical origin, 302
    phenolic pigment composition, 331
Tequila, 20
Terpenoid group, 71
Tetrasaccharides, 116, 123
N,N-Tetramethylurea, 63
Textural changes, 236
Theaflavin, 92
Thermal ionization mass spectrometry (TIMS), 283
Thermoluminescence, 8
Thioglycoll, 138
Thomson's gazelle, *261*
Time domain, 38
Time from slaughter, 186
TIMS, *271*
Tin, 278
Titanium, 273
TMS (tetramethylsilane), 38
Tobacco blends, 293
Toluene, 28
Tomato proteins, 234
Tomatoes, *20*, 25, 27
Topi, *261*
Total invert sugar, 98, 103, 109
Trace element analysis, 270
Trace elements in orange juice, 278
Trace metals, 65
Traditional techniques, 7
Transfer system, 295
Transgenic animals, 200
Transgenic fruit, 200
Transgenic vegetables, 200
Transglycosylation, 101

Transmission, 83
Transpiration, 24, 26
Trehalose, 52
Triglyceride(s), 37, 55, 56, *143*
    profiles, 52
Trimethylamine, 59
Trimethylamine oxide, 58
Triple pulsed amperometry, 107
Trisaccharide(s), 116, 123, 128
Triterpenic alcohols, 58
Triticum durum, 88
'True' cross reactivity, 255
TSP-d_4, (sodium salt of 3-(trimethylsilyl)propionic-2,2,3,3-d_4 acid, 38
Tunas, 188, 196
Turbidimetry, 245
Turkey, 182, 183

UHT milk, 157
Ultra trace determinations, 272
Umbelliferyl, 222
Ungulates, 195
Univariate statistics, 308
Unsaponifiable material, 58
Unsaponifiable matter, 57
Unsaturated oil, **54**
Unsupervised pattern recognition, 319
Urea, *142*, 164, 212
Urease, *142*

V.SMOW (Vienna Standard Mean Ocean Water), 63
Valencia or non-Valencia varieties, 281
Valladolid wine, 277
Valine, **43**
Vanadium, 273
Vanilla Planifolia, 70
Vanilla, 14, 59, 70
    extract, 21
Vanillic acid, 22
Vanillin, 14, 21, 63, 70
Variance, s2, 309
Varieties of rice, 197
Variety, 182
    identification, 233, 235
Vegetable fat(s), 91, 298, 302
Vegetable oil(s), 21, 53, 59
Vegetable products, 147
Vegetables, 222
Venetian wines, 277
Very high-field magnets, 36
Vibrational spectroscopy, 76
Vienna sausage, 186
Vinegar, 70, 298
Vinegar fruit origin, 303
Vintages, 51
Virgin oil, 58

Virgin olive oil, 56, 71, 90
Vitamins, 237

Walnut oil, 90
Water signal, 40
Water suppression, 39
Waterbuck, *261*
Watermelon, *20*
Weather conditions, 7
Weighting factors, 42
Western Blotting, 227
Whale meat, 191
Wheat, 18
  blends, 278
  hardness, 88
  variety, 88
Wheatflour – type for purpose, 88
Wheatmeal biscuit, 298
Whisky, 298, 300
White wines, 334
Wild mussels, 278
Wildebeest, *261*

Wildlife, 182
Wine(s), 20, 51, 59, 64, 70, 146, 147, 152, 154, 158, 160, 161, 287, 298, 301
  country of origin, 302
  region, 276
Wine vinegars, 21, 161
  production process, 322
Wood lignin, 21

Xanthan, 91
Xanthine, 137
Xylose, 52, 86
Xylose isomerase, 101

Year of production of wine, 276
Yellowfin tuna, 183, 192
Yellowfin, 188
Yogurt, 71, 146

Zero-order kinetics, 138
Zinc, 276, 277, 282
  as a reference element, 280
Zirconium, 273